Topics in
Current Physics

8

## Topics in Current Physics    Founded by Helmut K. V. Lotsch

Volume 1 **Beam-Foil Spectroscopy**
Editor: S. Bashkin

Volume 2 **Modern Three-Hadron Physics**
Editor: A. W. Thomas

Volume 3 **Dynamics of Solids and Liquids by Neutron Scattering**
Editors: S. W. Lovesey and T. Springer

Volume 4 **Electron Spectroscopy for Surface Analysis**
Editor: H. Ibach

Volume 5 **Structure and Collisions of Ions and Atoms**
Editor: I. A. Sellin

Volume 6 **Neutron Diffraction**
Editor: H. Dachs

Volume 7 **Monte Carlo Methods** in Statistical Physics
Editor: K. Binder

Volume 8 **Ocean Acoustics**
Editor: J. DeSanto

Volume 9 **Inverse Source Problems** in Optics
Editor: H. P. Baltes

Volume 10 **Synchrotron Radiation** Techniques and Applications
Editor: C. Kunz

Volume 11 **Raman Spectroscopy** of Gases and Liquids
Editor: A. Weber

Volume 12 **Positrons in Solids**
Editor: P. Hautojärvi

Volume 13 **Computer Processing of Electron Microscope Images**
Editor: P. W. Hawkes

Volume 14 **Excitons**
Editor: K. Cho

Volume 15 Physics of **Superionic Conductors**
Editor: M. B. Salamon

# Ocean Acoustics

Edited by J. A. DeSanto

With Contributions by
N. Bleistein J. K. Cohen R. L. Deavenport
J. A. DeSanto F. R. DiNapoli J. P. Dugan
R. P. Porter J. G. Zornig

With 109 Figures



Springer-Verlag Berlin Heidelberg New York 1979

Dr. John A. DeSanto

Naval Resarch Laboratory, Washington, DC 20375, USA



ISBN 3-540-09148-3 Springer-Verlag Berlin Heidelberg New York
ISBN 0-387-09148-3 Springer-Verlag New York Heidelberg Berlin

Library of Congress Cataloging in Publication Data. Main entry under title: Ocean acoustics. (Topics in current physics; v. 8), Bibliography: p. Includes index. 1. Underwater acoustics. I. Bleistein, Norman. II. DeSanto, John A., 1941–. III. Series. QC242.2.023 551.4′601 78-25859



Offset printing and bookbinding: Konrad Triltsch, Graphischer Betrieb, Würzburg.
2153/3130-543210

# Preface

This Topics volume is devoted to a study of sound propagation in the ocean. The effect of the interior of the ocean on underwater sound is analogous to the effect of a lens on light. The oceanic lens is related, as in light propagation, to the index of refraction of the medium. The latter is given by the ratio of the sound frequency to the speed of sound in water, typically about $1500\,\mathrm{m\,s^{-1}}$. It is the variation of the sound speed due to changing temperature, density, salinity, and pressure in the complex ocean environment which creates the lens effect. Many oceanic processes such as currents, tides, eddies (circulating, translating regions of water), and internal waves (the wave-like structure of the oceanic density variability) contribute in turn to the changes in sound speed. The net effect of the ocean lens is to trap and guide sound waves in a channel created by the lens. The trapped sound can then propagate thousands of miles in this oceanic waveguide.

In addition to the propagation in the interior of the ocean, sound can propagate into and back out of the ocean bottom as well as scatter from the ocean surface. Just as the sound produced by a loudspeaker in a room is affected by the walls of the room, so the ocean boundaries and the material properties below the ocean bottom are essential ingredients in the problem.

Several techniques are used to study underwater sound propagation. These include experiments at sea, controlled simulation experiments in a water tank, and modeling the phenomenon using mathematical techniques and computer simulation. The experimental methods use a sound source such as an explosive (at sea) or a mechanical device which radiates sound (like a loudspeaker) and a hydrophone receiver (like a microphone). The different paths which the sound waves take result in different modifications of the initial signal. The received signals are recorded and processed using computers to yield the desired measurement. Mathematical methods model the physical situation, the resulting equations describing the model are solved using computers, and the results compared to the experimental data.

All the above methods for the study of underwater sound propagation are discussed in the articles in this book. They are treated in such a way as to provide a rapid introduction to the field, as well as a thoroughly referenced review of the state-of-the-art experimental, theoretical, and computational methods in use.

Washington, D.C.
November 1978

*John A. DeSanto*

# Contents

# List of Contributors


BLEISTEIN, NORMAN

Mathematics Department, University of Denver, Denver CO 80208, USA

COHEN, JACK K.

Mathematics Department, University of Denver, Denver CO 80208, USA

DEAVENPORT, ROY L.

Naval Underwater Systems Center, New London CT 06320, USA

DESANTO, JOHN A.

Naval Research Laboratory, Washington, DC 20375, USA

DINAPOLI, FREDERICK R.

Naval Underwater Systems Center, New London CT 06320, USA

DUGAN, JOHN P.

Naval Underwater Systems Center, New London CT 06320, USA

PORTER, ROBERT P.

Schlumberger-Doll Research Center, Ridgefield, CT 06977, USA

ZORNIG, JOHN G.

Department of Engineering and Applied Science, Yale University,
15 Prospect St., New Haven, CT 06520, USA

# 1. Introduction

J. A. DeSanto

Our purpose in this Topics volume is to discuss the correlation between oceanic variability and acoustics. There are two aspects of ocean acoustics. The first is the study of effects that oceanic variability have on acoustic propagation. For example, a dynamic process in the ocean, such as internal gravity waves, can alter the speed at which an acoustic wave propagates. The variation in local sound speed induces modifications of the acoustic field, and the relationship can be expressed mathematically. The second and more recent aspect is concerned with the use of acoustics as a tool to prove the oceanic structure. For example, acoustics can be used to monitor ocean currents in the sense that characteristic modifications of the acoustic field infer local changes in the ocean current.

These two aspects of ocean acoustics can be viewed as direct or inverse ("remote sensing") problems, but I intentionally omit the word "respectively" since its use would only serve to characterize whether its user was an underwater acoustician or an oceanographer. In this book we view both aspects as complementary rather than separate, and this sums up the spirit of the book probably better than any other we might devise. The chapters of this book, however, concentrate more on examining the effect that oceanic variability has on acoustic propagation, than on the opposite situation. This is simply because it has been more extensively developed.

In attempting to understand any subject it is helpful to know its boundaries. As a function of frequency, ocean acoustics is bounded on one side by ultrasonics and on the other side by seismic phenomena. The frequency range of the human ear extends roughly from 30 Hz to 20 kHz. Ocean acoustics exceeds this range by more than an order of magnitude in both directions, from approximately 1 Hz to about 300 kHz. An acoustic signal propagates with the speed of sound in water approximately equal to 1500 $\mathrm{m\,s^{-1}}$. This means that acoustic wavelengths extending from 1500 m down to $5 \cdot 10^{-3}$ m can be present, a range of almost six orders of magnitude. In principle then any obstacle or process in the ocean within these dimensions can be probed using acoustics. In practice, however, high frequency sound is rapidly attenuated in the ocean, and if one treats long-range propagation problems, and implicitly includes the limitations of the tools necessary to study them, the frequency range is considerably narrower. It extends from roughly a few Hertz to several hundred Hertz. However, for short-range propagation problems, several kilohertz is often used. In addition, just as in seismic propagation problems, as one experiments with lower acoustic frequencies, sound can propagate into and back out of the bottom of the ocean, as well as scatter from the ocean surface. Thus the ocean boundaries and the material properties below the ocean bottom become essential parts of the problem. The latter is a characteristic difference between acoustic propagation in the ocean and in the atmosphere, where the effects of boundaries are generally ignored. For low-frequency ocean acoustic propagation it is the connection with seismic problems which matters. Our discussion in this book includes surface scattering, as a model of scattering from the ocean surface and bottom, but does not include any specific discussion of the effect of material properties below the ocean bottom on acoustic propagation. We limit our treatment to what happens within the oceanic waveguide.

As sound propagates through the ocean it stores information on the ocean conditions in its path. There are many processes which contribute to this information. They include effects of the surface and bottom as well as volume inhomogeneities. The latter include temperature and salinity structure in the ocean, density gradients, and marine organisms. Noise, both ambient noise due to shipping, rain, seismic activity and biological processes, and self-noise from the measuring equipment, limits the ability to measure in the ocean, particularly at low frequencies. The environmental background is thus quite complicated, and the complexity manifests itself in the difficulty and high cost of doing ocean acoustics experiments, in the difficulty in isolating characteristic oceanic effects, and in the interpretation of the resulting data.

Temperature, salinity, and density (or pressure) properties contribute to forming the sound speed at which an acoustic wave travels in the ocean. The resulting sound speed structure has the same effect on acoustic waves as a lens has on light. It can trap or channel sound energy so that very long range acoustic propagation in the ocean is possible. Hence the amount of information the acoustic signal can store can be very large. Indeed, it can even become saturated with information. It is then very difficult to unravel the various effects in the data. It is helpful, therefore, to consider either shorter propagation paths or alternative frequency regimes where a combination of both deterministic and stochastic modeling is more effective, and, in general, where the variability in the environment or the measuring apparatus is such that characteristic effects can be isolated. Model experiments in tanks are valuable in determining useful range and frequency combinations and in isolating the effects of individual oceanic processes. In order to attempt this isolation it is necessary to understand the spatial and temporal scales of oceanic processes and, consequently, the parameter domains of their coherence or lack thereof. Our present understanding of these quantities leads to the conclusion that a model must combine both deterministic and stochastic variations in the ocean to be appropriate in describing sound propagation. Just as in any other scientific enterprise, one needs to control the experimental environment in order to develop and sharpen one's analytical tools.

## 1.1 A Brief History

Many of these remarks are not new. However, many of the details associated with them in this book are comparatively new. The basic idea of attempting to correlate oceanic variation with acoustic signal properties appears prominently in the first two volumes of *Physics of Sound in the Sea* [1.1] which "presents the essential results obtained in the studies of underwater sound up to the middle of 1945". Introductory sections in these volumes provide an excellent intuitive background to the study of acoustic propagation. An introduction to the "usefullness of underwater sound in solving oceanographic problems" can be found in the text by CLAY and MEDWIN [1.2]. Historical developments in ocean acoustics are reviewed by LINDSAY [1.3], LASKY [1.4], and HERSEY [1.5], and a compilation of papers which traces the historical foundations of underwater sound can be found in [1.6]. These authors point out that the speed of sound in water was already known in 1826 to within a percent of its presently accepted value following measurements by Daniel Colladon in Lake Geneva. By 1912, underwater sound was in use as a communication tool to a range of about 30 miles. By 1945, it had been discovered by Ewing that a sound speed minimum existed due to temperature and pressure properties in the ocean. This was called the SOFAR (sound fixing and ranging) channel and, using it, propagation ranges of a thousand miles were possible. Today, using advanced signal processing techniques, this distance can be increased by over an order of magnitude, with underwater-sound propagation possible half way around the globe.

The early work in the field concentrated on the physics of the medium. Theoretical models used to explain the effects of the medium on acoustics were intensively studied in the 1940's. Two fundamental methods were studied by BERGMANN [1.7], who investigated the use of ray theory to describe acoustic signal fluctuations, and PEKERIS [1.8] who applied normal mode theory to the study of underwater acoustic propagation in a waveguide. Both methods are still in use today. Computational methods developed in the 1950's and 60's have tremendously increased the usefullness of these ocean acoustic models and others developed more recently. This has been pointed out by BERMAN and GUTHRIE [1.9] in their review paper on the ocean medium, as viewed by acousticians. These authors also review developments in the field to about 1968 and discuss attenuation, thermal microstructure, and scattering from the surface and bottom of the ocean. TOLSTOY and CLAY [1.10], in an evaluation of ocean acoustics in 1966, pointed out the necessity of including the random aspects of the ocean medium, in addition to its deterministic background structure. They discussed methods of data analysis which are in common use today. The ocean medium from the oceanographic viewpoint is very well presented in the books by PHILLIPS [1.11], SVERDRUP et al. [1.12], and DEFANT [1.13].

We have only very briefly touched on these historical developments and documentation since they are fully treated in the references we mentioned. What is clear from studying the above references, however, are the three areas of research we consider in this book. We discuss them next.

## 1.2 Outline of the Book

The book is divided into three parts. The physics of the ocean and its relation to acoustic propagation is presented in Chaps.5 and 7, the theoretical models in Chaps.2 and 6, and some aspects of the simulation of the ocean from both numerical and controlled-tank experimental points of view is presented in Chaps.3 and 4, respectively.

In Chapter 2, I present a review of theoretical methods which are being used in ocean acoustic propagation and scattering. I show how combined acoustic-internal wave equations can be derived from first principles, and where the acoustic sound speed enters in the derivation. Properties of equations on the acoustic field amplitudes are then studied for deterministic propagation models, and equation properties on the mean (coherent) field and second moment of the field (coherence function) are discussed for the case where stochastic effects are present. A similar breakdown into deterministic and stochastic equations and their properties is also presented for the problem of sound scattering from rough-surface models of the surface and bottom of the ocean.

In Chapter 3, F.R. DiNapoli and R.L. Deavenport present many results on the current state-of-the-art in numerical simulation methods for deterministic propagation. The presentation is very detailed and covers many different models, their comparison, and a critical examination of their respective domains of validity. A general formulation, derived in terms of the matrizant method, is provided for obtaining the depth-dependent Green's function when the source and receiver are located at any depth in a medium which is arbitrarily stratified in depth. The majority of the work is confined to purely depth-dependent sound speeds, but there are also sections in which range-dependent models are discussed. The chapter is one of the most comprehensive and detailed assessments of computer models in underwater acoustic propagation.

In Chapter 4, J.G. Zornig discusses certain aspects of the alternative analogue simulation method for ocean acoustics, that of experimental measurements in a water tank. He treats the relationship of the method to ocean experimentation, the various facilities which have been used, and possible targets such as wind driven, mechanically excited , and solid surfaces, solid targets, and volume effects. He also presents instrumentation and signal processing methodology, and several of his results from this model experimentation.

In Chapter 5, J.P. Dugan investigates the use of oceanography in acoustics in the sense of explaining how characteristic sound speed profiles arise from fundamental ocean measurements. In addition, he studies the spatial scales of ocean variability from the largest scale size of the ocean basin (ocean "climate"), through that of ocean "weather" (e.g. eddies as storms in the ocean) on a scale of roughly 50 to 500 km, down to very small (oceanic) scale variability such as internal waves and fine structure. These latter have horizontal scales of perhaps 100 m to 10 km, and vertical scales about two orders of magnitude less. Although the acoustic scale sizes we discuss are considerably smaller than, say, those of eddies, the latter strongly influence long range acoustic propagation by changing the vertical position of the sound speed minimum.

N. Bleistein and J.K. Cohen study the topic of acoustic probing in Chap.6. They present some of their recent work on mapping high-contrast interfaces in the seabed using physical-optics far-field inverse scattering (POFFIS) methods, and show that it is possible to probe sound speed profiles in media with small variations in sound speed. Extensions to regions of small inhomogeneities are also possible. The methods described are useful for both band-limited and aspect-angle limited data sets, and they operate directly on these data sets. Their practical utility is demonstrated with several examples.

In Chapter 7, R.P. Porter discusses progress in the use of acoustics to probe the ocean. Recent experiments for measuring ocean currents have used neutrally buoyant floats that use acoustic pulses to signal their location. Another method uses reciprocal transmissions between fixed transmitters and receivers to measure ocean currents. Many experiments have been conducted during the last twenty years to measure the effect of tides, internal waves, and ocean circulation on propagating acoustic signals. The results of these experiments are discussed and theories presented to quantity the relationship between measured sound and ocean dynamics. Experiments discussed either use short pulses propagated over single paths or narrow-band transmissions for which multipath fluctuations are important. It is shown that multipath phase fluctuations are related to ocean dynamics with time scales of several hours or longer. At shorter time scales, phase fluctuations of continuous tone signals depend on the statistics of amplitude fluctuations. The chapter concludes with some observations about experimental ocean probes of the future.

Finally, these chapters have concentrated on the physics of the ocean, its mathematical representation, and its numerical and controlled-experimental simulation. We have tried to present in as coherent a manner as possible the status of research in these areas, while simultaneously giving the reader an introduction to several important areas of research in ocean acoustics.

## References


1.1 *Physics of Sound in the Sea: I. Transmission*, ed. by P.G. Bergmann, A. Yaspan; *II. Reverberation*, ed. by E. Gerjuoy, A. Yaspan (Gordon and Breach, New York 1968)
1.2 C.S. Clay, H. Medwin: *Acoustical Oceanography: Principles and Applications* (Wiley-Interscience, New York 1977)
1.3 R.B. Lindsay: *Acoustics-Historical and Philosophical Development* (Douden, Hutchinson and Ross, Stroudsburg, PA. 1973)
1.4 M. Lasky: J. Acous. Soc. Am. *61*, 283-297 (1977)
1.5 J.B. Hersey: Oceanus *20*, 8-21 (1977)
1.6 V.M. Albers (ed.): *Underwater Sound* (Douden, Hutchinson, and Ross, Stroudsburg, PA. 1972)
1.7 P.G. Bergmann: Phys. Rev. *70*, 486-492 (1946)
1.8 C.L. Pekeris: "Theory of Propagation of Explosive Sound in Shallow Water", Memoir 27, Geological Society of America (1948)
1.9 A. Berman, A.N. Guthrie: J. Acoust. Soc. Am. *51*, 994-1009 (1972)
1.10 I. Tolstoy, C.S. Clay: Rev. Geophys. *4*, 33-40 (1966)
1.11 O.M. Phillips: *The Dynamics of the Upper Ocean* (Cambridge University Press, Cambridge 1969)
1.12 H.U. Sverdrup, M.W. Johnson, R.H. Fleming: *The Oceans, Their Physics, Chemistry, and General Biology* (Prentice-Hall, Englewood Cliffs, N.J. 1942)
1.13 A. Defant: *Physical Oceanography*, in 2 vols. (Pergamon Press, New York 1961)

# 2. Theoretical Methods in Ocean Acoustics

J. A. DeSanto

With 24 Figures

In this chapter we mathematically describe how sound propagates through the ocean. This is the study of sound in a fluid waveguide, i.e., a fluid with bounding walls, the surface and bottom of the ocean. Sound is a small amplitude wavelike disturbance on the hydrodynamic background and thus we must first describe the properties of the fluid equations. We derive equations for these properties in Sect.2.1 from first principles, but briefly, and then obtain the equations governing the sound field which we discuss in more detail in the remaining parts of this chapter.

As the sound waves propagate through the fluid they experience various effects that are familiar from other fields of physics. They are reflected, transmitted or scattered by the surface and bottom of the ocean, and they refract in a manner analogous to light propagating through a lens with a variable index of refraction. In our case the lens is created by the variation of sound speed in the ocean, which changes are due to density, salinity and temperature variations in the water. These effects can be semi-empirically combined to yield sound speed values. This is discussed in Sect.2.1 and elaborated in the remainder of the book. The depth variation of sound speed creates an effective interior waveguide important in long-range propagation. Sound waves may also diffract out of the plane because of horizontal variations of sound speed or by encountering an obstacle.

Other effects modify the sound wave as it propagates. Refraction effects can be complicated by mesoscale ocean phenomena such as currents and eddies. The latter induce modifications of the sound speed field, so that what is created is a complicated three-dimensional refractive medium. They also blur the distinction between refraction and diffraction. Internal waves related to density variations of the medium contribute further modifications which are generally treated as purely random variations and which can be thought of as a scattering phenomenon. The sound speed is thus modelled as a large deterministic refraction (due to depth variations), further three-dimensional deterministic modifications due to mesoscale phenomena, and random modifications in space and time due to smaller scale phenomena such as internal waves. We treat the effects of the sound speed functional dependence in detail in this chapter, and remark on them here to point out the complementary deterministic and random aspects of the propagation model discussed in Sects.2.2 and 2.3.

Analogous complementary models contribute to surface and bottom scattering treated in Sect.2.4. The ocean surface can be modelled as having deterministic periodic components as well as superimposed random variations, both of which contribute characteristic effects to sound scattering and hence modify the sound field at other points in the waveguide. For long-range propagation, multiple scatter effects become important. We argue that, even in short-range propagation, the scattering treatment should include multiple scattering. We show how this becomes particularly important when the surface is very rough by demonstrating how the multiple scatter theory found by solution of an integral equation yields results more comparable to those observed than that predicted by single scatter theories.

The main difficulty in the mathematical development is in treating the deterministic and random variations in combination. For deterministic variations alone simple scattering and propagation models exist. For example, normal mode theory in Sect.2.2.1 provides a useful propagation model when the sound speed variation is only one-dimensional and the surface and bottom are flat. Scattering from deterministic periodic surfaces provides much information on surface scatter models (Sect.2.4). Both are computationally feasible and have been experimentally verified, so they provide certain canonical benchmarks to measure more complicated models against. But both are of limited applicability. The situation with purely random models is less well developed. Random propagation models borrowed from atmospheric optics are of limited use since they generally do not include refraction as a first-order effect, and neglect the boundary waveguide nature of the propagation. Both simplifications make them more tractable mathematically but they are not generally applicable to acoustic propagation. Some results of ours on scattering from a random surface (Sect.2.4.4) as well as a general formalism for propagation in a flat waveguide with both deterministic and random sound speed variations (Sect.2.3.4) are presented from which some conclusions can be drawn, but only partial results are available for the general problem. We do present a general theoretical formalism for treating scattering from an arbitrary surface but much remains to be done computationally on this problem.

Our presentation of the propagation and scattering models is very general from the standpoint of theoretical formalism. We present some numerical results and point out where they are experimentally important but we stress the generality of what we do and the interplay between deterministic and random problems. We point out active areas of research, some inactive areas where more activity appears necessary, and try to make some sense of a collage of models of both propagation and scattering. Further development and elaboration on many of the topics we treat will be found in the successive chapters in this volume as well as in articles referenced in our discussion.

## 2.1 Conservation Laws, Fluid and Acoustic Equations

In this section we derive the fluid equations from general conservation laws of density and momentum, and state then in a parabolic equation form. Standard perturbation methods are used to yield the zeroth- and first-order equations and from the latter combined multidimensional acoustic internal wave equations are derived. A sound speed model is then presented which relates to subsequent sections of this chapter.

### 2.1.1 Conservation Laws and Fluid Equations

The simplest way to derive the fluid equations is to first derive a general conservation law for an arbitrary fluid quantity and then apply it to special cases [2.1]. Define a fluid volume V independent of fluid position and of time, and surrounded by a surface S. Differential volume and surface elements are also defined. Given that the function $\psi$ is some general conserved quantity per unit volume with $\iiint \psi\, dV$ the total conserved quantity, and given the function q as the external source of $\psi$ per unit volume with $\iiint q\, dV$ the total source of $\psi$. Then the flow of $\psi$ through a differential area $d\underline{S}$ per unit time can be written as $\psi\underline{u} \cdot d\underline{S}$, where $\underline{u}$ is the vector fluid velocity. The total outflow of $\psi$ is, using Green's theorem

$$-\iint \psi\underline{u} \cdot d\underline{S} = -\iiint \underline{\nabla} \cdot (\psi\underline{u})dV \quad ,$$

and we can state the general conservation law as

$$\partial_t \iiint \psi dV = \iiint q\, dV - \iiint \underline{\nabla} \cdot (\psi\underline{u})dV \quad . \tag{2.1}$$

Equation (2.1) can also be written as a differential statement in index notation

$$\partial_t\psi + \partial_j(\psi u_j) = q \quad , \tag{2.2}$$

where $\partial_j = \partial/\partial x_j$, $\partial_t = \partial/\partial t$, $\underline{x} = (x_1, x_2, x_3) = (x, y, z) = (x_\perp, z)$ and repeated subscripts are summed from 1 to 3 (summation convention).

If we let $\psi = \rho$, the mass density, and $q = 0$ in (2.2) we get the continuity equation for mass density conservation

$$\partial_t\rho + \partial_j(\rho u_j) = 0 \quad . \tag{2.3}$$

If we let $\psi = \rho u_j$, the momentum density, and $q = F_j$ for the external force density, insert the results into (2.2) and use (2.3) we get

$$\rho(\partial_t u_j + u_i \partial_i u_j) = F_j \quad , \tag{2.4}$$

which are the Euler equations, the fluid version of Newton's second law. We express the force density as

$$F_j = -\partial_j p + g\rho\delta_{j3} - 2\rho\varepsilon_{jmn}\Omega_m u_n \quad , \tag{2.5}$$

where $-\partial_j p$ is the force on the fluid due to the local water pressure, $g\rho\delta_{j3}$ is the gravitational force density with $\delta_{j3}$ the Kronecker delta function (=1 for j=3 and 0 otherwise) and the z-direction points into the Earth. Here g is approximately equal to 978 cm $s^{-2}$ and we neglect any variation in g at different positions on the Earth. These gravity forces when acting on changes from the mean density in the medium yield motions which have a wavelike character. These are called internal waves. They are discussed further below and in Chaps.5 and 7. The term

$$F_j^c = -2\rho\varepsilon_{jmn}\Omega_m u_n \quad , \tag{2.6}$$

is the Coriolis force density, $\varepsilon_{jmn}$ the antisymmetric third rank tensor

$$\varepsilon_{jmn} = \begin{cases} 1 & j,m,n = 1,2,3 \text{ or cyclic permutation,} \\ -1 & \text{anti-cyclic,} \\ 0 & \text{any two subscripts equal,} \end{cases}$$

and $\Omega$ is the angular velocity of the Earth whose approximate magnitude is $\Omega = 0.73 \times 10^{-4}\ s^{-1}$. The geometry is illustrated in Fig.2.1. We thus have that $\underline{\Omega} = \Omega(0,\cos\theta, -\sin\theta)$ where $\theta$ is the angle of latitude. If we write $\underline{u} = (u,v,w)$ then

$$\underline{F}^c = -2\rho\Omega(w\cos\theta + v\sin\theta, -u\sin\theta, -u\cos\theta) \quad . \tag{2.7}$$

Physically, the vertical motion is small with respect to the other velocities so that $w \ll v$ and $w \ll u$. In addition, the vertical (z) component of the Coriolis force is much less than the gravitational force. Thus the only appreciable component of the

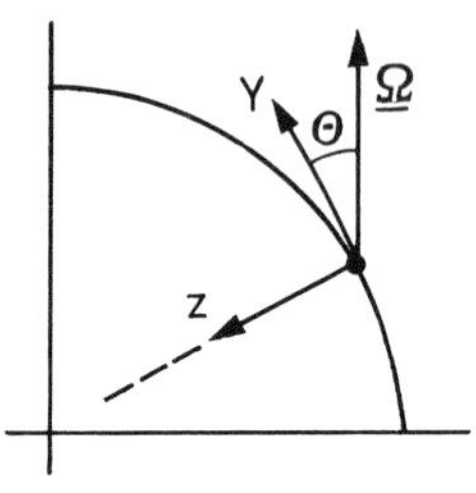


Fig. 2.1. Coordinate system with z directed into the earth, y north, $\Omega$ the angular velocity of the Earth, and $\theta$ the angle of latitude

Earth's rotation is the local vertical and we can approximately write (for $\theta$ in the neighborhood of $30^{\circ}$)

$$F_j^c = \rho f \varepsilon_{j3n} u_n \quad , \tag{2.8}$$

where $f = 2\Omega_3$. If f is constant, the result is called the f-plane approximation. A first-order approximation to the North-South gradient of vertical rotation can be introduced if $df/dy = \beta$ is constant. This is called the $\beta$-plane approximation, and $f = f_0 + \beta y$. Combining these results, (2.4) is

$$\rho(\partial_t u_j + u_i \partial_i u_j) = -\partial_j p + g\rho\delta_{j3} + \rho f \varepsilon_{j3n} u_n \quad . \tag{2.9}$$

Equations (2.3) and (2.9) are four equations, but we have five unknowns, $u_j$, p, and $\rho$. We add a fifth equation, a state equation of the simple form

$$p = p(\rho) \quad , \tag{2.10}$$

excluding the specific functional dependence of the pressure on salinity and temperature, although in Sect.2.1.4 we introduce a sound speed term in a specific representation of the state equation and show how the sound speed is determined empirically through its dependence on salinity and temperature. More fundamental oceanographic studies (Chap.5) describe salinity and temperature diffusion and their interrelationship, but our purpose here is to derive only those quantities necessary to arrive at the acoustic results. Other papers [2.2] as well as books [2.1,3] treat this question in more detail.

### 2.1.2 A Parabolic Equation

There is an interesting alternative form of the fluid equations. We can directly relate the solution of the fluid equations to the solution of an equivalent parabolic partial differential equation. To derive the latter assume only the continuity equation (2.3). Then decompose the vector velocity $u_j$ in terms of real scalar ($\phi$) and vector ($B_j$) potentials as

$$u_j = \partial_j \phi + B_j \quad , \tag{2.11}$$

where $B_j$ is divergence-free, i.e., $\partial_j B_j = 0$. We assume the hydrodynamic pressure p can be written in terms of $\phi$ as

$$p = -\rho \partial_t \phi \quad . \tag{2.12}$$

Next, define the function $\psi$ by

$$\psi = \rho^{1/2} \exp(i\alpha\phi) \quad , \tag{2.13}$$

where $\alpha$ is a constant having the dimensions of time per length squared. From the above equations we derive the following relations

$$\rho = \psi\psi^* \quad , \tag{2.14}$$
$$u_j = (2i\rho\alpha)^{-1}(\psi^*\partial_j\psi - \psi\partial_j\psi^*) + B_j \quad , \tag{2.15}$$
$$p = -(2i\alpha)^{-1}(\psi^*\partial_t\psi - \psi\partial_t\psi^*) \quad , \tag{2.16}$$

and from (2.14) and (2.15) derive $\partial_j(\rho u_j)$, substitute it and $\partial_t\rho$ from (2.14) into (2.3) and write the result as

$$\psi^*T = -\psi T^* \quad , \tag{2.17}$$

where

$$T = \partial_t\psi + (2i\alpha)^{-1}\nabla^2\psi + B_j\partial_j\psi \quad . \tag{2.18}$$

Thus $\psi^*T$ is pure imaginary and for (2.17) to hold for *any* $\psi$ we require that

$$T = i\alpha V\psi \quad , \tag{2.19}$$

where $V$ is a real function. Equations (2.18) and (2.19) combine to yield a parabolic partial differential equation for $\psi$

$$2i\alpha\partial_t\psi + \nabla^2\psi + 2\alpha^2V\psi + 2i\alpha B_j\partial_j\psi = 0 \quad , \tag{2.20}$$

analogous to the Schrödinger equation in quantum mechanics, where simple examples of $B_j$ allow $\psi$ to describe the motion of a particle in various electric and magnetic fields [2.4]. This general hydrodynamic formulation of quantum mechanics is well known [2.5]. Note that (2.20) was derived by assuming only the real continuity equation [2.6] whereas (2.20) is complex.

To clarify this latter remark, substitute (2.13) into (2.20) and equate real and imaginary parts of the result separately to zero. The imaginary part just yields the continuity equation (2.3). The divergence of the real part yields with (2.11) the result

$$\partial_t u_m + u_\ell\partial_\ell u_m = \partial_m\left[V + (2\alpha^2)^{-1}R + \tfrac{1}{2}B_j^2\right] + \partial_t B_m + u_\ell\partial_\ell B_m - u_j\partial_m B_j \quad , \tag{2.21}$$

where

$$R = \rho^{-1/2}\nabla^2(\rho^{1/2}) \tag{2.22}$$

Equation (2.21) with $B_j = 0$ is just WONG's [2.7] form of the Euler equations. To compare (2.21) with (2.9) first identify $B_j$ with f via

$$\varepsilon_{mjn}\partial_j B_n = f\delta_{m3} \quad ,$$

which has a solution, in the β-plane approximation, of

$$B_1 = -\frac{1}{2}\left[f_0 y+(x^2+y^2)\beta/2\right] \quad ,$$

$$B_2 = (f_0+\beta y)x/2 \quad ,$$

and

$$B_3 = 0 \quad .$$

The remainder of the comparison between (2.21) and (2.9) follows by equating

$$\partial_m\left[V+(2\alpha^2)^{-1}R+\frac{1}{2}B_j^2\right] = g\delta_{m3} - \rho^{-1}\partial_m p \quad , \tag{2.23}$$

which has a particularly simple solution (up to a constant) if $\rho$ is constant,

$$V = -\rho^{-1}p + gz - \frac{1}{2}B_j^2 \quad . \tag{2.24}$$

Equation (2.24) relates the potential directly to the hydrodynamic pressure. Alternatively the solution of the linear Schrödinger equation for the potential, the potential inversion problem of (2.20), yields the hydrodynamic pressure.

We have included this section as a digression since nothing more will be done with it. We consider it interesting, however, since one can go by perturbation methods of the above to acoustics (we use the standard methods in the next paragraph), numerical codes exist for solving parabolic equations rapidly, and the direct solution of the nonlinear fluid equations is coupled to the inverse solution of a linear equation. This latter idea has been extensively exploited in nonlinear-equation methods [2.8]. We remark that the parabolic equation used as an approximation in acoustic propagation in Sect.2.2.2 is parabolic in range rather than time. Also, the potential V is related to sound speed, and an iversion method for the sound speed is discussed in Chap.6.

### 2.1.3 Perturbation Method

We use the method of perturbation to derive the equations describing acoustic propagation. Assume $\rho$, p, and $u_j$ in Sect.2.1.1 are replaced by

$$\rho = \rho^{(0)}(\underline{x},t) + \varepsilon\rho^{(1)}(\underline{x},t) \quad , \tag{2.25}$$

$$p = p^{(0)}(\underline{x},t) + \varepsilon p^{(1)}(\underline{x},t) \quad , \tag{2.26}$$

and

$$u_j = u_j^{(0)}(\underline{x},t) + \varepsilon u_j^{(1)}(\underline{x},t) \quad , \tag{2.27}$$

where the 0-superscript terms refer to the hydrodynamic background, $\varepsilon$ is a small parameter, and the 1-superscript terms refer to acoustic quantities, i.e., to a small amplitude disturbance in a compressible ($\partial_j u_j \neq 0$) fluid. For a justification of this procedure in terms of various scale sizes of the different parameters and a thorough study of the brief discussion below there is a rich literature [2.1-3,9].

Substitute (2.25-27) into (2.3) and (2.9) and equate successive powers of $\varepsilon$ to zero. The zeroth-order equations are just (2.3) and (2.10) with zero superscripts. For zero hydrodynamic velocity, $u_j^{(0)} = 0$, these equations yield the hydrostatic result $\rho^{(0)} = \rho^{(0)}(z)$, $p^{(0)} = p^{(0)}(z)$ and

$$p^{(0)'} = g\rho^{(0)} \quad . \tag{2.28}$$

The prime indicates differentiation. If we include a state equation of the form $p^{(0)} = A(\rho^{(0)})$, and differentiate it

$$p^{(0)'} = (dA/d\rho^{(0)})\rho^{(0)'} = c_0^2(z)\rho^{(0)'} \quad , \tag{2.29}$$

where $c_0(z)$ is the lowest-order refractive sound speed (Sect.2.1.5). Combining (2.28) and (2.29) yields

$$c_0^2(z)\rho^{(0)'} = g\rho^{(0)} \quad , \tag{2.30}$$

a perturbed example of which is used in the next section.

The first-order equations are now written as (we drop the superscript 1 with the understanding that for the rest of the chapter we only deal with first-order quantities) from (2.3)

$$\partial_t\rho + u_j^{(0)}\partial_j\rho + \rho\partial_i u_i^{(0)} + u_i\partial_i\rho^{(0)} + \rho^{(0)}\partial_i u_i = 0 \quad , \tag{2.31}$$

and from (2.9)

$$\rho^{(0)}\partial_t u_j + \rho\partial_t u_j^{(0)} + \rho^{(0)}u_\ell^{(0)}\partial_\ell u_j + \rho^{(0)}u_\ell\partial_\ell u_j^{(0)} + \rho u_\ell^{(0)}\partial_\ell u_j^{(0)}$$
$$= -\partial_j p + g\rho\delta_{j3} + \rho^{(0)}f\varepsilon_{j3m}u_m + \rho f\varepsilon_{j3m}u_m^{(0)} \quad . \tag{2.32}$$

Next we derive coupled acoustic-internal wave equations from these latter equations and a particular choice of the state equation for first-order quantities.

## 2.1.4 Combined Acoustic-Internal Wave Equations

The combined acoustic-internal wave equations are derived from (2.31) and (2.32) assuming no background fluid velocity. Then the continuity equation is

$$\partial_t\rho + u_3\rho^{(0)'} + \rho^{(0)}\partial_i u_i = 0 \quad , \tag{2.33}$$

and the Euler equations are

$$\rho^{(0)}\partial_3 u_j = -\partial_j p + g\rho\delta_{j3} + \rho^{(0)}f\varepsilon_{j3m}u_m \quad , \tag{2.34}$$

which for $g = f = 0$ is called the first field equation of acoustics (analogous to the Maxwell equations of electromagnetic theory). Since we are dealing with multidimensional isentropic processes we make the assumption that the linearized state equation can be written as [2.2,10]

$$p = c^2(\underline{x},t)\rho \quad . \tag{2.35}$$

The sound speed c is a function of all the coordinates and time. The dependence of the state equation on temperature, salinity, etc. is through the sound speed and will be discussed later.

Differentiate (2.35) with respect to time and use the (now) approximate result from hydrostatics

$$c^2\rho^{(0)'} \simeq g\rho^{(0)} \quad , \tag{2.36}$$

to derive the equation

$$c^2\partial_t(p/c^2) + \rho^{(0)}c^2\partial_i u_i + g\rho^{(0)}u_3 = 0 \quad , \tag{2.37}$$

which is a generalized version of the second field equation of acoustics. The approximation (2.36) can be described in a perturbation sense if we weakly perturb the hydrostatic density. We discuss our sound speed field perturbation model in the next section.

The derivation of the coupled equations uses (2.33,34,37) in the $\beta$-plane approximation. If we define the generalized Väisälä frequency N via

$$N^2(\underline{x},t) = g\rho^{(0)'}/\rho^{(0)} - g^2/c^2(\underline{x},t) \quad , \tag{2.38}$$

and the operators $L = \partial_t^2 + f^2$, $L^* = \partial_t^2 - f^2$, and $R = \partial_t^2 + N^2$, and neglect $LN^2$ terms, we get the result [2.11]

$$\begin{aligned} R^2\Lambda &- gRL^2\partial_t\left[\partial_t^2(p/c^2)+g\partial_3(p/c^2)-\partial_3^2 p\right] \\ &= RN^2L^2\partial_t\partial_3 p - g\left(\partial_3 N^2\right)L^2\partial_t\left(gp/c^2-\partial_3 p\right) \quad , \end{aligned} \tag{2.39}$$

where

$$\Lambda = gL\partial_t\nabla_\perp^2 p - 2\beta g f\partial_t\partial_2 p + \beta g L^*\partial_1 p \quad , \tag{2.40}$$

and $\nabla_\perp^2 = \partial_x^2 + \partial_y^2$ is the transverse or horizontal Laplacian operator. Except for the neglect of $LN^2$ terms, (2.39) is exact but essentially useless because of its complexity. It becomes more useful and certainly more familiar for the special cases treated below.

If the sound speed is only a function of the spatial coordinates,

$$c(\underline{x},t) = c_2(\underline{x}) \quad , \tag{2.41}$$

then we can introduce the temporal Fourier transform

$$\tilde{p}(\underline{x},\omega) = \int \exp(i\omega t)p(\underline{x},t)dt \quad , \tag{2.42}$$

where $\omega$ is the circular frequency of sound, and from (2.39) derive

$$\begin{aligned} &\rho^{(0)}\partial_3\left[(\rho^{(0)})^{-1}\partial_3\tilde{p}\right] + (\omega^2-N^2)(\omega^2-f^2)^{-1}\nabla_\perp^2\tilde{p} + (\omega^2-N^2)^{-1}(\partial_3 N^2)\partial_3\tilde{p} \\ &+ 2\beta f(\omega^2-N^2)(\omega^2-f^2)^{-2}\partial_2\tilde{p} + i\beta\omega^{-1}(\omega^2-N^2)(\omega^2+f^2)(\omega^2-f^2)^{-2}\partial_1\tilde{p} \\ &+ \left[\omega^2 c_2^{-2} + 2gc_2^{-3}\partial_3 c_2 + gc_2^{-2}(\omega^2-N^2)^{-1}\partial_3 N^2\right]\tilde{p} = 0 \quad . \end{aligned} \tag{2.43}$$

For $\beta = g = 0$ and $\omega >> f_0$ this reduces to the familiar three-dimensional Helmholtz equation describing the propagation of acoustic waves

$$\rho^{(0)}\partial_3\left[(\rho^{(0)})^{-1}\partial_3\tilde{p}\right] + \nabla_\perp^2\tilde{p} + \omega^2 c_2^{-2}\tilde{p} = 0 \quad , \tag{2.44}$$

which for constant density is just

$$\left(\nabla^2+\omega^2 c_2^{-2}\right)\tilde{p} = 0 \quad . \tag{2.45}$$

If we further assume the sound speed to be only a function of depth

$$c(\underline{x},t) = c_0(z) \quad , \tag{2.46}$$

we can introduce an additional Fourier transform in x and y

$$\bar{p}(z,k_\perp,\omega) = \int \exp(-ik_\perp \cdot x_\perp)\tilde{p}(\underline{x},\omega)dx_\perp \quad , \tag{2.47}$$

where $k_\perp$ is the horizontal wavenumber (transverse to the z-direction). The resulting one-dimensional equation follows from (2.43). For $\beta = 0$ it is

$$\rho^{(0)}\left[(\rho^{(0)})^{-1}\bar{p}'\right]' + (\omega^2-N^2)^{-1}2NN'\bar{p}' + \left[\omega^2 c_0^{-2}+2gc_0^{-3}c_0' + k_\perp^2(N^2-\omega^2)\left(\omega^2-f_0^2\right)^{-1}-2gNN'c_0^{-2}(\omega^2-N^2)\right]\bar{p} = 0 \quad , \tag{2.48}$$

and is analogous to an equation on displacements derived and thoroughly discussed by TOLSTOY [2.12], but we have retained several additional terms. Again for $g = 0$ and $\omega >> f_0$ we get the one-dimensional Helmholtz equation

$$\rho^{(0)}\left[(\rho^{(0)})^{-1}\bar{p}'\right]' + \left(\omega^2 c_0^{-2}-k_\perp^2\right)\bar{p} = 0 \quad , \tag{2.49}$$

which, for constant density is just

$$\bar{p}'' + \left(\omega^2 c_0^{-2}-k_\perp^2\right)\bar{p} = 0 \quad , \tag{2.50}$$

whose solution and analytic properties as a function of $k_\perp$ are discussed in Sect.2.2.1.

As a final example, the internal wave equation follows in three-dimensions with the assumption of an incompressible fluid ($\partial_i u_i=0$) in (2.37). This yields (with a Fourier transform in time)

$$\tilde{p} = -ig\omega^{-1}\rho^{(0)}\tilde{u}_3 \quad , \tag{2.51}$$

where $\omega$ is now the internal gravity wave frequency parameter. Dropping the $c_2$ terms in (2.43), substituting (2.51) into the result, and defining a further Fourier transform in $x_\perp$ yields (for $\beta$=0) the one-dimensional equation

$$(\rho^{(0)})^{-1}\left(\rho^{(0)}\bar{u}_3'\right)' + g\left[(\rho^{(0)})^{-1}\rho^{(0)\prime}\right]'(\omega^2-N^2)^{-1}\bar{u}_3' + \left\{k_\perp^2(N^2-\omega^2)\left(\omega^2-f_0^2\right)^{-1} + \left[(\rho^{(0)})^{-1}\rho^{(0)\prime}\right]' + 2NN'\left[(\rho^{(0)})^{-1}\rho^{(0)\prime}\right](\omega^2-N^2)^{-1}\right\}\bar{u}_3 = 0 \quad , \tag{2.52}$$

where we have explicitly written the Väisälä frequency term in the first derivative of $\bar{u}_3$. Replacing the explicit density terms in (2.52) by a constant (equivalent to the Boussinesq qpproximation) yields the usual internal-wave equation (Sect.5.3.3, [2.13])

$$\bar{u}_3'' + k_\perp^2(N^2-\omega^2)\left(\omega^2-f_0^2\right)^{-1}\bar{u}_3 = 0 \quad . \tag{2.53}$$

Solutions of (2.53) have been used in acoustic propagation studies in the following way. The Helmholtz equation for the acoustic pressure field is given by a two-dimensional (cylindrical range r and depth z) version of (2.45). The parabolic approximation is then used (Sect.2.2.2) and the sound speed is composed of two parts, a deterministic refractive part, and a random part $\delta c$. The latter is related to the vertical displacement of the fluid and hence to $\bar{u}_3$. More precisely, it is given by a superposition of the eigenfunctions of (2.53) calculated from a boundary-value problem. The coefficients of the expansion are random variables. The ensemble of realizations of $\delta c$ is found by letting each mode progress harmonically in time with its corresponding eigenfrequency $\omega$ of (2.53), specific realizations being labeled by values of the time. It can be shown that these internal-waves induced fluctuations of the sound speed cause significant variation in sound intensity, and that for fixed source and receiver, they are the principle source of fluctuations within a frequency interval of cycles per day and cycles per hour [2.14]. This is discussed further in Chap.7.

We have ignored the other fluid equations on temperature, entropy, or salinity in directly deriving equations on acoustic quantities. We essentially consider these properties to be self-consistent and assume that their effects can be empirically included in our equation, via the sound speed. In practice this is what is done and the sound speed is derived from polynomial equations of the general form [2.15]

$$c(\underline{x},t) = c(T,S,\rho) \quad , \tag{2.54}$$

where T is temperature and S salinity. Their oceanographic properties are discussed further in Chap.5.

### 2.1.5 Sound Speed

We have already encountered many different "sound speeds", from purely depth-refractive to those depending on all the spatial coordinates and time. We also mentioned the empirical method of calculating sound speed from temperature and salinity. It is the functional dependence on space and time which concerns us. Our general model of the sound speed field can be written as

$$c(\underline{x},t) = c_0(z) + c_1(\underline{x}) + c_2(\underline{x},t) \quad , \tag{2.55}$$

where $c_0(z)$ is the deterministic refractive term, $c_1(\underline{x})$ a sound speed correction due to mesoscale phenomena such as eddies and fronts, which is modelled deterministically, and $c_2(\underline{x},t)$ the random component used, for example, to model internal wave phenomena, as we previously noted. The approximate size scales are that $c_0 \simeq 1500$ m s$^{-1}$, $c_1$ smaller than $c_0$ by a factor $10^{-2}$, and $c_2$ smaller than $c_0$ by a factor $10^{-4}$. Further remarks on the spatial scales involved in this sound speed breakup can be found in Chap.5.

Problems involving only $c_0(z)$ are found in Sect.2.2.1, extensions to sound speeds which are functions of two dimensions in Sect.2.2.2, with remarks extending to three dimensions in Sect.2.2.3, and examples including the random case in Sect.2.3. Computational methods for one- and two-dimensional problems are in Chap.3.

## 2.2 Propagation in Deterministic Media

In this section we discuss the solution and properties of the partial differential equations for the sound field where the medium properties and boundary conditions are deterministic. The examples we present are simple in order to illustrate as many general results as possible, and they are listed in terms of the coordinate dimensionality of the sound speed. For one-dimension, the Pekeris waveguide (a two-layer fluid) is used to illustrate the discrete and continuous spectral content of the boundary value problems encountered in mode theory, and the different representations available for the solution. We further include a method of finding the solution of the problem for certain functional sound speeds, and for solving a simple inversion problem.

In two dimensions, we illustrate a well-known high frequency method, ray theory, and some solvable examples of it, as well as a much more general method, the parabolic approximation, and a correction to it. It is also possible to find exact solutions in two dimensions using conformal mapping. In terms of routine use, two-dimensional problems are the state of the art. Only a brief mention of the interesting current research in three-dimensional models is given.

### 2.2.1 One-Dimensional Problems

Here we discuss the solution of propagation problems when the sound speed is a function of only the depth dimension. The Green's function G (or velocity potential) in cylindrical coordinates $\underline{r} = (r,z)$ for a point source at $(0,z')$ satisfies the equation

$$r^{-1}\partial_r(r\partial_r G) + \partial_z^2 G + k^2n^2(z)G = -\delta(r)\delta(z-z')/2\pi r \quad , \tag{2.56}$$

where $n(z) = c/c_0(z)$ is the dimensionless index of refraction, c an arbitrary sound speed, $c_0(z)$ the sound speed at depth z, and $k = \omega/c$ where $\omega$ is the circular frequency of sound. The harmonic time dependence $\exp(-i\omega t)$ is suppressed throughout.

Since n is only a function of z, a separable solution of (2.56) can be written in the form [2.16,17]

$$G(\underline{r},\underline{r}') = k^2(2\pi i)^{-1} \int_{C_j} G_1(r;\lambda)G_2(z,z';\lambda)d\lambda \quad , \tag{2.57}$$

where the $C_j$ contour encloses all the singularities of the $G_j$ (j=1,2) on the proper sheet of the complex $\lambda$-plane, $C_1$ in a clockwise manner and $C_2$ counterclockwise. Here $G_1$ and $G_2$ satisfy the equations

$$r^{-1}(rG_1')' = -k^2\lambda G_1 - \delta(r)/2\pi r \quad , \tag{2.58}$$

and

$$G_2'' + k^2n^2(z)G_2 = k^2\lambda G_2 - \delta(z-z') \quad , \tag{2.59}$$

where $\lambda$ is a separation constant. We assume that as $r \to \infty$, $G_1$ behaves like an outgoing wave so that the solution of (2.58) is the outgoing Hankel function

$$G_1(r;\lambda) = (i/4)H_0^{(1)}(k\xi r) \quad , \quad \lambda = \xi^2 \quad , \tag{2.60}$$

with $0 < \arg\lambda < 2\pi$ or $0 < \arg\xi < \pi$. On the complex $\xi$-plane this has a branch point at $\xi = 0$. The restriction $0 < \arg\xi < \pi$ is the radial radiation condition, $\mathrm{Im}\{\xi\} > 0$. We now discuss the analytic properties of $G_2$ in a simple canonical model.

### *a) The Pekeris Waveguide*

Consider the problem of the two-layer liquid half-space illustrated in Fig.2.2 and originally discussed by PEKERIS [2.18]. We wish to solve (2.59) where $\lambda = \xi^2$ and

$$\begin{aligned} n(z) &= n_1 = c/c_1 \quad 0 \leq z \leq d \quad , \\ &= n_2 = c/c_2 \quad z > d \quad . \end{aligned} \tag{2.61}$$

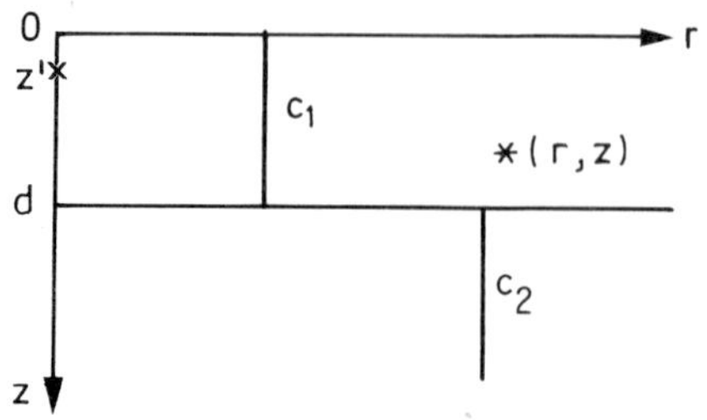


Fig. 2.2. Pekeris waveguide with constant sound speeds ($c_1<c_2$), a point source at (0,z'), receiver at (r,z) and the bottom extending to infinity

The $n_j$ are (dimensionless) constants and $n_2 < n_1$ or $c_2 > c_1$. In the interior (i) of the waveguide ($0 \leq z \leq d$) the solution is

$$G_2^i(z,z';\xi^2) = \alpha\left(z_<,\xi^2\right)\left[k\mu_1\Delta(\xi)\right]^{-1} \cdot \left[\mu_1\beta(d-z_>,\xi^2)-i\mu_2\alpha(d-z_>,\xi^2)\right] \quad , \tag{2.62}$$

where $\alpha(z,\xi^2) = \sin(k\mu_1 z)$, $\beta(z,\xi^2) = \cos(k\mu_1 z)$, $\mu_j = \left(n_j^2-\xi^2\right)^{1/2}$, and $z_<(z_>)$ are the lesser (greater) of z and z'. Here

$$\Delta(\xi) = \mu_1\beta(d,\xi^2) - i\mu_2\alpha(d,\xi^2) \quad , \tag{2.63}$$

and we assume a soft surface, $G_2^i = 0$ at $z = 0$. In the exterior (e) of the waveguide ($z>d$) the solution is

$$G_2^e(z,z';\xi^2) = \alpha(z',\xi^2)[k\Delta(\xi)]^{-1} \exp[ik(z-d)\mu_2] \quad . \tag{2.64}$$

Note that for $\mu_2 = \mu_{2r} + i\mu_{2i}$, both $\mu_{2r} > 0$ and $\mu_{2i} > 0$ in order to satisfy the radiation condition. Equations (2.62) and (2.64) were derived using continuity of $G_2$ and its normal derivative at the interface $z = d$. We can thus write the integral representation (2.57) as

$$G(\underline{r},\underline{r}') = (k^2/4\pi) \int_C H_0^{(1)}(k\xi r)G_2(z,z';\xi^2)\xi\, d\xi \quad , \tag{2.65}$$

where the contour C is chosen to enclose (in a counter-clockwise direction) all the singularities of $G_2$ on the proper Riemann sheet of the complex $\xi$-plane (i.e., satisfying the radiation condition). The singularities are found from the denominator function $\Delta(\xi)$ in (2.63). It has branch points at $\xi = \pm n_2$ when $\mu_2 = 0$ and poles at points $\xi_j$ when $\Delta(\xi_j) = 0$. The curves on which the branches lie can be found if we give $n_2$ a small positive imaginary part, i.e., let $n_2 \to n_2 + i\varepsilon$. It can then be easily seen that $\mu_{2i} = 0$ when $\xi_r\xi_i = n_2$ which defines the hyperbola shown in Fig.2.3. The branch cuts start at $\pm(n_2+i\varepsilon)$ and must be asymptotic to $\mu_{2r} > 0$ such that we have a true outgoing wave. The asymptotes are the positive and negative parts of the imaginary axis. The original integral must lie within the shaded region. By choosing the branch cuts along the curve $\mu_{2i} = 0$ we ensure that on the top (physical) sheet $\mu_{2i} > 0$ and on the second (unphysical) sheet $\mu_{2i} < 0$.

We define the sheet structure more carefully as follows. Define angles $\theta$ and $\phi$ by

$$n_2 - \xi = |n_2-\xi| \exp(i\theta) \quad , \tag{2.66}$$

$$n_2 + \xi = |n_2+\xi| \exp(i\phi) \quad , \tag{2.67}$$

so that

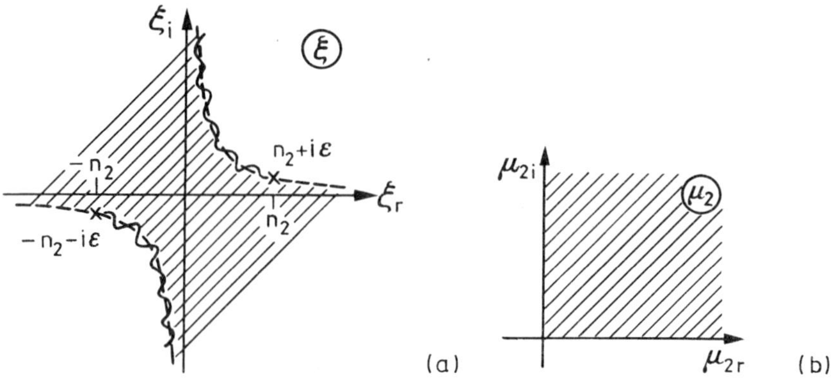


Fig. 2.3. (a) Branch cuts (wavy lines) and the hyperbola (dashed lines) on which they lie when $n_2$ has a small positive imaginary part $\varepsilon$, and the shaded region ($\mu_{2r}>0$, $\mu_{2i}>0$) in (a) the complex $\xi$-plane and (b) the complex $\mu_2$-plane

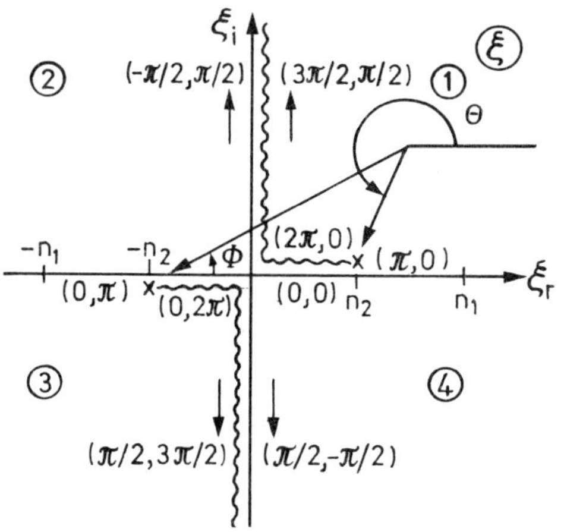


Fig. 2.4. Definition of $\mu_2=(n_2^2-\xi^2)^{1/2}$ in the complex $\xi$-plane illustrating the sheet where $\mu_{2i}>0$, various $(\theta,\phi)$ combinations, and slightly offsetting the branch cuts (wavy lines) from the axes for clarity

$$\mu_2 = |\mu_2|\{\cos[(\theta+\phi)/2] + i\sin[(\theta+\phi)/2]\} \quad . \tag{2.68}$$

The proper Riemann sheet is defined by the criterion that $\mu_{2i} \sim \sin[(\theta+\phi)/2] > 0$. This can be done [2.19] by using the definition of $\mu_2$ illustrated in Fig.2.4.

The poles of $G_2$ are the points $\xi_j$ where $\Delta = 0$. From (2.63) and (2.68) it can easily be seen that the only poles on the top sheet occur on the real axis when $n_2^2 < \xi^2 < n_1^2$. In this region $\theta + \phi = \pi$ and $\Delta = 0$ is given by

$$\cot(kd\mu_1) = -|\mu_2|\mu_1^{-1} \quad . \tag{2.69}$$

The two sides of (2.69) are graphed in Fig.2.5 where, as an example, three points of intersection are shown. It can be shown that there are only a finite (though possibly large) number of intersections. Since $\mu_{2i} > 0$ (and the $\xi_j$ also have a small

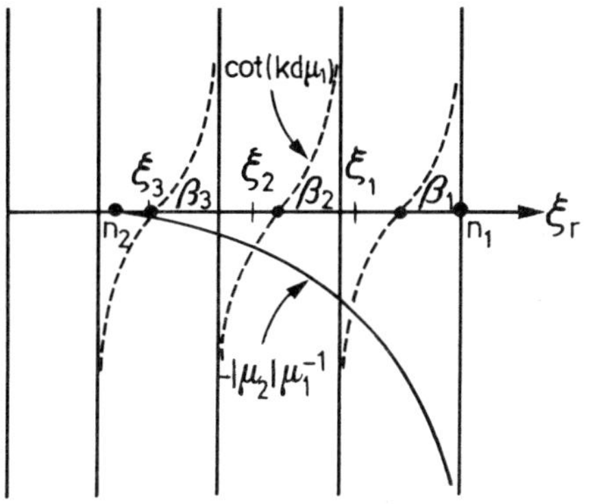


Fig. 2.5. Plot of the two sides of (2.69) for $\xi_r > 0$ showing three points of intersection $\xi_j$ and the zeroes of $\cos(kd\mu_1)$ given by $\beta_j$

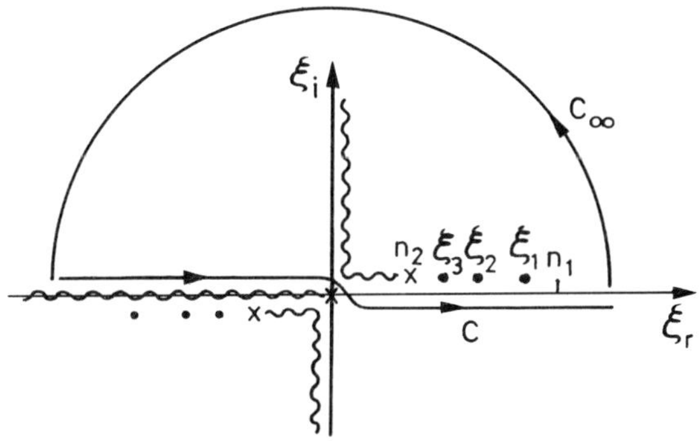


Fig. 2.6. Contour C closed in the upper half plane ($C_\infty$) and enclosing the $n_2$ branch and poles at $\xi_j$ with the branch at $\xi = 0$ due to the Hankel function

positive imaginary part) their residue contributions contain a finite amount of energy. They are thus referred to as proper modes. The low order modes are those close to $n_1$, the high order modes close to $n_2$. The points $\beta_j$ (j=1,2,3) shown in Fig.2.5 are the zeroes of $\cos(kd\mu_1)$ given by

$$\beta_j = \left\{n_1^2-[(2j+1)\pi/2kd]^2\right\}^{1/2} \quad . \tag{2.70}$$

We have illustrated parameter values in depth and frequency which yield three real $\beta_j$. Clearly, the larger d, or the higher the frequency, the more real $\beta_j$ occur and hence the more proper modes occur. We ignore the poles at $-\xi_j$ since these yield incoming radial waves.

Using these results the contour C of (2.65) is shown in Fig.2.6. It proceeds through the shaded region of Fig.2.3 and can be closed in the upper half plane ($C_\infty$) enclosing the poles at $\xi_j$ and the branch cut at $n_2$. In this region $\mu_{2i} > 0$ (decaying waves in depth) and $\xi_i > 0$ (outgoing radial waves). The contribution due to $C_\infty$ vanishes by the radiation condition.

The contour $C + C_\infty$ is equivalent to the contour shown in Fig.2.7. Thus (2.65) can be written as

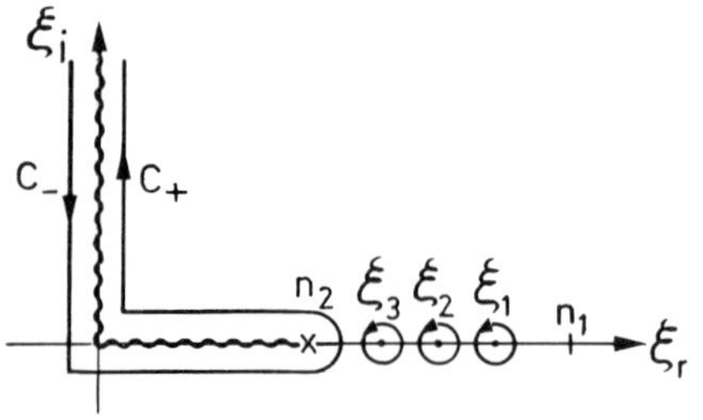


Fig. 2.7. Contour equivalent to $C+C_\infty$ in Fig.2.6

$$G(\underline{r},\underline{r}') = (k^2/4\pi)\left\{\int_C H_0^{(1)}(k\xi r)\mathrm{DISC}\left[G_2(z,z';\xi^2)\right]\xi\, d\xi \right.$$
$$\left. +2\pi i \sum_{j=1}^{3} H_0^{(1)}(k\xi_j r)\xi_j \mathrm{RES}\left[G_2(z,z';\xi^2)\right]\right\} \quad , \tag{2.71}$$

where $\mathrm{DISC}[G_2] = G_2|_{C_-} - G_2|_{C_+}$ and $\mathrm{RES}[G_2]$ is the residue of $G_2$ at the $\xi_j$ poles. Explicitly, in the interior of the waveguide, we get

$$\mathrm{DISC}\left[G_2^i\right] = 2i\mu_2[k\Delta(\xi)\Delta^*(\xi)]^{-1}\alpha\left(z_<,\xi^2\right)\alpha\left(z_>,\xi^2\right) \quad , \tag{2.72}$$

and at $\xi = \xi_j$

$$\mathrm{RES}\left[G_2^i\right] = |\mu_2|\alpha\left(z_<,\xi_j^2\right)\alpha\left(z_>,\xi_j^2\right)\left[k\xi_j(1+kd|\mu_2|)\right]^{-1} \quad . \tag{2.73}$$

Analogous formulae can be found for $z > d$.

We have thus been able to express the solution of our problem via (2.71) in terms of a discrete set of proper modes (PM) and a continuous spectral part corresponding to the branch line integral (BLI). The normal mode (NM) terminology corresponds to a spectral decomposition which is complete, and hence must include both discrete and continuous contributions. It can be shown [2.20] that the discrete spectral content corresponds in a ray picture to rays which undergo either total internal refraction or if they strike the bottom do so at angles greater than the critical angle $\theta_c = \sin^{-1}(c_1/c_2)$. The continuous spectral part corresponds to rays striking the bottom at angles less than critical (closer to the normal) so that they lose more energy into the bottom. The latter contribute a fine structure to the acoustic field which diminishes in range faster than $r^{-1/2}$. For long range propagation only the discrete modes are important. If there were no bottom region, i.e., if the interface at $z = d$ were perfectly reflecting, no branch point contribution would occur, the complete representation being given exactly by the discrete spectrum. Physically the continuous spectrum means that energy can propagate to infinity out of the waveguide.

Our method of drawing the cut corresponds to that of EWING, JARDETZKY, and PRESS (EJP) [2.20]; see also STICKLER [2.21]. This has the advantage that the entire top

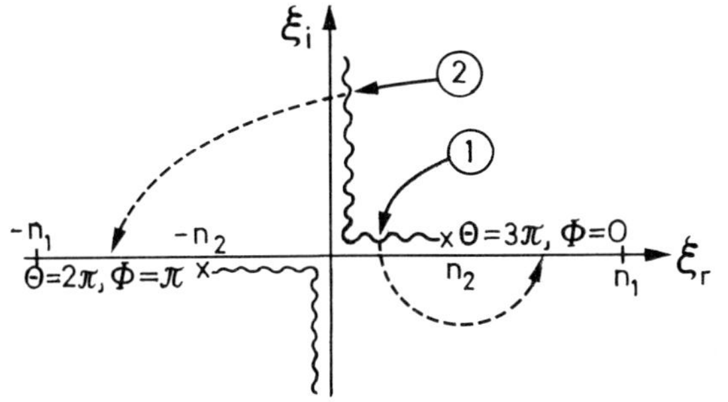


Fig. 2.8. Two possible analytic continuation paths from the first sheet (solid line) to pole regions where $\theta + \phi = 3\pi$ on the second sheet (dashed lines)

sheet in the complex plane corresponds to the physical region, i.e., to solutions which satisfy the radiation condition. PEKERIS (P) [2.18] originally drew the cut vertically upwards from $n_2$. Rotating the EJP cut into the P cut has the effect of exposing some of our second sheet to the closed contour integral we used. To determine the consequences of this we must ask whether there are any singularities on the second sheet which are "close" to the physical region, and which can appear in our contour integral when we rotate the EJP cut. The word "close" is defined in the sense of analytic continuation and the answer is yes. There can occur complex-valued poles on the second sheet but not on the first. We next show where these poles occur.

We saw from (2.69) that $\theta + \phi = \pi$ yielded the proper modes. The angular combination $\theta + \phi = 3\pi$ also yields solutions of $\Delta = 0$, but these are on the second sheet. The equation is

$$\cot(kd\mu_1) = |\mu_2|\mu_1^{-1} \quad , \tag{2.74}$$

and two continuation paths to these poles are illustrated in Fig.2.8. From Fig.2.5 we see that the plot of (2.74) gives pole positions near those of the proper modes but of course on the second sheet. Poles in the region $n_2 < \xi < n_1$ in Fig.2.8 are near the physical region especially if they are also near cutoff (near $n_2$).

The asymptotic value of the poles on the second sheet can be found by writing $\Delta = 0$ from (2.63) as an exponential. This has periodic solutions which, for large values of $\xi$, are (n=1,2,3,...)

$$\pm\xi_n = \left(\frac{1}{2} + n\pi i\right)(kd)^{-1} \quad . \tag{2.75}$$

These poles occur in the first and third quadrants. We have that on the second sheet $\mu_{2i} < 0$, and in the first quadrant $\mu_{2r} > 0$, opposite in sign to $\mu_{2r}$ on the first sheet, and $\mathrm{Im}\{\xi_n\} > 0$. Hence these pole contributions yield residue terms which decay exponentially in range, grow exponentially in depth, and result in a modal contribution with infinite energy. They are thus called improper or leaky modes (LM), the latter terminology due to the fact that they leak energy out of the waveguide. They occur

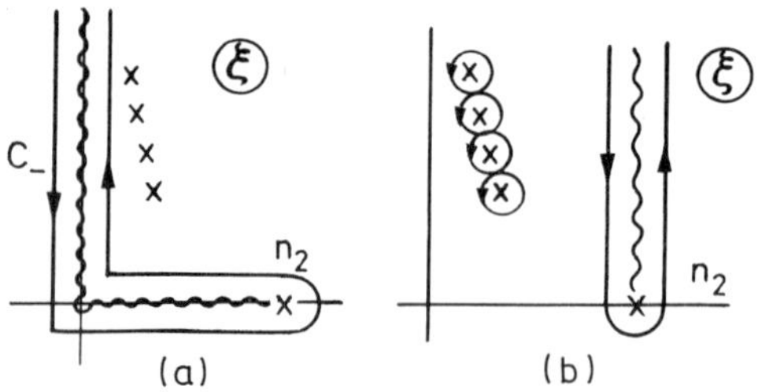


Fig. 2.9a and b. Contour integrals around the Ewing, Jardetzky, Press cut (a) and the Pekeris cut (b) with the leaky poles (x) on the second sheet in (a) and contributing to the field representation in (b)

close to the physical region (i.e., the top sheet) in the sense that the analytic continuation distance from the $C_-$ integral over the continuous spectrum is small. The EJP and P contour integrals are shown in Fig.2.9. Shifting the branch cut thus yields the relation

$$\int(\mathrm{EJP}) = \int(\mathrm{P}) + \sum(\mathrm{LM}) \quad . \tag{2.76}$$

In an example BARTBERGER [2.22] has shown that at 50 Hz the leaky modes contribute very little, and the EJP and P branch line integrals are essentially equal. At 100 Hz the P branch line integral contributes very little and the EJP branch line integral is approximated by a sum of leaky poles. The solutions based on the two different branch cuts yield the same results.

The leaky-mode terminology is consistent with that of others [2.21,23]. Several authors refer to these poles as virtual modes [2.24,25]. The latter agrees with the corresponding virtual or antibound state terminology in quantum mechanics only for a very weak potential [2.26]. In general, the quantum mechanical virtual states have different analytic properties than the virtual modes, in particular they do not exhibit the resonance behavior common to the latter and to our leaky modes. The analytic treatment we describe has been extended to the Furry duct model [2.24] where the sound speed is bilinear and results in sound being propagated in a surface duct.

Finally, we have discussed a static solution, where kd is constant. As we indicated, if kd increases, more proper modes could occur. Analytically, as kd increases, the leaky-wave poles migrate from the second sheet, through the branch point threshold at $n_2$ and onto the first sheet as proper modes.

We conclude that there are several alternative representations for the solution. The most common are summarized next.

### *b) Alternative Representations*

There are several alternative representations for the Green's function. The NM (normal mode) and PM (proper mode) representations follow directly from (2.71-73). Inside the waveguide they can be written as

$$G_{NM}(\underline{r},\underline{r}') = (i/4) \int_C H_0^{(1)}(k\xi r)\tau\left(z_>,\xi^2\right)\tau\left(z_<,\xi^2\right)\xi\, d\xi + G_{PM}(\underline{r},\underline{r}') \quad , \tag{2.77}$$

where

$$\tau(z,\xi^2) = \left[2k\mu_2/\pi\Delta(\xi)\Delta^*(\xi)\right]^{1/2}\alpha(z,\xi^2) \quad , \tag{2.78}$$

and

$$G_{PM}(\underline{r},\underline{r}') = (i/4) \sum_{j=1}^{N} H_0^{(1)}(k\xi_j r)\psi_j\left(z_>,\xi_j^2\right)\psi_j\left(z_<,\xi_j^2\right) \quad , \tag{2.79}$$

written for an arbitrary number N of PMs and with

$$\psi_j(z,\xi^2) = \left[2k|\mu_2|/(1+kd|\mu_2|)\right]^{1/2}\alpha(z,\xi^2) \quad . \tag{2.80}$$

The NM representation is the complete spectral representation, and the PM is the discrete part. The PM sum is also called the residue series representation by others [2.24] but since the leaky poles can be interpreted as residue contributions depending on the choice of cut we prefer the terminology "proper modes". Details on how different mode programs treat these spectral contributions can be found in Chap.3 and in [2.21,27].

The Fourier-Bessel (FB) representation follows directly from (2.65) by folding the integral from $-\infty$ to $\infty$ into an integral from 0 to $\infty$ or by choosing the clockwise contour $C_1$ in the original representation (2.57). The result is

$$G_{FB}(\underline{r},\underline{r}') - (k^2/2\pi) \int_0^\infty J_0(k\xi r)G_2(z,z';\xi^2)\xi\, d\xi \quad , \tag{2.81}$$

where $J_0$ is the zero-order Bessel function. Equation (2.81) is the basis for the fast field program discussed in Chap.3.

The above three representations are the most common. We have written them in the plane of the longitudinal wave number $\xi$. If in (2.59) we replace $\lambda$ by $1-\kappa^2/k^2$ then the differential equation for $G_2$ is

$$G'' + \left\{\kappa^2 - k^2[1-n^2(z)]\right\}G = -\delta(z-z') \quad ,$$

which is like the s-wave Schrödinger equation in non-relativistic quantum mechanics. Its analytical properties as a function of the quantum mechanical momentum ($\kappa$) and energy ($\kappa^2$) have been extensively studied [2.26,28]. This has also been done in acoustics in the transverse wave number or $\kappa$-plane [2.29] and in the $\kappa^2$-plane [2.24] and these lead to further representations of the acoustic field. As in quantum mechanics where the asymptotic properties of the potential govern the representation, here the asymptotic properties of the index of refraction (the "termination of the wave guide") are crucial to the specific form of the representation.

Finally, in practice the BLI must be evaluated numerically using discrete methods or approximated by a finite sum of leaky poles. The latter has been generalized to an expansion of the Mittag-Loeffler type [2.25,26,29].

### *c) Solvable Profiles*

There is a general method of finding one-dimensional sound speed profiles which are solvable, i.e., profiles for which the differential equation describing the velocity potential $\phi$ has known solutions in terms of special functions. In one dimension, $\phi$ satisfies

$$\phi''(z) + K(z)\phi = 0 \quad , \qquad (2.82)$$

where

$$K(z) = \omega^2 c_0^{-2}(z) - k_\perp^2 \quad . \qquad (2.83)$$

This is similar to (2.50). The question is: for which $K(z)$ and hence $c_0(z)$ can (2.82) be solved in terms of special functions?

Begin with a general second-order linear differential equation

$$u'' + m(r)u' + \ell(r)u(r) = 0 \quad , \qquad (2.84)$$

where r is a dummy variable. Transform both independent and dependent variables

$$r = f(z) \quad , \quad u(r) = g(z)\phi(z) \quad , \qquad (2.85)$$

where $g(z) \neq 0$, and the resulting equation is

$$\phi'' + A(z)\phi + B(z)\phi = 0 \quad , \qquad (2.86)$$

where

$$A(z) = 2g'/g + M(z)f' - f''/f' \quad , \qquad (2.87)$$

$$B(z) = g''/g + L(z)(f')^2 + [M(z)f'-f''/f']g'/g \quad , \qquad (2.88)$$

$$M(z) = m[f(z)] \quad , \qquad (2.89)$$

and

$$L(z) = \ell[f(z)] \quad . \qquad (2.90)$$

If we identify (2.84) with a canonical-differential equation whose special-function solutions are known, for example the hypergeometric equation, so that

$$m(r) = [d-(a+b+1)r]/r(1-r) \quad ,$$

and

$$\ell(r) = -ab/r(1-r) \quad , \tag{2.91}$$

where a, b, and d are constants, then identify (2.86) with (2.82) so that

$$A(z) = 0 \quad , \quad B(z) = K(z) \quad , \tag{2.92}$$

we can then integrate (2.87) to get

$$f' = \alpha g^2(z) f^d(z)[1-f(z)]^{a+b-d+1} \quad , \tag{2.93}$$

where $\alpha$ is an integration constant. Using (2.92), (2.87,88) become

$$K(z) = g''/g - ab(f')^2/f(1-f) - 2(g'/g)^2 \quad . \tag{2.94}$$

We thus have the sound-speed profile given by two unknown functions f and g, and the functions related by (2.93). We must choose one of the functions. For example, if we choose

$$f(z) = \frac{1}{2}[1+\tanh(z/2)] \quad , \tag{2.95}$$

we can find g(z) from (2.93) and get

$$K(z) = a_1 + a_2 \tanh(z/2) + a_3 \operatorname{sech}^2(z/2) \quad , \tag{2.96}$$

where $a_1$, $a_2$, and $a_3$ are constants. Equation (2.96) is called the Epstein profile, and its solution, in terms of hypergeometric functions, has been discussed along with general methods of solution of other solvable profiles by DEAVENPORT [2.17]. Other solvable profiles involving hypergeometric, confluent hypergeometric, Bessel, and Mathieu functions, and solutions to Hill and Lame differential equations have been extensively tabulated as solvable potentials for the Schrödinger equation in quantum mechanics [2.30] and solvable profiles in electromagnetic propagation problems [2.31].

*d) Inverse Propagation*

We have been discussing the direct propagation problem, e.g. given the differential equation for the pressure field (including the sound speed) and the boundary conditions, find the field in the waveguide. There are several types of inversion problems [2.32]. One of them is finding the sound speed from other data discussed in Chap.6. Another is finding the "near" field from the "far" field, an example of which we present here [2.33]. Assume the velocity potential $\phi$ satisfies the differential equation (in Cartesian coordinates)

$$\phi_{xx} + \phi_{zz} + k^2 n^2(z)\phi = 0 \quad . \tag{2.97}$$

The waveguide we describe is shown in Fig.2.10. The field is known on the plane $x = x_2$ and is to be found on the plane $x = x_1$. Direct propagation proceeds from $x_1$ to $x_2$.

Equation (2.97) has a separable solution and is assumed given by an infinite discrete spectral representation

$$\phi(x,z) = \sum_{j=0}^{\infty} A_j \psi_j(z) \exp(ikxm_j) \quad , \tag{2.98}$$

where $\psi_j$ satisfies the equation

$$\psi_j'' + k^2[\mu_j - q(z)]\psi_j = 0 \quad , \tag{2.99}$$

with $q(z) = 1 - n^2(z)$, and $m_j = (1-\mu_j)^{1/2}$ with $\mathrm{Im}\{m_j\} \geq 0$. Using orthogonality we can write $\phi$ as the sum of the two terms

$$\phi(x_1,z_1) = \phi_1(x_1,z_1) + \phi_2(x_1,z_1) \quad , \tag{2.100}$$

where (m=1,2)

$$\phi_m(x_1,z_1) = \int K_m(x_1,z_1;x_2,z_2)\phi(x_2,z_2)dz_2 \quad . \tag{2.101}$$

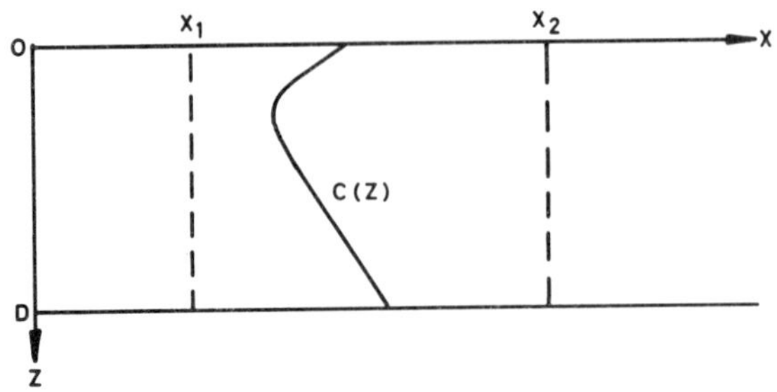


Fig. 2.10. Waveguide example for inverse sound propagation with the field known on the $x = x_2$ plane and unknown on the $x = x_1$ plane

The kernels $K_m$ can be written as

$$K_1(x_1,z_1;x_2,z_2) = -\partial G(x_1,z_1;x_2,z_2)/\partial x_2 \quad , \tag{2.102}$$

where G is the Green's function

$$G(x_1,z_1;x_2,z_2) = \sum_{j=0}^{\infty} \psi_j(z_1)\psi_j(z_2)(2ikm_j^*)^{-1} \exp(-ikm_j^*|x_1-x_2|) \quad , \tag{2.103}$$

and

$$K_2(x_1,z_1;x_2,z_2) = 2\sum_{j=J+1}^{\infty} \psi_j(z_1)\psi_j(z_2) \sinh\left[k(\mu_j-1)^{1/2}(x_2-x_1)\right] \quad , \tag{2.104}$$

where $\mu_J < 1$ and $\mu_{J+1} > 1$. The kernel $K_1$ is bounded and expresses the results at $x_1$ of waves at $x_2$ reversing both their direction of propagation and decay. Its relation to the Green's function shows that $\phi_1$ is the inverse of a Rayleigh-type diffraction formula. The kernel $K_2$, on the other hand, is singular since the summation goes to $\infty$, and the full problem is thus ill-posed because a small change in the initial condition $\phi(x_2,z_2)$ can produce a large change in $\phi_2(x_1,z_1)$ by (2.101). Physically these are the evanescent wave terms which now grow exponentially as their direction is reversed.

Neglect of $\phi_2$ means neglect of large wavenumber, short wavelength terms and hence an inability to gather information on an obstacle or process with a characteristic length smaller than $k^{-1}(\mu_{J+1}-1)^{1/2}$. Since high-frequency sound is rapidly attenuated, neglect of $\phi_2$ probably does not affect propagation results very much. Further remarks on inversion problems are in Chap.6.

### 2.2.2 Two-Dimensional Problems

In this section we present methods capable of treating problems where the sound speed is a deterministic function of two variables. Whereas the one-dimensional problems we discussed were amenable to exact analytic treatment, the methods presented here are approximate. We first discuss ray theory and then a general class of profiles for which it is exactly solvable. Next we discuss the parabolic approximation, its meaning, and how its solution is related to the solution of the Helmholtz equation by an integral transform. A numerical approximation derived from the stationary phase approximation of the transform is compared to a normal mode solution and shown to be an improvement on the parabolic approximation. Lastly, and briefly, there is a class of analytically solvable models using conformal mapping which although synthetic, might be useful in checking approximation methods in two-dimensions. These latter models yield a modal solution for particular range-dependent profiles. We omit a discussion of general range-dependent normal-mode theory [2.34]. Other range-dependent models can be found in Sect.3.2.

*a) Ray Theory*

In two-dimensional cylindrical coordinates the Helmholtz equation is

$$\partial^2\psi/\partial r^2 + r^{-1}\partial\psi/\partial r + \partial^2\psi/\partial z^2 + k^2n^2(r,z)\psi = 0 \quad . \tag{2.105}$$

Assume a solution

$$\psi(r,z) = A(r,z)\exp[ik\theta(r,z)] \quad , \tag{2.106}$$

where the amplitude A and phase $\theta$ are real. Substitute (2.106) into (2.105) and write the result in inverse powers of k. Neglecting diffractive terms [2.35] gives the eikonal equation for the phase

$$|\underline{\nabla}\theta|^2 = n^2(r,z) \quad , \tag{2.107}$$

and a transport equation for the amplitude

$$A\nabla^2\theta + 2\underline{\nabla}A \cdot \underline{\nabla}\theta = 0 \quad . \tag{2.108}$$

The coefficient of $\underline{\nabla}A$ perpendicular to $\underline{\nabla}\theta$ has been neglected (diffraction), and large changes in amplitude along $\underline{\nabla}\theta$ may violate (2.108) if $\nabla^2\theta$ does not change as much. This latter occurs in regions near shadows, edges of obstacles, caustics or focal points. Neglect of the diffractive term is invalid near local sources. In general, (2.107) and (2.108) are valid for large k (high frequency) and in fact the wavelength should be much less than any significant length scale of the problem. For example, taking the gradient of (2.107) in regions where $\nabla^2\theta \ll 1$ yields $(kn)^{-1}|\underline{\nabla}n| \ll 1$ so that the fractional change in the index of refraction is small in a wavelength. This is violated near the ocean surface and bottom.

The advantage of ray theory is that the rays can be found from (2.107) independent of the amplitude, and the amplitude found by finding the intensity ($\sim A^2$) along each ray from (2.108) neglecting the coupling (diffraction) between rays. The intensity $I_j$ is defined as the energy per unit time crossing a unit surface element $d\sigma_j$ perpendicular to the ray. Energy conservation is defined as

$$I_i d\sigma_i = I_j d\sigma_j \quad . \tag{2.109}$$

For rays emitted from a point source or for rays crossing at a point (caustic) the above definition leads to infinite intensity since some $d\sigma_j$ is zero. It is possible to generate equations directly on the ray intensity [2.36], but the caustic problem remains.

Thus the lack of coupling between rays, advantageous in the tracing of intensity, leads to fundamental difficulties if the rays bunch together. Caustic corrections can be calculated using uniform asymptotic expansions [2.36-38], or modified ray theory using Airy functions [2.39,124], all of which methods still deal fundamentally with rays, or by going to wave theory. The relation of ray and wave theory is the same as the relation of classical and quantum mechanics [2.40] and the ray and wave mode solutions can be related via the Poisson summation (Sect.3.1.6). Ray effects in wave theory are also well understood [2.41].

A differential equation for the rays can easily be derived from the eikonal equation. It is

$$\frac{d}{ds}\left(n\frac{d\underline{r}}{ds}\right) = \underline{\nabla}n \quad , \tag{2.110}$$

where $\underline{r}$ is an arbitrary point on a ray, and s the arc length of the ray. Many simple solutions exist for a one-dimensional index of refraction. For example, the rays are straight lines for $n = \text{const.}$, ellipses for $n = z^{-1}$ and circles for $n = (a^2-z^2)^{1/2}$. The latter is the "harmonic oscillator" case.

There is also a method of solving (2.110) in two-dimensions with n having any of the following forms [2.42], $f(ar+bz+d)$, $f[(ar+b)/(dz+e)]$, and $f[(ar+b)(dz+e)]$ where f is an arbitrary function and a, b, d, and e are constant. Complex function theory can also be used to generate solvable two-dimensional models [2.43].

In addition, in one-dimension simple graphical methods can be used [2.44] to define ray angles relative to the horizontal, ray tracing has been done using power series expansions of the sound speed [2.45] and spline fits to it [2.46], with sound speeds having two turning points [2.47] and multidimensionally in complicated regions using ray invariants [2.48]. Ray theory has been derived using integral transform methods [2.49] and more fundamentally, ray equations have been derived in a general Riemannian space [2.50]. Also, we have been discussing examples where A and $\theta$ are real, but ray tracing can also be done using a complex form of Snell's law [2.51].

Our remarks on ray theory have been brief since the method has been amply treated in textbooks [2.35] and conference proceedings [2.52]. Further remarks and references can be found in Chap.3.

*b) Corrected Parabolic Approximation*

As we have noted, sound propagation in a two-dimensional waveguide is described by the solution of the Helmholtz equation

$$\psi_{rr} + (2a/r)\psi_r + \psi_{zz} + k^2K(r,z)\psi = 0 \quad , \tag{2.111}$$

where $\psi$ is the velocity potential, $k = \omega/c$, $\omega$ the frequency, c a reference sound speed and $K = [c/c(r,z)]^2$ with $c(r,z)$ the range and depth dependent sound speed. For $a = 0$, r and z are Cartesian coordinates, for $a = 1/2$ cylindrical coordinates, and $a = 1$, spherical coordinates.

Assume

$$\psi(r,z) = (kr)^{-a}\phi(r,z)\exp(ikr) \quad , \tag{2.112}$$

and substitute this in (2.111). The equation for $\phi$ is

$$\phi_{rr} + 2ik\phi_r + \phi_{zz} + k^2[K-1+a(1-a)(kr)^{-2}]\phi = 0 \quad . \tag{2.113}$$

The parabolic approximation (PA) consists in assuming the terms $\phi_{rr}(2ik\phi_r)^{-1}$ and $\phi/r^2$ to be small. Thus $\phi$ is approximated by a function p satisfying the parabolic partial differential equation

$$2ikp_r + p_{zz} + k^2(K-1)p = 0 \quad . \tag{2.114}$$

A history of the approximation, the justification for its validity, and the areas of current research are discussed by TAPPERT [2.53] who first introduced the method to ocean acoustics, and others [2.54-57]. The main advantage of the method is a numerical one, since (2.111) is an elliptic partial differential equation which must be numerically solved in the entire two-dimensional region simultaneously, while (2.114) is a parabolic partial differential equation which can be marched in the range dimension.

We illustrate some of the interpretations of the PA with some simple examples. If K is only a function of z, (2.113) is separable, $\phi = F(r)G(z)$, and the radial equation is

$$F'' + 2ikF' + \left(k_n^2-k^2\right)F = 0 \quad , \tag{2.115}$$

where $k_n$ is an eigenvalue of the depth equation. For each mode a solution can be written in the form $\exp(i\alpha r)$ where $\alpha_\pm = -k \pm k_n$. Computing the neglected term in the PA yields

$$\left|\phi_{rr}(2ik\phi_r)^{-1}\right| = \left|-\frac{1}{2} \pm k_n(2k)^{-1}\right| \ll 1 \quad . \tag{2.116}$$

For outgoing waves ($\alpha_+$) the inequality holds provided $k_n \simeq k$, i.e., in some narrow spectral region centered about k (which is arbitrary). For backscatter ($\alpha_-$) the inequality does not hold. Thus the PA is a narrow spectral approximation which neglects backscatter. It is also easy to illustrate the narrow angle interpretation of the PA for the free field case. If we assume that the K-1 term can be neglected, the Riemann or fundamental solution of (2.114) is

$$p^{(0)}(r,z) = r^{-1/2} \exp(ikz^2/r) \quad . \tag{2.117}$$

Again, computing the neglected term

$$p_{rr}^{(0)}\left(2ikp_r^{(0)}\right)^{-1} = -\frac{1}{4}(z/r)^2 - (4ikr)^{-1} \cdot \left[1+2(1+2ikz^2/r)(1+ikz^2/r)^{-1}\right] \quad . \tag{2.118}$$

which is small at long ranges provided the tangent of the angle, $\tan\theta = z/r$, is small. Note that the spectral method yielded a result dependent on k which could be freely chosen and related to an equivalent ray angle so as to give a narrow spread about any chosen direction, whereas the Riemann solution yielded a narrow angle approximation defined solely from the horizontal.

It is possible to exactly relate the solutions $\psi$ and p using an integral transform due to DESANTO [2.57]. It has the form

$$\psi(r,z) = A_0 r^{1-2a} \int_0^\infty p(t,z)R(r,t,z) \cdot \exp[(ik/2t)(r^2+t^2)]t^{a-3/2}dt \quad , \tag{2.119}$$

where $A_0$ is a constant and R satisfies the parabolic equation

$$\begin{aligned} &R_{rr} + 2[(1-a)r^{-1}+ikr/t]R_r + 2ikR_t + R_{zz} \\ &\quad + 2R_z[\partial_z p(t,z)]p^{-1}(t,z) = k^2[K(t,z)-K(r,z)]R \quad . \end{aligned} \tag{2.120}$$

The solution of (2.120) is formidable since it has higher dimensionality than (2.114) and also contains the solution of (2.114) as a coefficient. We do notice that the phase term of the integrand of (2.119) has a stationary point at $t = r$, and that at $t = r$ the right hand side of (2.120) vanishes provided R is finite at this point. The solution of the resulting equation is just a constant and since we have an unspecified constant $A_0$ we can choose $R = 1$ at $t = r$. We also note that for range-independent profiles, the right hand side of (2.120) vanishes exactly and $R = 1$ exactly. It thus appears plausible to look for an expansion of R about the stationary phase point since it is related to the modal solution.

We first discuss the modal solution. If $K(r,z) = K_1(z)$, then (2.114) is separable. Its solution is

$$p(t,z) = \sum_j p_j N_j(z) \exp\left[i\left(k_j^2-k^2\right)t/2k\right] \quad , \tag{2.121}$$

where the eigenfunctions $N_j$ satisfy the ordinary differential equation

$$N_j'' + \left[k^2K_1(z)-k_j^2\right]N_j = 0 \quad , \tag{2.122}$$

and boundary conditions which we do not need to specify. Substitute (2.121) into (2.119) with $R=1$ and note that the resulting integrals are Hankel functions. The result is

$$\psi(r,z) = i\pi A_0 \exp[i\pi(1-2a)/4] \cdot \sum_j p_j N_j(z)(rk_j/k)^{1/2-a} H^{(1)}_{1/2-a}(k_j r) \quad , \tag{2.123}$$

which is the eigenmodal solution of (2.111). For $a=0$ the integral transform representation was noted by POLYANSKII [2.58].

For the general case where K is a smooth function of both depth and range we can expand $p(t,z)$, $R(r,t,z)$ and $K(t,z)$ in a Taylor series about the stationary phase point $t=-r$, and provided we keep the depth dependence of the profile, simultaneously about the eigenmodal solution $R=1$. The resulting integrals are again Hankel functions and asymptotically, we can relate $\psi$ to an expansion in p and its range derivatives. The result yields the same lowest-order terms as the purely stationary phase argument in [2.57]. It was found numerically [2.59] from the general results in [2.57] that we could define an improvement to the PA, called the corrected parabolic approximation (CPA) of the form $\left(a=\frac{1}{2}\right)$

$$\psi_{CPA}(r,z) = (kr)^{-1/2}[p(r,z)+(ir/2k)p_{rr}(r,z)]\exp(ikr) \quad ; \tag{2.124}$$

where the second-range derivative $p_{rr}$ is calculated from a discrete approximation and included as a correction term at each range step. We compared the PA and CPA results quantitatively to a normal mode (NM) example using the phase-inclusive error function

$$E(r) = \left\{ J^{-1} \sum_{j=1}^{J} \left| \frac{\psi_{NM}(r,z_j)-\psi(r,z_j)}{\psi_{NM}(r,z_j)} \right|^2 \right\}^{1/2} \quad , \tag{2.125}$$

which is a depth averaged value as a function of range, where J is the total number j of discrete points $z_j$ in depth. Here $\psi$ is either the PA with the result shown in Fig.2.11a or the CPA shown in Fig.2.11b. It is known that the PA does not approximate the phase of the acoustic field well and it is seen that the CPA is a considerable improvement. The results shown were characteristic of other depth dependent sound speeds. Further details on the CPA can be found in [2.59]. Other methods of improving the accuracy of the PA are also available [2.60].

*c) Conformal Mapping*

Using a conformal-mapping technique it is possible to relate the solutions of the Helmholtz equation with sound speed $c(x,y)$, where x is range and y is depth to the solution of a parabolic equation whose sound speed varies in only the mapped depth coordinate [2.61]. Assume the velocity potential satisfies the Helmholtz equation

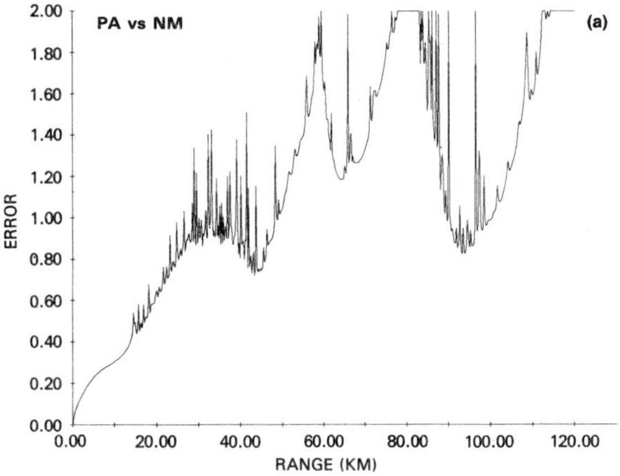


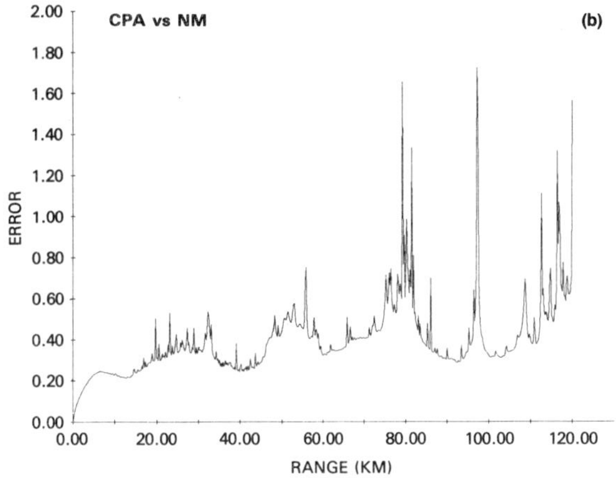


Fig. 2.11a and b. Error E as a function of range for (a) the parabolic approximation (PA) versus a normal mode (NM) example and (b) the corrected parabolic approximation (CPA) versus NM

$$\psi_{xx} + \psi_{yy} + k^2K(x,y)\psi = 0 \quad , \tag{2.126}$$

in the waveguide $x \geq 0$ and $0 \leq y \leq L$, and where $K(x,y) - [c/c(x,y)]^2$. Define the coordinate transformation $\xi = u(x,y)$ and $\eta = v(x,y)$ and use the functional definition $\chi(\xi,\eta) = \psi(x,y)$. Assume the transformation is conformal so that we can write $f(z)=u+iv$ with $z = x + iy$ where u and v satisfy the Cauchy-Riemann equations, f is a regular analytic function and $f'(z) \neq 0$ in the waveguide. Then $\chi$ satisfies

$$\chi_{\xi\xi} + \chi_{\eta\eta} + k^2K_2(\xi,\eta)\chi = 0 \quad , \tag{2.127}$$

where

$$K_2(\xi,\eta) = K(x,y)|f'(z)|^{-2} \quad . \tag{2.128}$$

Next, assume p satisfies the parabolic equation

$$\alpha p_\xi + p_{\eta\eta} + k^2 K_1(\eta) p = 0 \quad , \tag{2.129}$$

where $\alpha$ is a constant and $K_1$ is only a function of the transformed depth coordinate. Then $\psi$ and p can be related as

$$\psi(x,y) = A \int_0^\infty p[\alpha(4it)^{-1},\eta] \exp(it\xi^2) t^{-3/2} dt \quad , \tag{2.130}$$

where A is a constant. This is a combined Fourier-Mellin inversion transformation on the parabolic solution, and is a variation of the transform discussed in the previous section. The respective sound speeds of $\psi$ and p are related via

$$K(x,y) = K_1(\eta)|f'(z)|^2 \quad . \tag{2.131}$$

Equation (2.129) is separable, and, for a discrete set of modes its solution is

$$p(\xi,\eta) = \sum_j p_j N_j(\eta) \exp\left(-k_j^2 \xi/\alpha\right) \quad , \tag{2.132}$$

where the $p_j$ are constants and the $N_j$ satisfy (2.122) in $\eta$. Substituting (2.132) into (2.130) yields a separable solution for $\psi$

$$\psi(x,y) = A\pi i \xi^{1/2} \sum_j p_j N_j(\eta) H_{1/2}^{(1)}(k_j \xi) \quad . \tag{2.133}$$

Requiring each mode to satisfy a radiation condition as $x \to \infty$ implies that asymptotically $\xi \sim x$, so that the transformation can be written as

$$f(z) = z + f_1(z) \quad , \tag{2.134}$$

where $f_1(z) \to 0$ as $z \to \infty$ in the waveguide. As an example choose $K_1 = 1$ and

$$f(z) = z + \sum_{m=1}^{M} f_m \exp(-m\pi z/L) \quad , \tag{2.135}$$

so that $\psi(x,0) = 0$ and $(\partial\psi/\partial y)(x,L) = 0$, that is, we ensure that we have a soft surface and a hard bottom. This yields an M-parameter family of curves for (2.131), each having M-1 turning points. An example for $M = 2$ is illustrated in Fig.2.12. The profile is sinusoidal in depth, exponentially decreasing in range and has one turning point. The solution of the Helmholtz equation (2.126) is given exactly by the separable form (2.133). The details can be found in [2.61].

Additional transformations are possible. Thus $f_m$ could be finite polynomials (splines) or oscillatory functions. So long as $f_1(z)$ has no singularities in the

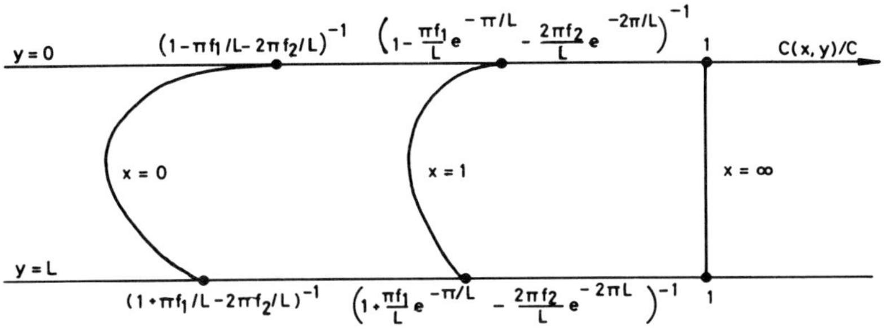


Fig. 2.12. An example of a two-parameter family of depth- and range-dependent sound speed profiles for which the Helmholtz equation is exactly solvable

waveguide and vanishes as $z \to \infty$ it can be quite arbitrary. This admits polynomial ratios (Pade approximants) as well as functions with more complicated singularities outside the waveguide. Further examples have been considered by others [2.62].

### 2.2.3 Multi-Dimensional Problems

We only briefly mention the work being done on multidimensional problems since it is in an active area of research and not yet appropriate for review. There is a method of combining ray theory in a horizontal plane with mode theory in depth [2.63]. The method assumes a ray series expansion in a scale length parameter $\varepsilon$, reflecting the fact that the horizontal variations are smaller than the vertical ones. For example, the velocity potential is expanded as

$$\phi(\underline{x},\varepsilon) \simeq \exp[\theta(x,y)/i\varepsilon] \cdot \sum_{\ell=0}^{\infty} (i\varepsilon)^{\ell} \sum_{m=0}^{\ell} a_{\ell m}(x,y)\psi_m(x,y;z) \quad ,$$

where $\theta$ satisfies an eikonal equation, the $\psi_m$ are normal mode eigenfunctions found via the solution of a one-dimensional Helmholtz equation in z, and the $a_{\ell m}$ satisfy a recursive system of transport equations. Modal eigenfunctions and eigenvalues depend parametrically on the horizontal coordinates x and y. Further remarks on this method are in Chap.3.

BOJARSKI [2.64] has developed a method of solving the exact integral equations arising from the Helmholtz equation by using an iterative scheme based on the fast Fourier transform, and locality in the transform space (or, equivalently, a constitutive relation). In principle, the method will work for an arbitrary number of dimensions. In practice it is being tried in two dimensions first, with a desired extension to three dimensions.

Finally, there is an extension of the two-dimensional parabolic method to three dimensions, range, depth, and angle, being done by BAER [2.65]. Here the crucial theoretical (and numerical) questions are the methods of treating the angular variation of the field and its sampling, since only a finite angular sector can be numerically handled, and interpolative algorithms on sound speed data in three dimensions. Results on this method should be available at the time of printing of this book.

All the methods described face the formidable problem of handling an enormous amount of data.

## 2.3 Wave Propagation in a Random Medium

Wave propagation in a random medium reduces to the study of the linear partial differential equations satisfied by the field and whose coefficients are random functions. For a general discussion of stochastic equations and random media propagation there are the excellent review papers of FRISCH [2.66], BARABANENKOV et al. [2.67], and several books [2.68-70]. In our case it is the sound speed which contributes a random coefficient although if random boundaries are present (Sect.2.4.4) these also contribute. There are two ways to study these equations [2.71], either "honestly" or "dishonestly". In the honest approach first one solves the unaveraged field equation exactly, as a series for example, and presents the solution as a known functional of the random variable. *Then* one performs the averaging process on the field (to yield the coherent part of the field) or first multiplies the field by itself (possibly evaluated at different points) and *then* averages to yield the higher-order statistical moments. In the initial solution of the equation the randomness plays no essential role.

In the dishonest approach the randomness is used ab initio before the solution is obtained. For example, the stochastic differential equation for the field is "averaged". In our example this is valid for the differential part of the equations which has constant coefficients but, since the sound speed term multiplies the field, and both are dependent random functions, the average of this term does not yield a coherent field term unless some approximation is made. The latter is the dishonest contributor to the method. We explain these remarks on what is termed the hierarchy problem and some of the different methods of breaking the hierarchy using a simple example due to RICHARDSON [2.72] from ordinary differential equations. We also include a simple example which shows the effect of hierarchy breaking on the wave solutions of the equations. The latter two subsections treat the second moment of the acoustic field, the coherence function. We show how the equations governing the propagation of the coherence function are related to other work in stochastic wave theory, and what form the equations assume when we treat propagation in a waveguide with a multidimensional sound speed structure of the form (2.55).

## 2.3.1 The Hierarchy Problem

Consider the following one-dimensional second-order differential equation

$$L\psi(x) + u(x)\psi(x) = 0 \quad , \tag{2.136}$$

where L is the differential operator defined by $L\psi(x) = \psi''(x) + \psi(x)$. The solution has (deterministic) boundary conditions $\psi(0) = \alpha$ and $\psi'(0) = \beta$, and $u(x)$ is a zero-mean stationary random process, $\langle u(x)\rangle = 0$, where the brackets represent the mathematical average or mathematical expectation over the ensemble of realizations of u. Thus $\psi(x)$ is a stationary random process but does not necessarily have zero mean. The mean of $\psi(x)$, $\langle\psi(x)\rangle$, is called the coherent field. Introduce a perturbation parameter $\varepsilon$ in (2.136) as

$$L\psi(x) + \varepsilon u(x)\psi(x) = 0 \quad . \tag{2.137}$$

The solution now depends on two parameters, x and $\varepsilon$. A review of the various types of perturbation methods for general multidimensional stochastic equations is available [2.73] and for our purposes we use the Born approximation

$$\psi(x,\varepsilon) = \sum_{n=0}^{\infty} \psi_n(x)\varepsilon^n \quad . \tag{2.138}$$

Substituting (2.138) in (2.137) and equating coefficients of successive powers of $\varepsilon$ to zero yields ($\psi_{-1} \equiv 0$)

$$L\psi_n = -u\psi_{n-1} \quad , \tag{2.139}$$

whose solution can be written as

$$\psi_n(x) = -\int_0^x \sin(x-x')u(x')\psi_{n-1}(x')dx' \quad . \tag{2.140}$$

Since the average operator commutes with the differential operator we get from (2.139) and the boundary conditions

$$\langle\psi_0(x)\rangle = \psi_0(x) = A \sin(x+a) \quad , \tag{2.141}$$

where $A = (\alpha^2+\beta^2)^{1/2}$ and $a = \tan^{-1}(\alpha/\beta)$. From (2.140) we get $\langle\psi_1(x)\rangle = 0$ and

$$\langle\psi_2(x)\rangle = (A/4)\int_0^x C(s)\sin(s)[\cos(x-s+a)-(1+2x-2s)\cos(x-s-a)]ds \quad , \tag{2.142}$$

where

$$C(s) = \langle u(x)u(x+s)\rangle \quad , \tag{2.143}$$

is the two-point correlation function which, for homogeneous statistics, only depends on the difference of the arguments of the u terms. Higher-order terms can also be calculated. The asymptotic properties of $\psi$ depend on the existence and sign of integrals of the form $\int_0^\infty s^n C(s)ds$. If the correlation function does not decrease fast enough the perturbation solution yields secular terms. For example, if u is not a function of x, so that C(s) is a constant, it can be seen from (2.142) that $\langle\psi_2(x)\rangle \sim x^2$. But if the series converges we can first write the full solution of the problem term by term, and then apply the average operator. In this case we are honest, and the computations are straightforward, but secularity prevents us from going very far from the initial point. On the other hand, the non-convergence of the unaveraged series does not necessarily imply that its average is divergent, and renormalization methods similar to those in quantum field theory can sometimes be applied to give a convergent averaged solution [2.74]. A more general connections between random classical fields and quantized fields has also been demonstrated [2.75].

Now define the following two hierarchies [2.72], the unaveraged hierarchy

$$\begin{aligned} &[L+u(x)]\psi(x) = 0 \quad , \\ &[L+u(x)]u(x_1)\psi(x) = 0 \quad , \\ &[L+u(x)]u(x_1)u(x_2)\psi(x) = 0 \quad , \ldots , \end{aligned} \tag{2.144}$$

with the number of independent variables increasing with hierarchy order, and the averaged hierarchy

$$\begin{aligned} &L\langle\psi(x)\rangle + \langle u(x)\psi(x)\rangle = 0 \quad , \\ &L\langle u(x_1)\psi(x)\rangle + \langle u(x)u(x_1)\psi(x)\rangle = 0 \quad , \\ &L\langle u(x_1)u(x_2)\psi(x)\rangle + \langle u(x)u(x_1)u(x_2)\psi(x)\rangle = 0 \quad , \ldots , \end{aligned} \tag{2.145}$$

where the averaging operator is applied directly to the differential equations for $\psi(x)$. It is obvious that the averaged hierarchy grows, e.g. to calculate $\langle\psi(x)\rangle$ one needs the next term $\langle u(x)\psi(x)\rangle$, etc. So methods of truncation become necessary. Two methods which are used on the averaged hierarchy are the correlation discard and the cumulant discard.

At the $n^{th}$ step of the hierarchy assume that $\psi(x) = \langle\psi(x)\rangle - \Delta\psi(x)$. Neglecting the $\Delta\psi(x)$ term (neglect of the incoherent field contribution) yields the correlation discard result which thus approximates the $n^{th}$ hierarchy member as

$$\langle u(x_1)\ldots u(x_{n-1})u(x)\psi(x)\rangle \simeq \langle u(x_1)\ldots u(x_{n-1})u(x)\rangle\langle\psi(x)\rangle \quad . \tag{2.146}$$

If instead we approximate the $n^{th}$ hierarchy member as a linear combination of lower quantities, $<\psi(x)>$, $<u(x_1)\psi(x)>,\ldots,$ $<u(x_1)\ldots u(x_{n-1})\psi(x)>$, we get the cumulant discard result. In the lowest two orders both methods are identical.

A third method, and an example which is used on the unaveraged hierarchy, is the least mean square error method. For example, in lowest order of (2.144) we write $u(x)\psi(x)$ as a linear functional of $\psi(x)$, via

$$u(x)\psi(x) = w(x) + \int_0^x K(x,x')\psi(x')dx' + \eta(x) \quad , \tag{2.147}$$

then minimize $\eta^2$, with respect to the unknown functions w and K, in the least square sense. The hierarchy is broken by averaging (2.147) so that $<u(x)\psi(x)>$ is now a linear functional of only $<\psi(x)>$.

Examples of all the above methods are in [2.72] with comparison to the exact results. Each works well in different situations. As a simple example let u be independent of x. Then the exact right-going wave solution of (2.137) is $\psi(x) = \exp[ix(1\mp\varepsilon u)^{1/2}]$. Expanding the exponent in powers of $\varepsilon$, regrouping terms and further expanding the resulting exponentials yields, to order $\varepsilon^2$, the result $<\psi(x)> = \exp(ix)[1-\varepsilon^2Cx(x+i)/8]$, where $C=<u^2>$. The same problem using first-order correlation discard yields $<\psi(x)>=\exp(ix)$, and for second-order correlation discard yields two possible solutions $<\psi(x)>=\exp(ix)(1\mp\varepsilon^2Cix/2)$. The methods yield clearly different results, the Born results being more secular for example, and in general a great deal of care must be taken when applying any truncation method.

In the previous simple example for both cases we wrote an exact term in the phase as an expansion in a small parameter in the amplitude, i.e., in a Born series. Presumably the phase approximation should be a more accurate representation of a wave than an amplitude expansion. We illustrate the difference following KELLER [2.76], which shows in what sense the Rytov approximation is better.

Assume the exact solution of a one-dimensional problem is a wave travelling to the right, and depends on x and $\varepsilon$ as

$$\psi(x,\varepsilon) = \exp[ixk(\varepsilon)] \quad . \tag{2.148}$$

Then expand $k(\varepsilon)$ as

$$k(\varepsilon) = \sum_{j=0}^{\infty} k_j\varepsilon^j \quad . \tag{2.149}$$

For a Born (B) approximation, substitute (2.149) into (2.148), expand each of the resulting exponentials in a power series in $\varepsilon$ and recombine terms to yield

$$\psi^B(x,\varepsilon) = \exp(ikx) \sum_{p=0}^{\infty} \varepsilon^p \sum_{q=0}^{p} (ix)^q(q)^{-1} \sum k_{j\ell}\ldots k_{jq} \quad ,$$

where the latter summation is over all combinations such that $j_1 + \ldots + j_q = p$. For the $n^{th}$ Born approximation cut off the p-sum at $p = n$. The relative remainder can be written as

$$\left[\psi(x,\varepsilon)-\psi_n^B(x,\varepsilon)\right]\left[\psi(x,\varepsilon)\right]^{-1} = O(\varepsilon^{n+1}x^{n+1}) \quad ,$$

where O is the order symbol. The Born remainder is thus the same order in both parameters.

The Rytov (R) approximation [2.70] keeps the expansion in the phase, however, so that the $n^{th}$ Rytov approximation is

$$\psi_n^R(x,\varepsilon) = \exp\left(ix\sum_{j=0}^{n} k_j\varepsilon^j\right) \quad ,$$

and its remainder is

$$\left[\psi(x,\varepsilon)-\psi_n^R(x,\varepsilon)\right]\left[\psi(x,\varepsilon)\right]^{-1} = O(\varepsilon^{n+1}x) \quad ,$$

which shows that the Rytov approximation is the same order as the Born approximation in $\varepsilon$ but is more general since it is uniform in x [2.77]. Higher order Rytov approximations do not have as serious a secularity problem as do the Born approximations.

### 2.3.2 Coherent Waves

We continue the discussion of the coherent field by illustrating the effect of a simple hierarchy breaking scheme which is commonly used. We remain in one-dimension for simplicity but the results have wider validity. The first two unaveraged hierarchy equations are

$$L\psi(x) + k^2u(x)\psi(x) = 0 \quad ,$$

and

$$L\psi(x)u(x_1) + k^2u(x_1)u(x)\psi(x) = 0 \quad , \tag{2.150}$$

where now $L\psi = d^2\psi/dx^2 + k^2\psi$. These are the same equations as in (2.144) but in the latter we set $k = 1$ for simplicity. Next, write the first two equations of the averaged hierarchy and break the hierarchy using the correlation discard method at the second step. The result is

$$L\langle\psi(x)\rangle + k^2\langle u(x)\psi(x)\rangle = 0 \quad , \tag{2.151}$$

$$L\langle\psi(x)u(x_1)\rangle + k^2C(x_1-x)\langle\psi(x)\rangle = 0 \quad , \tag{2.152}$$

where the correlation function C is defined in (2.143). Using the one-dimensional Green's function G which satisfies the equation

$$LG(x,y) = -\delta(x-y) \quad ,$$

and is explicitly given by

$$G(x,y) = G(x-y) = (-1/2ik)\exp(ik|x-y|) \quad , \qquad (2.153)$$

we can write the solution of (2.152) as

$$\langle u(x_1)\psi(x)\rangle = I(x_1,x) + k^2 \int_{-\infty}^{\infty} G(x-y)C(x_1-y)\langle\psi(y)\rangle dy \quad , \qquad (2.154)$$

where I is related to some unknown incident field and $LI = 0$. In the limit as $x_1 \to x$ (2.154) becomes

$$\langle u(x)\psi(x)\rangle = I(x) + k^2 \int_{-\infty}^{\infty} M(x-y)\langle\psi(y)\rangle dy \quad , \qquad (2.155)$$

where $I(x) = I(x,x)$ and $M(x) = G(x)C(x)$. Substitute (2.155) into (2.151). The result is

$$L\langle\psi(x)\rangle + k^2 I(x) + k^4 \int_{-\infty}^{\infty} M(x-y)\langle\psi(y)\rangle dy = 0 \quad , \qquad (2.156)$$

which is a non-local wave equation of convolution form. Since it is over the whole space it can easily be solved using Fourier transforms. It can also be solved over a half space using Wiener-Hopf and other methods [2.78]. Introduce the Fourier transform notation

$$\langle\tilde{\psi}(p)\rangle = \int_{-\infty}^{\infty} \langle\psi(x)\rangle \exp(-ipx)dx \quad , \qquad (2.157)$$

and apply it to (2.156) to yield

$$\langle\tilde{\psi}(p)\rangle = k^2 I(p)/[p^2-k^2-k^4\tilde{M}(p)] \quad . \qquad (2.158)$$

The Fourier inversion of (2.158) can be computed using residue calculus methods provided its singularities are known, i.e., provided we know the zeroes of the dispersion relation $p^2 - k^2 - k^4\tilde{M}(p)$. As a simple example, consider the exponential correlation function $C(x) = c\exp(-\alpha|x|)$ where $\alpha^{-1}$ is the correlation distance. Then

$$\tilde{M}(p) = i(\alpha+ik)[p^2+(\alpha+ik)^2]^{-1}\, k^{-1} \quad ,$$

so that we can write (2.158) in the form

$$\langle\tilde{\psi}(p)\rangle = k^2[p^2+(\alpha+ik)^2]I(p)/(p^2+\beta^2)(p^2+\gamma^2) \quad , \tag{2.159}$$

where $\beta$ and $\gamma$ are complex solutions of the equations $\beta^2+\gamma^2=\alpha^2+2ik\alpha-2k^2$ and $\beta^2\gamma^2=-k^2(\alpha+ik)(\alpha+2ik)$. Fourier inversion of (2.159) is easily seen to lead to two waves of the form $\exp(-\beta|x|)$ and $\exp(-\gamma|x|)$ so that $\beta$ and $\gamma$ yield effective wave-numbers shifted from k for the coherent field $\langle\psi(x)\rangle$. This is a simple and often used result, particularly in phenomenological modelling where waves with more general effective wave numbers are assumed for the solution, and the effective wave number values are calculated by other means.

In this and the previous section we have considered the problem of the coherent wave. The full wave function solution of a problem also includes the fluctuating or incoherent contribution. Its neglect, and the consequent study of the coherent wave alone, is thus a smooth approximation to the total field, and is called just that, the smoothing approximation. Breaking the hierarchy as we did is equivalent to a first-order smoothing theory (FOST). As mentioned, this method is often used, but there are problems with it. LERCHE and PARKER [2.79] have solved the following two equations exactly (t is time)

$$\frac{\partial^2 B}{\partial x^2} - [1+\varepsilon v(t)]\frac{\partial B}{\partial t} = 0 \quad , \tag{2.160}$$

and

$$\frac{\partial^2 B}{\partial x^2} - [1+\varepsilon v(x)]\frac{\partial B}{\partial t} = 0 \quad , \tag{2.161}$$

where v is Gaussian distributed and has an exponential correlation function (Uhlenbeck-Ornstein process) and compared their results to the FOST results. What they found was agreement with FOST when the differential equation was first order in the variable on which the random function depends, i.e., (2.160), and disagreement when it is second order, i.e., (2.161). For the latter case the disagreement occurs at the first correction term ($\varepsilon^2$). However, they did find general validity for short correlation lengths and short times, i.e., in what they called the short-sudden approximation.

Finally, we note that our example and also the Uhlenbeck-Ornstein process used exponential correlation functions. They are quite simple for calculation purposes but also introduce problems of their own. It was shown by DOOB [2.80] that the first derivative of an exponential correlation function is everywhere discontinuous, and this is physically unacceptable. The general result, where the solution is in terms of waves with effective wave numbers, does, however, hold true for more acceptable correlation functions. Correlation function problems recur in our discussion of random rough surface scattering in Sect.2.4.4.

## 2.3.3 Coherence Function and Related Work

In this section we present a mathematical method of treating the propagation of the two-point moment of the acoustic field, the coherence function, in a random medium, and show how the derived equations are related to other stochastic differential equations. The method is approximate since it uses the parabolic approximation on the field (Sect.2.2.2). Here we discuss only unbounded media problems, and in Sect.2.3.4 treat the problem in a waveguide. For a more general treatment using the Bethe-Salpeter equation which results from the Helmholtz equation on the field, we refer to FRISCH [2.66].

We begin with the velocity potential $\psi$ satisfying the Helmholtz equation in cylindrical coordinates, $\underline{x} = (r,\theta,z)$

$$[\nabla^2+k^2K(\underline{x})]\psi(\underline{x}) = 0 \quad . \tag{2.162}$$

The function K is assumed to be real and to contain both deterministic and random parts so that the equation is stochastic. Assume the parabolic approximation

$$\psi(\underline{x}) \simeq (kr)^{-1/2} \exp(ikr)p(\underline{x}) \quad , \tag{2.163}$$

where p satisfies the stochastic parabolic partial differential equation

$$2ikp_r + p_{zz} + r^{-2}p_{\theta\theta} + k^2[K(r,\theta,z)-1]p = 0 \quad . \tag{2.164}$$

The coherence function $\Gamma$ is defined [2.35,81] as the two-point density

$$\Gamma(r;\theta,\bar{\theta};z,\bar{z}) = p(r,\theta,z)p^*(r,\bar{\theta},\bar{z}) \quad , \tag{2.165}$$

and is here restricted to a single plane. The * denotes complex conjugation. To form a differential equation on $\Gamma$, multiply (2.164) by p $(r,\bar{\theta},\bar{z})$ and subtract from it $p(r,\theta,z)$ times the complex conjugate of (2.164) for $(r,\bar{\theta},\bar{z})$. The equation for $\Gamma$ is then

$$2ik\Gamma_r + \Box_x^2\Gamma + k^2[K(r,\theta,z)-K(r,\bar{\theta},\bar{z})]\Gamma = 0 \quad , \tag{2.166}$$

where the d'Alembert-type operator using the four vector $x = (\theta,\bar{\theta},z,\bar{z})$ is defined by

$$\Box_x^2 = r^{-2}\left(\partial_\theta^2-\partial_{\bar{\theta}}^2\right) + \left(\partial_z^2-\partial_{\bar{z}}^2\right) \quad . \tag{2.167}$$

An equation analogous to (2.166) occurs in the description of particle motion in a randomly perturbed potential in quantum mechanics [2.82] and is called the von Neumann equation. In the latter our range variable r is replaced by time since the equation follows from the Schrödinger equation on the particle wave function.

Introduce sum ($\underline{\sigma}$) and difference ($\underline{\delta}$) coordinate two-vectors for the second $(\theta,\bar{\theta})$ and third $(z,\bar{z})$ coordinates via

$$\underline{\sigma} = \frac{1}{2}(\theta+\bar{\theta},z+\bar{z}) = (\sigma_2,\sigma_3) \quad ,$$

and

$$\underline{\delta} = (\theta-\bar{\theta},z-\bar{z}) = (\delta_2,\delta_3) \quad .$$

Then $\Gamma$ can be written as

$$\Gamma_2(r;\underline{\sigma};\underline{\delta}) = \Gamma\left(r;\sigma_2+\frac{1}{2}\delta_2,\sigma_2-\frac{1}{2}\delta_2,\sigma_3+\frac{1}{2}\delta_3,\sigma_3-\frac{1}{2}\delta_3\right) \quad , \tag{2.168}$$

where, using (2.167), $\Gamma_2$ then evolves according to the equation

$$2ik\frac{\partial\Gamma_2}{\partial r} + 2r^{-2}\frac{\partial^2\Gamma_2}{\partial\sigma_2\partial\delta_2} + 2\frac{\partial^2\Gamma_2}{\partial\sigma_3\partial\delta_3} + k^2\left[K\left(r;\underline{\sigma}+\frac{1}{2}\underline{\delta}\right)-K\left(r;\underline{\sigma}-\frac{1}{2}\underline{\delta}\right)\right]\Gamma_2 = 0 \quad . \tag{2.169}$$

Next, Fourier transform $\Gamma_2$ in the difference coordinate (Weyl transform)

$$f(r;\underline{\sigma};\underline{p}) = \iint d\underline{\delta}\,\exp(-i\underline{p}\cdot\underline{\delta})\Gamma_2(r;\underline{\sigma};\underline{\delta}) \quad . \tag{2.170}$$

The function f is a partial spectral density representation of the coherence function, and it can also be thought of as a phase space representation in the space $(\underline{\sigma},\underline{p})$. The corresponding quantum-mechanical function is called the Wigner distribution function. Applying the Weyl transform to (2.169) shows that f evolves according to the equation

$$k\frac{\partial f}{\partial r} + r^{-2}p_2\frac{\partial f}{\partial\sigma_2} + p_3\frac{\partial f}{\partial\sigma_3} = [Mf](r;\underline{\sigma};\underline{p}) \quad , \tag{2.171}$$

where

$$[Mf](r;\underline{\sigma};\underline{p}) = (k^2/8\pi^2 i)\iint d\underline{p}'M(r;\underline{\sigma};\underline{p}-\underline{p}')f(r;\underline{\sigma};\underline{p}') \quad , \tag{2.172}$$

and

$$M(r;\underline{\sigma},\underline{q}) = \iint d\underline{\delta}\,\exp(-i\underline{\delta}\cdot\underline{q})\left[K\left(r;\underline{\sigma}-\frac{1}{2}\underline{\delta}\right)-K\left(r;\underline{\sigma}+\frac{1}{2}\underline{\delta}\right)\right] \quad . \tag{2.173}$$

If we expand the K terms in the integrand of (2.173) about $\underline{\delta} = 0$ and keep only the leading order term, the resulting integration can be done exactly. Substituting this result in (2.172) and performing a partial integration yields the approximate representation

$$[Mf](r;\underline{\sigma};\underline{p}) \simeq -\frac{1}{2}k^2\left[\frac{\partial}{\partial\underline{\sigma}}K(r;\underline{\sigma})\right]\cdot\left[\frac{\partial}{\partial\underline{p}}f(r;\underline{\sigma};\underline{p})\right] \quad , \tag{2.174}$$

which, when substituted in (2.171), yields an equation in the form of a stochastic Liouville equation familiar from statistical mechanics [2.83]. There f describes the density of molecules in phase space and the Liouville equation is an expression of Liouville's theorem, $df/ds = 0$, where s is the path coordinate along the characteristic of this first-order partial differential equation. Hence the spectral density function moves in phase space like the density in an incompressible fluid (Sect.2.1).

As a simple example of the above consider the harmonic oscillator example mentioned in our ray theory treatment in Sect.2.2.2 and discussed for the coherence function in [2.84]. Let K be a quadratic function of z, $K = a+bz+cz^2$, where a, b, and c are constant. In this case the approximate solution (2.174) is also exact. The result is

$$[Mf](r;\underline{\sigma};\underline{p}) = -k^2(b/2+c\sigma_3)\partial f/\partial p_3 \quad ,$$

and the characteristics are given by the equation

$$p_3^2 - k^2c(\sigma_3+b/2c)^2 = A - k^2b^2/4c \quad ,$$

where A is an integration constant and which for $c > 0$, are hyperbolae, and for $c < 0$, ellipses. The latter case is the desirable one since then energy will be channeled, as it is for acoustic propagation in the ocean.

We have thus seen that the equation for the propagation of the mutual coherence function in an unbounded region is closely related to other equations from stochastic wave theory and statistical mechanics. There is a difficulty in applying the Weyl transform to a bounded region, however, and this problem has not yet been treated. In addition, it has also been shown [2.85] that the Weyl transform can only be used for statistically homogeneous wavefields. In the next section, we treat the problem of the propagation of the mutual coherence function in a bounded region, and even though our statistical treatment is homogeneous (but anisotropic) we confine our discussion to coordinate space. Other unbounded space methods have been developed [2.86] as well as extensive work on applying the Feynman path integral formalism to random media problems [2.87]. Further remarks on sound propagation in a random ocean can be found in Chap.7.

## 2.3.4 Propagation of the Coherence Function in a Waveguide

In previous sections on propagation in random media we confined our attention to problems in unbounded media. This is understandable if we recognize that virtually all the literature on underwater acoustic applications of this problem was adopted from atmospheric optics results where the approximations of unbounded media and no lowest-order refraction are appropriate. In addition, these approximations have permitted the analytic methods to be carried quite far. The question remains, however, of whether it is all really applicable to ocean acoustic problems. It is known, for example, that surface scattering can effect propagation and, for long-range propagation, the effects accumulate. It is also well known that the predominant effect of the ocean medium on acoustics is one of refraction, and this results from the spatial dependence of the sound speed. It is also known that random variations in the sound speed can cause large fluctuations in the acoustic field, and that, in this regard, the ocean is a statistically anisotropic medium. Thus it is appropriate that we attempt to address all these factors at the same time.

In this section we present a brief outline of our recent work [2.88] on propagation of the mutual coherence function in a waveguide having both spatially dependent deterministic as well as statistically anisotropic sound speed contributions. The method uses Green's function techniques and Feynman diagram methods. Here we consider only two point statistics on the fluctuating part of the sound speed. The result is a matrix set of Volterra integral equations on the ensemble average of the coherence function.

We begin with (2.170), the equation on the coherence function. First rewrite the terms containing the sound speeds. Decompose K into deterministic and random parts as

$$K(r,\theta,z) = K_2(r,\theta,z) + \mu(r,\theta,z) \quad , \tag{2.175}$$

where $K_2$ is deterministic and $\mu$ random with zero mean, $\langle\mu\rangle = 0$. The deterministic part can be further decomposed into depth-refractive and mesoscale components analogous to our decomposition of the sound speed in (2.55). Next, define the functions L, D, and R using the four-vector notation in Sect.2.3.3 and (2.175) as

$$L(r,x) = k^2[K(r,\theta,z)-K(r,\bar{\theta},\bar{z})] = D(r,x) + R(r,x) \quad , \tag{2.176}$$

where D and R represent the deterministic and random parts of L, respectively. Then (2.170) is

$$2ik\Gamma_r + \square_x^2\Gamma + L(r,x)\Gamma(r,x) = 0 \quad . \tag{2.177}$$

Choose the Green's function G to satisfy the "heat equation" part of (2.177), i.e.,

$$2ikG_r(r,x;r_0,x_0) + \square_x^2 G = -\delta(r-r_0)\delta(x-x_0) \quad . \tag{2.178}$$

For a flat waveguide, G has the representation

$$G(r,x;r_0,x_0) = -(2ik)^{-1}H(r-r_0)\sum_\alpha E_\alpha(r)E_\alpha^*(r_0)F_\alpha(x)F_\alpha^*(x_0) \quad , \tag{2.179}$$

where $\alpha$ is the discrete four-vector $(m,\bar{m},n,\bar{n})$ and $H(r)$ is a unit step function. The $F_\alpha$ are products of individual eigenfunctions

$$F_\alpha(x) = Z_m(z)Z_{\bar{m}}(\bar{z})\Theta_n(\theta)\Theta_{\bar{n}}(\bar{\theta}) \quad , \tag{2.180}$$

which satisfy the equations

$$Z_m''(z) = -k_m^2 Z_m \quad , \tag{2.181}$$

and

$$\Theta_n''(\theta) = -n^2\Theta_n \quad , \tag{2.182}$$

where m and n are discrete indices, and the usual orthonormality conditions. The $Z_m$ eigenfunctions satisfy boundary conditions at the surface (z=0) and bottom (z=D) of the waveguide, and the $\Theta$ functions satisfy continuity conditions in angle. The function $E_\alpha$ in (2.179) is

$$E_\alpha(r) = \exp\left[\left(k_m^2-k_{\bar{m}}^2\right)r/2ik-(n^2-\bar{n}^2)/2ikr\right] \quad . \tag{2.183}$$

Applying Green's theorem to (2.177) and (2.178) in the waveguide yields an integral equation on $\Gamma$

$$\Gamma(r,x) = \Gamma^0(r,x) + \iint G(r,x;r_0,x_0)L(r_0,x_0)\Gamma(r_0,x_0)dr_0dx_0 \quad , \tag{2.184}$$

where the incident field at $r = r_s$ yields

$$\Gamma^0(r,x) = -2ik\int G(r,x;r_s,x')\Gamma(r_s,x')dx' \quad , \tag{2.185}$$

and $\Gamma$ is assumed to satisfy the same boundary conditions as G. Expanding (2.184) in a Neumann series and resumming enables us to write

$$\Gamma(r,x) = \Gamma^0(r,x) + \iint M(r,x;r_0,x_0)\Gamma^0(r_0,x_0)dr_0dx_0 \quad , \tag{2.186}$$

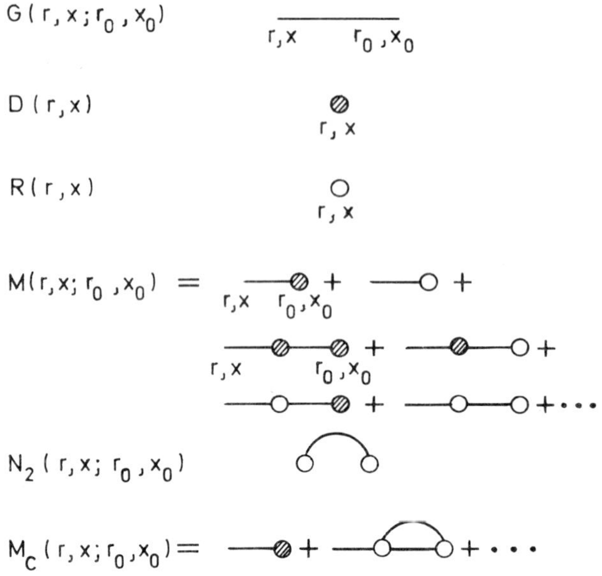


Fig. 2.13. Diagram notation for random medium propagation using the coherence function illustrating the propagator G, the deterministic (D) and random (R) sound speed contributions, the contribution to the two point correlation function $N_2$, and the stochastic (M) and connected ($M_c$) kernel functions

where the function M can be most easily defined by using the diagram notation in Fig.2.13 which is somewhat different from that in [2.66] for two-point statistical problems. Using this result, the integral equation for the ensemble average of $\Gamma$ is given by

$$\langle\Gamma(r,x)\rangle = \Gamma^0(r,x) + \iint M_c(r,x;r_0,x_0)\langle\Gamma(r_0,x_0)\rangle dr_0 dx_0 \quad , \tag{2.187}$$

with the first two terms of the connected kernel $M_c$ shown in the figure. Equation (2.186) is an integral *relation*, not an integral equation. Taking its ensemble average requires us to determine $\langle M\rangle$, which is the ensemble average of all the diagrams. Introducing the connected set of diagrams or connected kernel $M_c$ requires that (2.187) be an integral *equation*. This includes the remaining unconnected diagrams since they are products of connected diagrams and can be viewed as such through a Born expansion of (2.187). They are called unconnected since they can be broken by breaking only a propagator line [2.66]. The analytical representation of $M_c$ is given as an infinite series of terms, the first two of which are

$$M_c^{(1)}(r,x;r_0,x_0) = G(r,x;r_0,x_0)D(r_0,x_0) \quad , \tag{2.188}$$

and

$$M_c^{(2)}(r,x;r_0,x_0) = \iint G(r,x;r_1,x_1)G(r_1,x_1;r_0,x_0)N_2(r_1,x_1;r_0,x_0)dr_1dx_1 \quad . \quad (2.189)$$

Here $N_2$ is the contribution from the two-point correlation functions on the random sound speed terms, i.e.,

$$N_2(r,x;r_0,x_0) = \langle R(r,x)R(r_0,x_0)\rangle = k^4\langle[\mu(r,\theta,z)-\mu(r,\bar{\theta},\bar{z})] \cdot [\mu(r_0,\theta_0,z_0)-\mu(r_0,\bar{\theta}_0,\bar{z}_0)]\rangle \quad . \quad (2.190)$$

A simple form for the correlation function which preserves the statistically anisotropic nature of the ocean is the separable form

$$\langle\mu(r,\theta,z)\mu(r_0,\theta_0,z_0)\rangle = f(r-r_0)g(\theta-\theta_0)h(z-z_0) \quad , \quad (2.191)$$

where f, g, and h are separate correlation functions in range, angle, and depth. For further details on anisotropic media see [2.89].

Using these results it is possible to decompose $\langle\Gamma(r,x)\rangle$ and the incident field term into eigenfunctions as

$$\langle\Gamma(r,x)\rangle = H(r-r_s)\sum_\alpha \Gamma_\alpha(r)E_\alpha(r)F_\alpha(x) \quad , \quad (2.192)$$

and

$$\Gamma^0(r,x) = H(r-r_s)\sum_\alpha E_\alpha(r)E_\alpha^*(r_s)F_\alpha(x)\gamma_\alpha \quad , \quad (2.193)$$

where

$$\gamma_\alpha = \int F_\alpha(x')\,\Gamma(r_s,x')dx' \quad . \quad (2.194)$$

Substituting these results into (2.187) yield the coupled set of one-dimensional Volterra integral equations on $\Gamma_\alpha(r)$ given by

$$\Gamma_\alpha(r) = \gamma_\alpha E_\alpha^*(r_s) - (2ik)^{-1}\sum_{\alpha''}\int_{r_s}^{r} E_\alpha^*(r_0)E_{\alpha''}(r_0)D_{\alpha\alpha''}(r_0)\Gamma_{\alpha''}(r_0)dr_0$$
$$+ (k/2i)^2\sum_{\alpha'}\sum_{\alpha''} M_{\alpha\alpha'\alpha''}\int_{r_s}^{r} E_{\alpha\alpha'}(r,r_0)E_{\alpha'}^*(r_0)E_{\alpha''}(r_0)\Gamma_{\alpha''}(r_0)dr_0 \quad , \quad (2.195)$$

where the matrix elements are defined as

$$D_{\alpha\alpha''}(r) = \int F_\alpha^*(x)D(r,x)F_{\alpha''}(x)dx \quad ,$$
$$M_{\alpha\alpha'\alpha''} = \int M_{\alpha\alpha'}(x)F_{\alpha'}^*(x)F_{\alpha''}(x)dx \quad ,$$

$$M_{\alpha\alpha'}(x) = \delta_{\bar{m}\bar{m}'}\delta_{\bar{n}\bar{n}'}[G_{nn'}(\theta)H_{mm'}(z)-G_{nn'}(\bar{\theta})H_{mm'}(\bar{z})]$$
$$-(m\Leftrightarrow\bar{m},m'\Leftrightarrow\bar{m}',n\Leftrightarrow\bar{n},n'\Leftrightarrow\bar{n}') \quad ,$$

$$E_{\alpha\alpha'}(r,r_0) = \int_{r_0}^{r} E_\alpha^*(r_1)E_{\alpha'}(r_1)f(r_1-r_0)dr_1 \quad ,$$

$$G_{nn'}(\theta) = \int_{-\pi}^{\pi} \Theta_n^*(\theta_1)\Theta_{n'}(\theta_1)g(\theta_1-\theta)d\theta_1 \quad ,$$

and

$$H_{mm'}(z) = \int_0^D Z_m^*(z_1)Z_{m'}(z_1)h(z_1-z)dz_1 \quad .$$

The right-hand side of (2.195) is the sum of three terms resulting from the incident field, the deterministic sound speed contributions, and the random sound speed contributions, respectively. It can be shown by further analysis that the summations shown in the latter two terms can be reduced to two and three, respectively, but the equations are complicated by their matrix nature. Volterra equations have a computational advantage, however, in that iterative solutions are everywhere convergent and are rapidly convergent if the kernels are small. In our case this is so since the matrix elements depend on the difference in sound speed values (K-terms), an obvious advantage of equations on the coherence function which, however, does not seem to have been sufficiently exploited. Our method of exploitation of this fact was through the choice of the heat-equation-type Green's function whose eigenfunctions only depend on the boundary conditions and not on the particular form of K. Special cases of the above equations are under study as are the computational problems involved in the numerical solution of equations of this type. An alternate approach using equations on the coherent intensities formed on the stochastic amplitudes of a normal mode decomposition is also under study [2.90].

## 2.4 Scattering from Rough Surfaces

In our discussion of propagation models we restricted our attention to flat boundaries. If the boundaries are not flat, significant effects can be introduced into the propagation models due to boundary roughness. In this section we discuss the problem of scattering from rough boundaries per se and only indicate the methods of incorporating the effects in propagation models. The reason for this approach is the complexity of the scattering problem. Single (Born) scattering results are often misleading. Perturbation methods are better but work in a limited region. For large roughness, they are inadequate, as are the single scatter results. Shadowing functions involve some fundamental geometric principles of surfaces, but in practice amount to curve fitting of data.

It is the thesis of this section that it is necessary to use multiple scattering for any rough surface model, particularly for complicated and/or very rough surfaces. We illustrate this with our results of scattering from periodic surfaces and in the coherent specular scattering from a Gaussian distributed random surface. We also discuss the representation of the scattered field in the wells of and on a rough surface, i.e., the Rayleigh hypothesis, present a general integral equation satisfied by the scattering amplitude for scattering from an arbitrary deterministic surface, and show how it is related to other literature results.

### 2.4.1 The Rayleigh Hypothesis

Consider plane wave scattering from the two-dimensional periodic surface $h(x)$, illustrated later in Fig.2.20, where $h(x+L) = h(x)$ and $\max\{h(x)\} = 0$. The velocity potential $\psi$ is decomposed into plane wave incident field ($\psi_i$) plus scattered field ($\psi_{sc}$) which satisfies the two-dimensional Helmholtz equation

$$\left(\partial_x^2+\partial_z^2+k^2\right)\psi_{sc}(x,z) = 0 \quad , \tag{2.196}$$

and the boundary condition

$$a\partial_n\psi_{sc} + \psi_{sc} = f(u) \quad ,$$

on $z = h(x)$. Here $a$ is constant, $\partial_n$ is the normal derivative, the function $f$ is related to the incident field, and $u$ is the length measured along the boundary.

Define the set of eigenfunctions of (2.196) for $z \geqq 0$ as $(m=0,\pm 1,\dots)$

$$\phi_m(x,z) = \exp[ik(\alpha_m x+\beta_m z)] \quad ,$$

where $\alpha_m^2 + \beta_m^2 = 1$, and $\beta_m$ is real and positive, or positive imaginary. The $\alpha_m$ are given by the grating equation (2.200). RAYLEIGH [2.91] conjectured in 1903 that the set $\{\phi_m\}$ is complete for $z \geqq h(x)$, that is, any function defined on or above the surface could be expanded in the $\{\phi_m\}$ alone. For example, we could write

$$\psi_{sc}(x,z) = \sum_{m=-\infty}^{\infty} A_m\phi_m(x,z) \quad .$$

It has been shown [2.92,93] that the hypothesis is not always correct, after plausible objections were raised to it concerning the lack of downgoing waves in the $\{\phi_m\}$ and hence in the wells of the surface [2.94,95]. Several [2.96,97] discussions of the controversy which the hypothesis engendered are available. The success or failure of the hypothesis depends on the singularities of $\psi_{sc}$. These arise as follows.

$\psi_{sc}$ is a function of x and z (analytic in each), and a functional of h(x) and f(u). The validity of the Rayleigh hypothesis depends on the analytic continuation of $\psi_{sc}$ as a function of z, in particular continuation down into the wells of the surface. If h is analytic (e.g. no sharp corners) and f is analytic (no discontinuities) then, for a given x, $\psi_{sc}$ can be continued in z down into the wells until it reaches a singularity. As x varies, the envelope (convex hull) of singularities is exposed and this is just some surface g(x). Three cases can occur. To elucidate them it is helpful to again refer to Fig.2.20.

*Case 1:* $g(x) < -d$

Here the Rayleigh expansion is valid, the $A_m$ can be calculated from the boundary condition, and we have uniform convergence to the boundary values and to $\psi_{sc}$ in the wells.

*Case 2:* $g(x) < h(x)$

But there exists at least one $x_j$ such that $g(x_j) > -d$. For this case the Rayleigh hypothesis is invalid since the Rayleigh expansion can be continued only to that $z_m = \max\{g(x)\}$ and no further. That is, although the singularities are *outside* the wells, the continuation fails *in* the wells. However, for this case it is possible to define the Rayleigh hypothesis as valid in the following extended sense [2.96]. Define the sequence

$$\gamma_M(x,z) = \sum_{m=-M}^{M} A_m(M)\phi_m(x,z) \quad ,$$

where now the coefficients $A_m$ depend on sequence size. Then we can guarantee that the $\{\phi_m\}$ are complete for functions defined on h by choosing the $A_m(M)$ in a mean-square sense via

$$\lim_{M\to\infty} \int_h |f(u)-(1+a\partial_n)\gamma_M(x,z)|^2_{z=h(x)} du = 0 \quad ,$$

Hence $(1+a\partial_n)\gamma_M$ converges (on the boundary) to f(u) in the mean-square sense and *not* uniformly. Nevertheless, using these $A_m(M)$, $\gamma_M$ can be shown to converge uniformly to $\psi_{sc}$ for $z > h$. More precisely, in closed subsets of the region $z > h(x)$ we have that

$$\lim_{M\to\infty} |\psi_{sc}(x,z)-\gamma_M(x,z)| = 0 \quad ,$$

which is uniform convergence. The proof is easy. Alternatively, in principle, it is possible to construct from the $\{\phi_m\}$ a set $\{\psi_m\}$ which are complete and orthonormal on h [2.98] such that the sequence $\sum_{m=-M}^{M} B_m\psi_m(x,z)$ converges uniformly to f(u). Here the $B_m$ do not depend on M. Since it converges uniformly, it also converges in the mean (a weaker result) and, a fortiori, converges uniformly in the field by the same argument as that above. In practice, no one constructs this sequence, however. Thus we conclude that although the original hypothesis fails for this case, it is possible to find an extension of the hypothesis which holds so long as the only singularities which occur are outside the wells.

*Case 3*

At least one of the points of $g(x)$ occurs within the wells. Then both the Rayleigh hypothesis and its extension fail. No method of treating this case has been done, although we mention a possible method below.

Also, when h or f are not analytic the hypothesis in general is not valid.

As an example, consider the sinusoidal surface illustrated in Fig.2.20, $h(x) = -(d/2)[1+\cos(2\pi x/L)]$. It has been shown that the hypothesis fails when $\pi d/L > 0.448$. The quantity $\pi d/L$ is the maximum slope of the surface, and the number 0.448 arises as the solution of a transcendental equation analogous to that used in Sect.2.2.1 to find the modal solutions. Physically, these results mean that for a sufficiently deep well in the surface vertical eigenmodes or standing waves occur, at which point the Rayleigh hypothesis fails. Interestingly enough, the eigenmodes depend on only the geometric parameters of the surface, and are independent of the acoustic frequency. It would seem possible to use a Mittag-Löffler type expansion to take these singularities into account as has been done in the normal mode problem, but no one, apparently, has done this.

In the next section we discuss scattering from rectangular periodic surfaces which violate the Rayleigh hypothesis due to the non-analyticity of the surface, and a sinusoidal surface, which also violated the hypotheses since we put no restriction on the well depth. We show how to take into account the violation by correctly defining field expansions in the wells in two different ways. For surfaces which satisfy the conditions for the validity of the Rayleigh hypothesis, perturbation methods [2.99] have been extensively developed.

## 2.4.2 Periodic Surfaces

### *a) Rectangular Periodic Surfaces*

It is possible to solve exactly for the scattered field when a plane wave is incident on any of the periodic rectangularly corrugated surfaces illustrated in Fig.2.14. The results are due to DESANTO [2.100-103]. The method of solution uses a combination of complex function theory, the residue calculus method [2.104] and a rapidly convergent numerical iterative procedure. Results can be quoted for any values of the parameters of the problem, thus making the results, aside from their intrinsic interest, ideal as testing grounds for approximate theories. We briefly develop the results for the case illustrated in Fig.2.14a, and quote numerical results for all the surfaces.

For the surface illustrated in Fig.2.14a, the velocity potential satisfies the Helmholtz equation

$$\left(\partial_x^2+\partial_z^2+k^2\right)\psi(x,z) = 0 \quad , \tag{2.197}$$

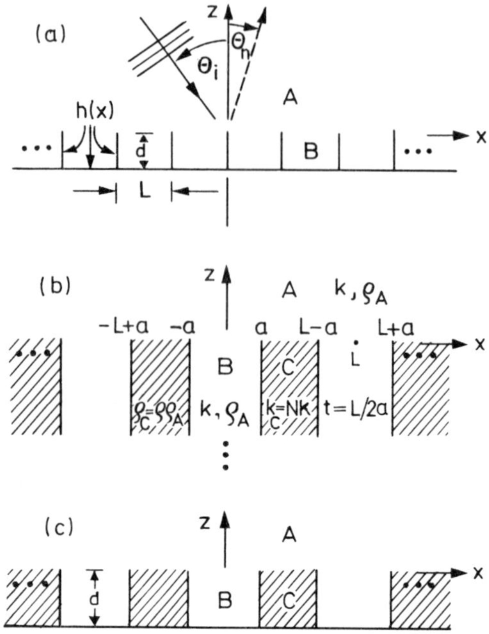


Fig. 2.14. Examples of periodic rectangularly corrugated surfaces for which the scattered field can be exactly calculated using the residue calculus method

where $k = 2\pi/\lambda$ is the acoustic wavenumber and $\lambda$ the wavelength. The surface has period L and is defined by

$$\begin{aligned} h(x) &= -d \quad , \quad x \neq 0, \pm L, \pm 2L, \ldots \quad , \\ &= 0 \quad , \quad x = 0, \pm L, \pm 2L, \ldots \quad . \end{aligned} \tag{2.198}$$

In addition, $\psi$ has the following restrictions:

(a) $\psi$ and $|\underline{\nabla}\psi|$ are finite in each subregion except at the sharp edges where $\psi = O(r^{1/2})$ and $|\underline{\nabla}\psi| = O(r^{-1/2})$, and where r is the radial distance from the edge and O the order symbol. This is the edge condition [2.105] and guarantees that the edges do not serve as sources of energy.

(b) $\psi$ and $\underline{\nabla}\psi$ are continuous in each subregion and across the $z = 0$ open interface.

(c) Apart from the incident wave, $\psi$ represents outgoing waves as $z \to \infty$. The following functions satisfy these restrictions. For $z \geqq 0$ (region A)

$$\psi_A(x,z) = \exp[ik(\alpha_0 x - \beta_0 z)] + \sum_{n=-\infty}^{\infty} A_n \exp[ik(\alpha_n x + \beta_n z)] \quad , \tag{2.199}$$

the first term being the incident plane wave ($\alpha_0 = \sin\theta_i$ and $\beta_0 = \cos\theta_i$), and the second term the scattered field ($\alpha_n = \sin\theta_n$ and $\beta_n = \cos\theta_n$). Periodicity of the surface implies the grating equation

$$\alpha_n = \alpha_0 + n\Lambda \quad , \quad \Lambda = \lambda/L \quad . \tag{2.200}$$

For $0 \geq z \geq -d$ (region B), $0 \leq x \leq L$, and the soft boundary condition $\psi(x,h(x)) = 0$, we have

$$\psi_B(x,z) = \sum_{j=1}^{\infty} B_j \sin(p_j kx) \sin[q_j k(z+d)] \quad , \tag{2.201}$$

with $p_j^2 + q_j^2 = 1$, and $p_j = j\Lambda/2$. Values of $\psi_B$ for other x-values can be found from the Floquet condition. Substituting (2.199) and (2.201) into the continuity equations

$$\psi_A(x,0) = \psi_B(x,0) \quad , \quad \partial_z\psi_A(x,0) = \partial_z\psi_B(x,0) \quad , \tag{2.202}$$

yields the set of linear equations

$$\sum_{n=-\infty}^{\infty} A_n\left[\frac{\exp(-iq_jkd)}{\beta_n - q_j} \pm \frac{\exp(iq_jkd)}{\beta_n + q_j}\right] - \left[\frac{\exp(-iq_jkd)}{\beta_0 + q_j} \pm \frac{\exp(iq_jkd)}{\beta_0 - q_j}\right]$$

$$= \begin{cases} (2\pi q_j/i\Lambda p_j)B_j\left[1-(-)^j \exp(2\pi i\alpha_0/\Lambda)\right]^{-1} \\ 0 \end{cases} \tag{2.203}$$

We can solve these equations by finding a meromorphic function f(w) with the following properties

(a) f(w) has simple poles at $w = \beta_n$, for n = 0, 1, 2,..., and a simple pole at $w = -\beta_0$,

(b) f(w) has simple zeroes at $w = \bar{q}_j$, for j = 1,2, determined from the condition $f(q_j) = f(-q_j) \exp(2iq_jkd)$,

(c) $f(w) \to 0$ as $|w| \to \infty$,

(d) f satisfies appropriate algebraic properties so that $\psi$ satisfies the edge condition,

and considering contour integrals of the form

$$\int_{C_s} \frac{f(w)}{w-q_j}\, dw \pm \exp(2ikdq_j) \int_{C_s} \frac{f(w)}{w+q_j}\, dw \quad , \tag{2.204}$$

where the contours $C_s$ enclose the points $w = \pm q_j$ for $j \leq s$, $-\beta_0$, and $\beta_n$ for $n \leq s$. The residue series from these integrals exactly match the linear equations (2.203) if $A_n = r(\beta_n)$, the residue of f at $w = \beta_n$, and the $B_j$ are appropriately related to $f(q_j)$. The problem then becomes one of constructing f(w) as a ratio of infinite products, and numerically finding the $\bar{q}_j$ from (b) above. The uniqueness of f follows from the edge condition. The $A_n$ and $B_j$ amplitudes are then determined. In addition, energy flux conservation gives the result

$$\sum_n R_n = 1 \quad , \quad R_n = (\beta_n/\beta_0)|A_n|^2 \quad , \tag{2.205}$$

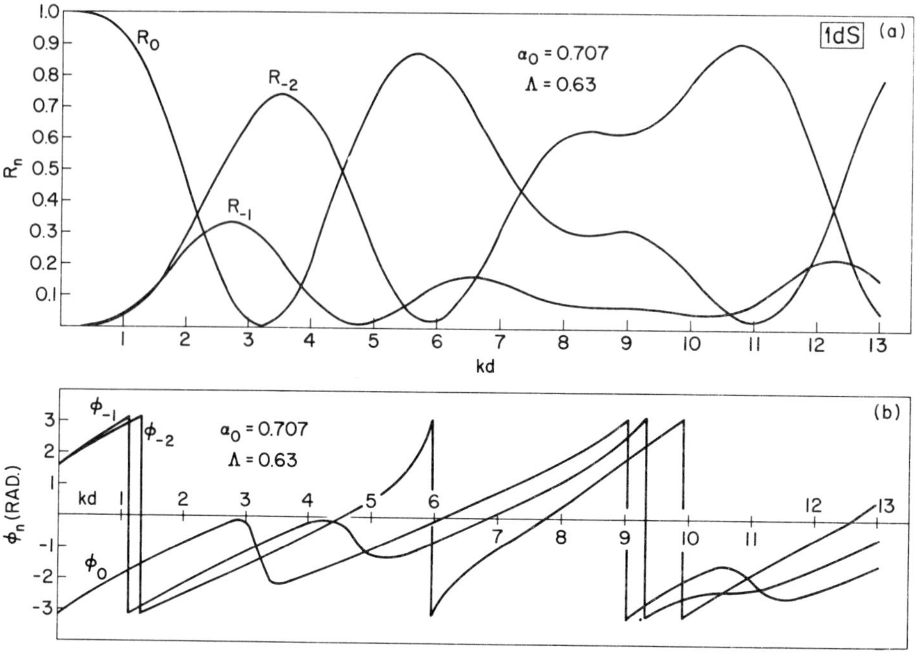


Fig. 2.15a and b. Reflection coefficients $R_n$ and amplitude phases $\phi_n$ plotted as a function of depth for fixed incident angle and surface period illustrating the Brewster angle anomaly

where the $R_n$ are the individual reflection coefficients and the sum extends over all n such that the $\beta_n$ are real (real scattering orders). We now illustrate some of the grating anomalies [2.106] which occur in the results for both $R_n$ and amplitudes phases $\phi_n$ defined by $A_n = |A_n| \exp(i\phi_n)$.

Figure 2.15 shows the Brewster angle anomaly [2.100] which occurs at parameter values which cause the specular scattering ($R_0$) to be extinguished. $R_{-1}$ and $R_{-2}$ are backscatter reflection coefficients.

Figure 2.16 shows the Wood P-anomaly [2.100], the rapid exchange of energy between specular and backscatter orders as one of the parameters is varied. For a hard surface the Wood S-anomaly is similar [2.101].

Figure 2.17 shows the Rayleigh or threshold anomaly, the cusp-like behavior of $R_0$ as either an evanescent wave becomes a real scattering wave ($R_{-2}$) or a real scattering order becomes evanescent ($R_1$). In either case the appropriate $\beta_n$ vanishes. The effect is visible in $R_0$ by energy conservation [2.100].

For surfaces of the types illustrated in Fig.2.14b and c, the shaded regions can be "loaded" with a material having different density and supporting waves with different sound speed than occur in the unshaded areas. Threshold anomalies similar to the Rayleigh anomaly occur at the point that the z-component of the wavenumber for waves in these regions vanishes. Fig.2.18 illustrates how the shape of one of these

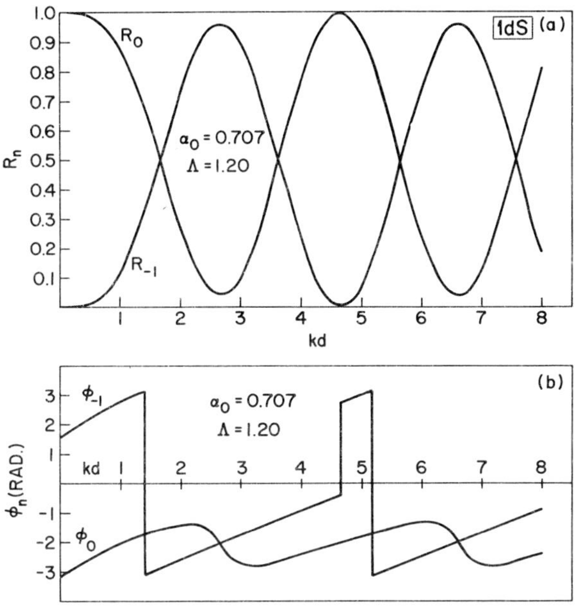


Fig. 2.16a and b. Reflection coefficients $R_n$ and amplitude phases $\phi_n$ plotted as a function of depth for fixed incident angle and surface period illustrating the Wood P-anomaly

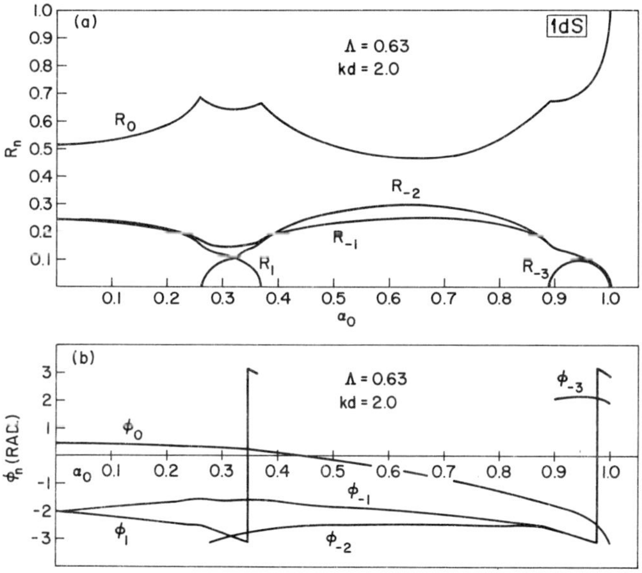


Fig. 2.17a and b. Reflection coefficients $R_n$ and amplitude phases $\phi_n$ plotted as a function of incident angle for fixed surface period and depth illustrating Rayleigh anomalies

anomalies changes at is "crosses" the Rayleigh anomaly at $\Lambda = 1$. Here, [2.102], the thickness parameter t of the shaded region increases from right to left for a fixed density ratio $\rho$ and sound speed ratio N. Finally, Fig.2.19 shows a resonance anomaly [2.103], the sharp rise in $R_0$ is due to a near zero in its denominator. The results are for a hard surface of the type shown in Fig.2.14c. In addition to the anomalous behavior, the figure also shows that nearly perfect specular reflection is possible

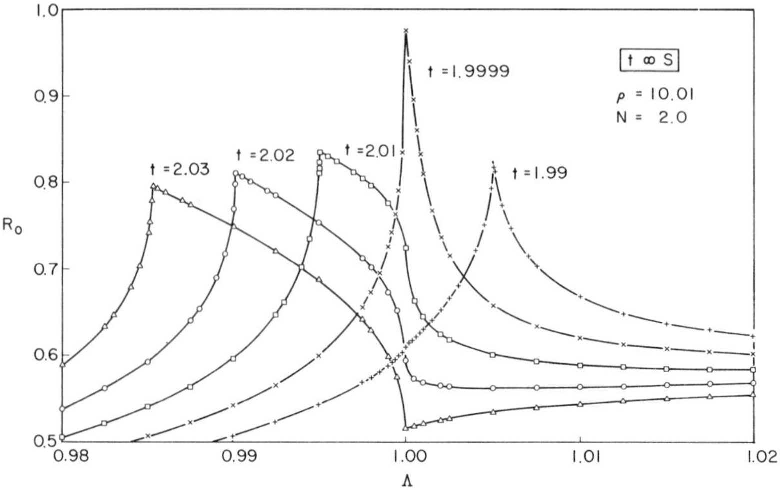


Fig. 2.18. Specular reflection coefficient $R_0$ plotted as a function of $\Lambda$, tracing a threshold anomaly as it crosses the Rayleigh anomaly

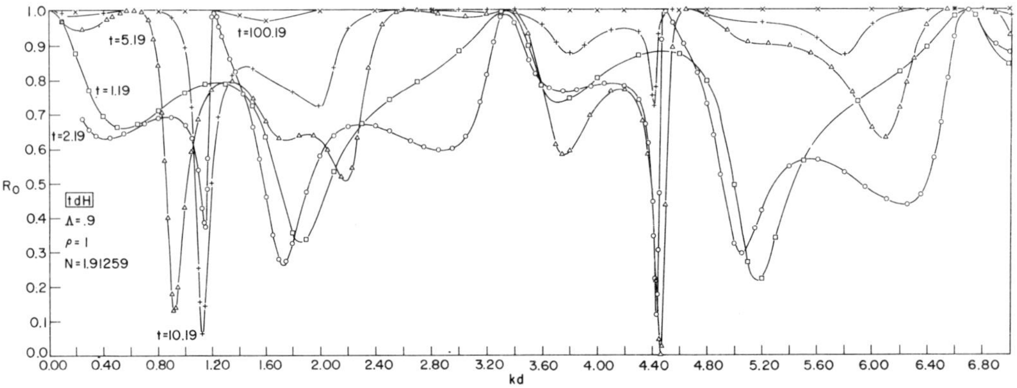


Fig. 2.19. Specular reflection coefficient $R_0$ plotted as a function of depth for fixed surface period, density ratio and sound speed ratio, illustrating a resonance anomaly and perfect specular blazing over a broad parameter domain

over a broad range of parameters. Perfect blazing of gratings in a particular diffraction order can also be illustrated [2.107] for these rectangular corrugated surfaces.

*b) The Sinusoidal Surface*

For plane wave incidence on the sinusoidal surface illustrated in Fig.2.20 a similar field expansion to (2.199) can be used [2.108]. Using the auxiliary functions

$$G^{\pm}(x,z) = \exp[ik(\pm\beta_m z-\alpha_m x)] \quad , \tag{2.206}$$

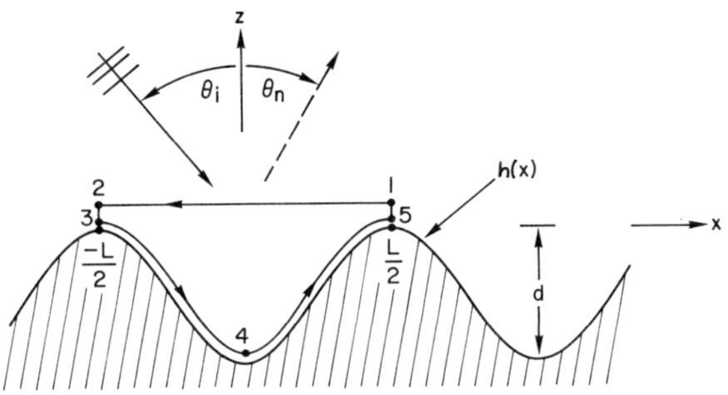


Fig. 2.20. Plane wave incidence and reflection from a sinusoidal surface illustrating the closed contour (123451) used in Green's theorem

which include both up and downgoing waves, in Green's theorem

$$\int_C (\psi \partial_n G^{\pm} - G^{\pm} \partial_n \psi) du = 0 \quad , \tag{2.207}$$

applied to the closed contour C = (123451) shown in the figure yields linear equations on the amplitudes. Since the (345) part of the contour is evaluated on the surface, we can expand the surface fields in Floquet (periodic) modes in the x-coordinate alone. For a soft (s) boundary we find the amplitude coefficients given by

$$A_m^S = -(\beta_0/\beta_m) i^m \exp\left(i\tau_m^+\right)\left[J_m\left(\tau_m^+\right) + \sum_{n \neq 0} a_n^S J_{m-n}\left(\tau_m^+\right)\right] \quad , \tag{2.208}$$

where the $a_n^S$ satisfy the equations

$$J_m\left(\tau_m^-\right) + \sum_{n \neq 0} a_n^S J_{m-n}\left(\tau_m^-\right) = 0 \quad , \tag{2.209}$$

with $\tau_m^{\pm} = kd(\beta_0 \pm \beta_m)/2$ and $J_m$ the Bessel function of order m. Only slightly more complicated equations result for the hard (H) surface [2.108]. They are

$$A_m^H = i^m \exp\left(i\tau_m^+\right)\left[(1+b_{m0})J_m\left(\tau_m^+\right) + \sum_{n \neq 0} a_n^H (1+b_{mn}) J_{m-n}\left(\tau_m^+\right)\right] \quad , \tag{2.210}$$

where the $a_n^H$ satisfy the equations

$$(1-\alpha_m\alpha_0-\beta_m\beta_0)J_M\left(\tau_m^-\right) + \sum_{n \neq 0} a_n^H (1-\alpha_m\alpha_n-\beta_m\beta_0)J_{m-n}\left(\tau_m^-\right) = 0 \quad , \tag{2.211}$$

and

$$b_{mn} = \alpha_m(\alpha_m-\alpha_n)/\beta_m(\beta_m+\beta_0) \quad . \tag{2.212}$$

The same equations were derived independently [2.109] and solved numerically. The results illustrate the same grating anomalies shown for the rectangularly corrugated surfaces. For a penetrable surface, matrix inversion methods were used [2.110] to illustrate several radiation patterns from a sinusoid.

### 2.4.3 Arbitrary Deterministic Surfaces

There are books [2.111,112] and many review articles [2.113-117] on rough surface scattering theory. What we present here is a theory of the scattering of an arbitrary incident field from a rough interface separating two fluids of different, but constant, density and sound speed. No restrictions are put on the parameters of the problem, and thus the theoretical results are quite general. For the time-independent case considered, the scattered field is found from the solution of an integral equation in Fourier transform space. The results are formal and complicated. To dispel the complexity, we show that, with certain parameter restrictions, the results reduce to familiar cases, and we introduce a diagram notation to give a graphic and interpretive account of the multiple scattering terms in the result. Results derived here are used to derive an integral equation for the random surface case in Sect.2.4.4 which is numerically solved. It is our belief that it is necessary to account for multiple scattering in order to understand the rough surface scatter problem.

#### *a) Green's Function Formalism*

Consider an arbitrary scalar wave incident on the rough surface $z = h(x_\perp)$ which separates two fluids. A projection of the surface is illustrated in Fig.2.21. Our notation for a vector is $\underline{x} = (x_1, x_2, x_3) = (x, y, z) = (x_\perp, z)$ and, for the vector evaluated

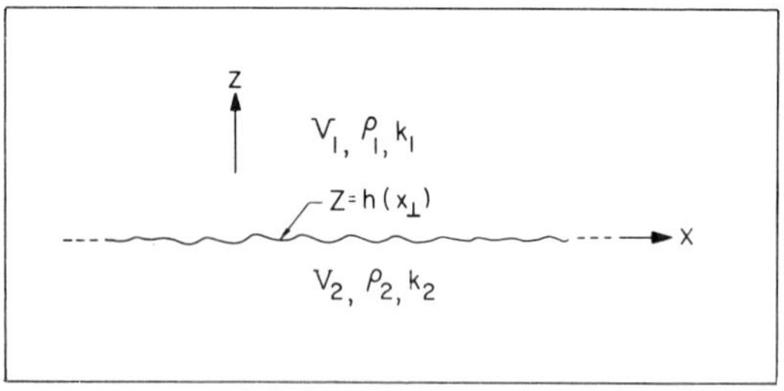


Fig. 2.21. Rough surface separating two semi-infinite fluids

on the surface, $\underline{x}_s = (x_\perp, h)$. The upper half space $V_1$, $z > h(x_\perp)$ is a fluid with density $\rho_1$ and supports waves with wavenumber $k_1 = \omega/c_1$, with $\omega$ the angular frequency and $c_1$ the sound speed. The lower half space $V_2$ is a fluid with density $\rho_2$ and wavenumber

$k_2 = \omega/c_2$. All $\rho_j$, $k_j$ and $c_j$ (j=1,2) are constant. The source is in $V_1$. The Green's functions $G_j$ in $V_j$ satisfy the Helmholtz equations

$$\left(\nabla^2+k_{1,2}^2\right)G_{1,2}(\underline{x},\underline{x}'') = -\delta(\underline{x}-\underline{x}'') \quad , \tag{2.213}$$

for $\underline{x} \in V_{1,2}$ and $\underline{x}'' \in V_1$, and continuity conditions at the interface

$$G_2(\underline{x}_s,\underline{x}'') = \rho^{-1}G_1(\underline{x}_s,\underline{x}'') \quad , \quad \rho = \rho_2/\rho_1 \quad , \tag{2.214}$$

and

$$\partial_m G_2(\underline{x}_s,\underline{x}'')n_m(x_\perp) = \partial_m G_1(\underline{x}_s,\underline{x}'')n_m(x_\perp) \quad , \tag{2.215}$$

with $\partial_m = \partial/\partial x_m$, $n_m = \delta_{m3} - \partial_{m\perp}h(x_\perp)$ a vector normal to the surface, and repeated subscripts are summed from 1 to 3 (summation convention). The continuity conditions correspond to the physical requirements of continuity of pressure and velocity normal at the surface. The free-space Green's functions satisfy (2.213) everywhere and are explictly given by

$$G_j^{0(\pm)}(\underline{x},\underline{x}') = (4\pi|\underline{x}-\underline{x}'|)^{-1}\exp(\pm ik_j|\underline{x}-\underline{x}'|) \quad , \tag{2.216}$$

corresponding to outgoing (+) and incoming (-) waves, respectively. Applying the Green's theorem in the two regions using only a single free-space Green's function yields two mixed surface-volume integral equations which can be combined using the continuity conditions [2.118,119]. The results can be interpreted in terms of nonlocal or extended boundary condition models [2.110,120]. We Fourier transform the result and restrict our attention to the part of the Green's function, T, scattered into $V_1$. The result is an integral equation of Lippmann-Schwinger form [2.26] analogous to that for the T-matrix of quantum scattering theory

$$T(\underline{k}',\underline{k}'') = V(\underline{k}',\underline{k}'')A(\underline{k}'-\underline{k}'') + \int V(\underline{k}',\underline{k})A(\underline{k}'-\underline{k})G_1^{0(+)}(k)T(\underline{k},\underline{k}'')d\underline{k} \quad . \tag{2.217}$$

Here $G_1^{0(+)}$ is the Fourier transform of the free-space Green's function and the vertex function V is defined as

$$V(\underline{k}',\underline{k}) = -2i(2\pi)^{-3}R(k_\perp)\left[k'_{m\perp}-\delta_{m3}P\left(K_1'^2/k_z'\right)\right](k_m'-k_m)(k_z'-k_z)^{-1} \quad , \tag{2.218}$$

where P stands for the Cauchy principle value, and R is the Fresnel reflection coefficient

$$R(k_\perp) = (\rho K_1-K_2)(\rho K_1+K_2)^{-1} \quad , \tag{2.219}$$

with $K_j = \left(k_j^2 - k_\perp^2\right)^{1/2}$ and where $K_j$ is real and positive, or positive imaginary. The function A in (2.217) is the phase modulated amplitude spectrum of the surface

$$A(\underline{k}) = \int_{-\infty}^{\infty} \exp(-i\underline{k}\cdot\underline{x}_s)dx_\perp \quad . \tag{2.220}$$

The hard surface limit ($\rho = \infty$) of (2.217) agrees with our previous results [2.121] and it can be shown [2.122,123] that (2.217) reduces to the correct flat surface limit [2.124] when $h(x_\perp) = 0$. The soft surface limit ($\rho = 0$) follows immediately. Also, for a sinusoidal surface it is obvious that (2.220) reduces to a sum over Bessel functions, and these results can be related to our previous discussion.

The fields can be expanded in plane waves (reduction technique) and the scattered intensity is then

$$I(k_\perp)\delta(k_\perp - k_\perp') = (\pi/K_1)^2 \int T(k_\perp, K_1; k_\perp'', -K_1'')\phi(k_\perp'')dk_\perp'' \cdot \int \left[T(k_\perp', K_1'; k_\perp''', -K_1''')\,\phi(k_\perp''')\right]^* dk_\perp''' \quad , \tag{2.221}$$

where $\phi$ is the spectrum of the incident field

$$\psi^i(\underline{x}) = \int \exp(i\underline{k}\cdot\underline{x})\phi(k_\perp)dk_\perp \quad , \quad k_z = -K_1 \quad . \tag{2.222}$$

For a single plane wave incidence, $\phi(k_\perp) = \delta\left(k_\perp - k_\perp^i\right)$, where $k_\perp^i$ is the incident transverse wavenumber. For an incident spherical wave $\phi(k_\perp) = \pi i(2\pi)^{-3}K_1^{-1/2}$, and an incident cylindrical wave $\phi(k_\perp) = \pi i(2\pi)^{-2}K_1^{-1/2}\delta(k_y)$.

### *b) Diagrams*

A convenient device to keep track of the terms in (2.217) is to associate with each term a diagram, as shown in Fig.2.22. This is analogous to the diagram notation in random volume scattering theory in Sect.2.3. The function $G_1^0(k)$ is called a propagator because it moves or "propagates" some field with momentum $\underline{k}$. The field then interacts both geometrically, the function A, and kinematically, the vertex function V, with a point on the surface. Single scattering corresponds to a term of the form VA, multiple scattering to a term of the form $VAG_1^0VAG_1^0VA\ldots$, the order of the multiple scattering being given by the number of VA combinations. Terms of this form arise from an iterative solution of (2.217). The diagram rules include a three-fold integration over any internal momentum line.

Diagram methods in surface scattering have also been used by us in problems involving elastic [2.125] and electromagnetic [2.126] media. They have also been used in the context of a perturbation expansion of h [2.112], and when the interaction term is related to a surface wave [2.127].

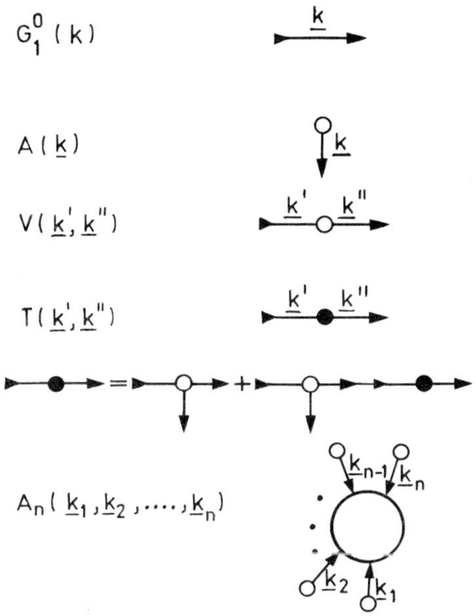


Fig. 2.22. Diagram notation for random rough surface scattering illustrating the propagator $G_1^0$, interaction term A, vertex function V, scattering amplitude T, the integral equation (2.217) and cluster terms $A_n$

## 2.4.4 Random Surfaces

### *a) Gaussian Surfaces*

We now consider $h(x_\perp)$ to be a Gaussian distributed random variable. It has a correlation function which for a homogeneous surface is

$$\gamma(x_\perp - x_\perp') = \langle h(x_\perp)h(x_\perp')\rangle \quad , \tag{2.223}$$

where the brackets represent the ensemble average. Here $\gamma(0) = \sigma^2$, where $\sigma$ is the root mean square height of the surface. If we assume that a Neumann series expansion of (2.217) is valid (note that this is not a perturbation expansion in powers of h), it can be written in operator notation as

$$T = VA + VAG_1^0VA + VAG_1^0VAG_1^0VA + \dots \quad ,$$

where, for each occurrence of $G_1^0$, a three-dimensional integration is assumed. Since only the A function involves the random variable, the ensemble average of T follows by knowing the ensemble average of products of A functions. It can be shown [2.121] that

$$\langle \prod_{j=1}^{n} A(\underline{k}_j)\rangle = (2\pi)^2 \prod_{j=1}^{n} C(k_{jz})\delta\left(\sum_{j=1}^{n} k_{j\perp}\right) I_n(\underline{k}_1,\dots,\underline{k}_n) \quad , \tag{2.224}$$

where

$$C(k_{jz}) = \exp\left(-\frac{1}{2}\sigma^2 k_{jz}^2\right) \quad . \tag{2.225}$$

The delta function is an expression of the two-dimensional translational invariance of the surface statistics and the first two $I_n$ are given by

$$I_1 = 1 \quad , \tag{2.226}$$

and

$$I_2(\underline{k}_1,\underline{k}_2) = \int \exp[-ik_{1\perp}\cdot\rho_\perp - \gamma(\rho_\perp)k_{1z}k_{2z}]d\rho_\perp \quad . \tag{2.227}$$

If we define functions $A_j$ via the cluster decomposition

$$\begin{aligned}
\langle A(\underline{k}_1)\rangle &= A_1(\underline{k}_1) = (2\pi)^2\delta(k_{1\perp})C(k_{1z}) \quad ,\\
\langle A(\underline{k}_1)A(\underline{k}_2)\rangle &= A_1(\underline{k}_1)A_1(\underline{k}_2) + A_2(\underline{k}_1,\underline{k}_2) \quad ,\\
\langle A(\underline{k}_1)A(\underline{k}_2)A(\underline{k}_3)\rangle &= A_1(\underline{k}_1)A_1(\underline{k}_2)A_1(\underline{k}_3) + A_1(\underline{k}_1)A_2(\underline{k}_2,\underline{k}_3) + A_1(\underline{k}_2)A_2(\underline{k}_1,\underline{k}_3)\\
&\quad + A_1(\underline{k}_3)A_2(\underline{k}_1,\underline{k}_2) + A_3(\underline{k}_1,\underline{k}_2,\underline{k}_3) \quad ,
\end{aligned} \tag{2.228}$$

and so forth we can write, in general,

$$A_n(\underline{k}_1,\dots,\underline{k}_n) = (2\pi)^2 \prod_{j=1}^{n} C(k_{jz})\delta\left(\sum_{j=1}^{n} k_{j\perp}\right) R_n(\underline{k}_1,\dots,\underline{k}_n) \quad , \tag{2.229}$$

where for $n = 1,2$

$$R_1 = 1 \quad , \tag{2.230}$$

and

$$R_2(\underline{k}_1,\underline{k}_2) = \int \exp(-ik_{1\perp}\cdot\rho_\perp)\left\{\exp\left[-\gamma(\rho_\perp)k_{1z}k_{2z}\right]-1\right\}d\rho_\perp \quad . \tag{2.231}$$

The $A_n$ are illustrated as the diagram term in Fig.2.22. The general term for all the above can be written [2.121], but here we have only retained the low-order interesting results. Note that if we expand the exponential involving the correlation function in $R_2$ in a power series, and keep only the first two terms, we find that $R_2$ is proportional to the Fourier transform of the correlation function, i.e., to the two point spectral function of the surface.

*b) Dyson Equation*

We can thus write an equation for the ensemble average (coherent part) of T, the Dyson equation, which is

$$\langle T(\underline{k}',\underline{k}'')\rangle = M(\underline{k}',\underline{k}'') + \int M(\underline{k}',\underline{k})G_1^0(k)\langle T(\underline{k},\underline{k}'')\rangle d\underline{k} \quad , \qquad (2.232)$$

where the function M is called the mass operator, and is an infinite series of terms

$$M(\underline{k}',\underline{k}'') = \sum_{j=1}^{\infty} M_j(\underline{k}',\underline{k}'') \quad , \qquad (2.233)$$

the diagrammatic representation of the first three terms of which is illustrated in Fig.2.23. The diagrams are all connected. The analytic representation of each diagram is calculated by applying the diagram rules to the pieces of each diagram, while integrating over internal momentum lines. For example,

$$M_2(\underline{k}',\underline{k}'') = \int V(\underline{k}',\underline{k})G_1^0(k)V(\underline{k},\underline{k}'')A_2(\underline{k}'-\underline{k},\underline{k}-\underline{k}'')d\underline{k} \quad . \qquad (2.234)$$

From the homogeneity of the statistics, it is possible to factor a two-dimensional delta function from each $M_j$ term in the mass operator, for example,

$$M_1(\underline{k}',\underline{k}'') = (K_1'^2/\pi i)P(1/k_z')R(k_\perp')C(k_z'-k_z'')\delta(k_\perp'-k_\perp'') \quad , \qquad (2.235)$$

Thus we can factor the T-matrix

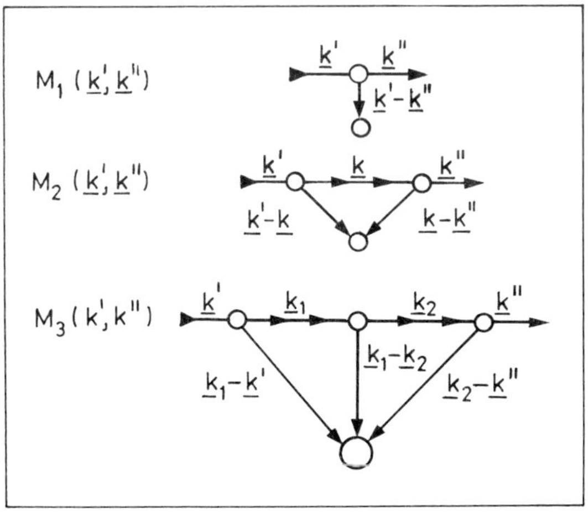


Fig. 2.23. Diagram notation for the first three terms in the mass operator

$$\langle T(\underline{k}',\underline{k}'')\rangle = \left(K_1'^2/\pi i\right)P(1/k_z')R(k_\perp')U(k_z',k_z'')\delta(k_\perp'-k_\perp'') \quad , \tag{2.236}$$

where we have suppressed the $k_\perp$ dependence in U. Substitute the results in (2.232), scale the resulting equation by the factor $K_1'$, viz., $k_z = K_1'\xi$, $k_z' = K_1'\xi'$ and $k_z'' = K_1'\xi''$, and introduce the notation [2.128] $T(\xi',\xi'') = U(K_1'\xi',K_1'\xi'')$. If we further subscript this T-function we can define a hierarchy of approximations satisfying one-dimensional singular integral equations of the form

$$T_{mn}(\xi',\xi'') = \sum_{i=1}^{m} u_i(\xi',\xi'') + (\pi i)^{-1}R(k_\perp')\int_{-\infty}^{\infty}\sum_{j=1}^{n} u_j(\xi',\xi)(\xi^2-1)^{-1}P(1/\xi) \cdot T_{mn}(\xi,\xi'')d\xi \quad . \tag{2.237}$$

Here for example

$$u_1(\xi',\xi'') = \exp\left[-\frac{1}{2}\Sigma^2(\xi'-\xi'')^2\right] \quad , \tag{2.238}$$

where $\Sigma = \sigma K_1'$ is the Rayleigh roughness parameter. The hierarchy is defined by taking successively more Born terms (m increases) or terms in the kernel (n increases). The notation $n = 0$ is used when we drop the integral term entirely. Unfortunately the equation rapidly becomes complicated since $M_2$ (and hence $u_2$) already involves an integration over the correlation function times complicated kinematical factors, but the numerical solution for the $T_{11}$ integral equation has been given by DESANTO and SHISHA [2.128] and is discussed next.

### *c) Coherent Specular Intensity*

For plane wave incidence at angle $\theta_i$ (defined from the normal) it can be shown that the ensemble average (coherent) intensity from (2.221) is given by

$$I_c(k_\perp) = |R|^2|T_{mn}(1,-1)|^2\delta\left(k_\perp-k_\perp^i\right) \quad , \tag{2.239}$$

with the delta function indicating specular return and where the reflection coefficient is

$$R = \frac{\rho K\cos(\theta_i) - \left[1-K^2\sin^2(\theta_i)\right]^{1/2}}{\rho K\cos(\theta_i) + \left[1-K^2\sin^2(\theta_i)\right]^{1/2}} \quad , \tag{2.240}$$

and $K = k_1/k_2 = c_2/c_1$. There are three T values available, the single scatter Kirchhoff approximation which is [2.111]

$$|T_{10}(1,-1)|^2 = \exp(-4\Sigma^2) \quad , \tag{2.241}$$

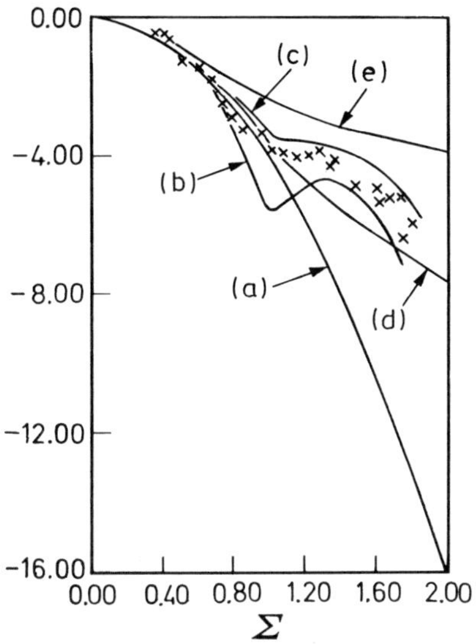


Fig. 2.24. Plot of the natural logarithm of the coherent specular intensity for random rough surface scattering versus the Rayleigh roughness parameter for data (×) and several theoretical models

the solution of the integral equation (2.237) for $m = n = 1$, which yields $|T_{11}(1,-1)|^2$, and is just Kirchhoff multiple scattering [2.128], and the eikonal (e) approximation [2.129] which is related to $T_{10}$ as

$$|T_e(1,-1)|^2 = I_0^2(2\Sigma^2)\exp(-4\Sigma^2) \quad , \tag{2.242}$$

where $I_0$ is the modified Bessel function of zeroth order. For these results, $\Sigma = k_1\sigma\cos\theta_i$.

For a soft surface $(\rho = 0, R = -1)$ these theoretical results as well as others explained below are plotted against experimental data from CLAY, MEDWIN, and WRIGHT (CMW) [2.130] and as a function of $\Sigma$ in Fig.2.24. The experimental surface has a probability density function (PDF) that was slightly non-Gaussian, and it is illustrated in Fig.4.1. In Fig.2.24 curve (a) is the result of the single scatter Kirchhoff approximation from (2.241). This was derived for a surface with a Gaussian PDF, and is essentially the magnitude squared of the Fourier transform of the PDF. The approximation works well until the surface starts to get rough ($\Sigma = 1$), but for very large roughness it fails badly. Curve (b) is also single scatter but is now the magnitude squared of the Fourier transform of the experimental PDF of the surface, $W_E$. There appears to be a small diffractive dip in the data near $\Sigma = 1$ but the theoretical results are poor in this region. For data showing a larger diffractive dip see Zornig's results in Sect.4.6. For large values of $\Sigma$ the results fit the data better than curve (a). Curves (a) and (b) illustrate the high sensitivity of single scatter results to a slight change in the PDF. Curve (c) is single scatter and is the plot of the magnitude squared of the Fourier transform of the shadow-modified PDF of the surface, $W_S$, for a large degree of shadowing. Shadowing [2.131] is a ray-optics formalism which treats the scattering from ensonified (or illuminated) portions of a rough surface different from those portions not ensonified. The result can be expressed as a multiplier of the PDF of the surface, and this yields an effective surface PDF. In this case the surface PDF was $W_E$ and the resulting shadowed PDF was $W_S$. Curve (b) was for no shadowing (shadowing parameter $B=0$, $W_S=W_E$) and curve (c) for large shadowing ($B=0.6$). This changes the effective interference in the scattered field due to different portions of the surface. CMW also show a curve ($B=0.3$) which covers the lower range of the data well. But there is a difficulty with these choices of values of B. The parameter B is related to the second derivative of the correlation function evaluated at zero, $\gamma''(0)$. It can be easily shown that $\gamma''$ is proportional to the correlation function of surface slopes. Its value at zero cannot be experimentally measured and must be extrapolated from measured correlation data at non-zero separations. The value so extrapolated is about two orders of magnitude lower than that value necessary to fit the data (i.e., a value of B between 0.3 and 0.6). One possible conclusion (CMW) is that the extrapolation procedure is questionable [2.132]. Another possible conclusion, and the one we believe, is that the use of shadowing functions to fit experimental data amounts to curve fitting with the free parameter $\gamma''(0)$. The reason, the extrapolated value of $\gamma''$ fails to give the correct results, is due to the neglect of multiple scattering effects.

In Fig.2.24, curve (d) illustrates Kirchhoff multiple scattering resulting from the numerical solution of the $T_{11}$ integral equation [2.128]. It shows a much larger coherent return than the single scatter Kirchhoff in (a). The results fall in the same region as the single scatter shadow results. They contain no correlation effects (which would be in $T_{22}$ for example) and, unlike the shadow results, are parameter free. Essentially in $T_{11}$ we have a Gaussian distribution of surfaces with infinite correlation distance, i.e., a Gaussian distribution of flat surfaces. The multiple scatter results thus illustrate an interference effect analogous to a random phase screen scatter model [2.133]. They still fall below the data, however. There are two reasons for this. The first is that the experimental surface was slightly non-Gaussian. Although this is not as strong a perturbation as in single scattering, there is some effect. It appears from the CMW data and other data [2.134] from an even more non-Gaussian surface that the purely Gaussian surface gives the lowest coherent return due to interference effects. Other results from random facet models [2.135] appear to corroborate some of this. Secondly, the smooth multiple scatter curve does not reproduce the oscillatory data because of the neglect of correlation effects. Finally, in Fig.2.24, curve (e) is the single scatter eikonal approximation from (2.242) for a Gaussian PDF. For large roughness it is much higher than the CMW data, although it fits other scatter data [2.134] but not the corresponding surface PDF results.

We have spent considerable time on coherent specular scattering. Scattering into other angles rigorously requires the solution of the Bethe-Salpeter equation [2.121].

This is very involved and has not been done. It is our belief that any theory must be able to predict the coherent specular return for large roughness or it will fail when applied to more complicated situations.

*d) Remarks*

We briefly mention some of the topics in surface scattering which we have neglected. There are papers which treat the relation between the Kirchhoff and perturbation expansion in powers of h [2.136,137] and the use of perturbation methods [2.138] or Monte-Carlo surface scattering [2.139] in propagation problems. Backscattering theories [2.140,141] and composite surface models [2.142] which treat the surface scatter problem as two superimposed rough surfaces with different scales, are sometimes combined. It is known that Kirchhoff scattering is not self-consistent [2.143], and questions arise as to its relation to other scatter models, most notably that due to ECKART [2.144] which is both self-consistent and satisfies reciprocity. In Chap.4 many of the interesting results from controlled tank experiments are discussed and include time-dependent scattering results which we have ignored. We mentioned an interesting experiment now being done by NUMRICH [2.145] which can be used to check some of our theoretical results. Finally, we mentioned the comprehensive thesis by ROSICH [2.146] which contains a thorough discussion of perturbation methods, many results on the form and position of grating anomalies in the presence of surface impedance boundary conditions, and includes an excellent historical introduction to rough surface scattering theory. Many other references and models are contained in the review articles [2.113-117] and in Chap.4. As one will see from even a brief perusal, they constitute a quite vast array.

## References


2.1 W. Krauss: *Methods and Results of Theoretical Oceanography I. Dynamics of the Homogeneous and Quasihomogeneous Ocean* (Gebrüder Bornträger, Berlin 1973)

2.2 F.V. Hunt: J. Acoust. Soc. Am. *27*, 1019-1039 (1955)
P.G. Bergmann: J. Acoust. Soc. Am. *17*, 329-333 (1946)

2.3 L.D. Landau, E.M. Lifshitz: *Fluid Mechanics* (Addison-Wesley, Reading, Mass. 1959)

2.4 K.M. Case: "Burger's Equation, Generalizations and Solutions", in *Statistical Models and Turbulence*, ed. by M. Rosenblatt, C. Van Atta, Lecture Notes in Physics, Vol.12 (Springer, Berlin, Heidelberg, New York 1972)

2.5 M. Jammer: *The Philosophy of Quantum Mechanics* (Wiley and Sons, New York 1974)

2.6 R.E. Collins: Found. Phys. *7*, 475-494 (1977)

2.7 C.Y. Wong: J. Math. Phys. *17*, 1008-1010 (1976)

2.8 A.C. Newell (ed.): *Nonlinear Wave Motion*, Lectures in Applied Mathematics, Vol.15 (American Mathematical Society, Providence, RI 1974)

2.9 C. Eckart: Phys. Rev. *73*, 68-76 (1948)
S. Hanish: "Nonlinear Acoustics"; Rpt. NRL 7772 (Naval Research Laboratory, Washington, DC 1974)

2.10 J.J. Markham: Phys. Rev. *86*, 710-711 (1952)
2.11 J.A. DeSanto: Submitted to J. Acoust. Soc. Am.
2.12 I. Tolstoy: *Wave Propagation* (McGraw-Hill, New York 1973)
2.13 O.M. Phillips: *The Dynamics of the Upper Ocean* (Cambridge University Press, Cambridge 1969);
J. Roberts: *Internal Gravity Waves in the Ocean* (Marcel Dekker, New York 1975);
R.P. Mied, J.P. Dugan: J. Phys. Ocean. *4*, 493-498 (1974)
2.14 W.H. Munk, F. Zachariasen: J. Acoust. Soc. Am. *59*, 818-838 (1976)
S.M. Flatte, F.D. Tappert: J. Acoust. Soc. Am. *58*, 1151-1159 (1975)
2.15 W.D. Wilson: J. Acoust. Soc. Am. *32*, 1357 (1960)
V.A. Del Grosso: J. Acoust. Soc. Am. *47*, 947-949 (1970)
2.16 I. Stakgold: *Boundary Value Problems of Mathematical Physics*, Vol.I (Macmillan, New York 1967)
2.17 R.L. Deavenport: Radio. Sci. *1*, 709-724 (1966)
2.18 C.L. Pekeris: *Theory of Propagation of Explosive Sound in Shallow Water*, Geol. Soc. Am. Mem., Vol.27 (1948)
2.19 L.B. Felsen, N. Marcuvitz: *Radiation and Scattering of Waves* (Prentice-Hall, Englewood Cliffs, NJ 1973)
2.20 W.M. Ewing, W.S. Jardetzky, F. Press: *Elastic Waves in Layered Media* (McGraw-Hill, New York 1957)
2.21 D.C. Stickler: J. Acoust. Soc. Am. *57*, 856-861 (1975); and "Normal Modes in Ocean Acoustics", in *International Workshop on Low-Frequency Propagation and Noise* (U.S. Government Printing Office, Washington, DC 1977) pp.125-154
2.22 C. Bartberger: J. Acoust. Soc. Am. *61*, 1643 (1977)
2.23 T. Tamir, A.A. Oliner: Proc. IEEE *51*, 317-332 (1963)
2.24 F.M. Labianca: J. Acoust. Soc. Am. *53*, 1137-1147 (1973)
2.25 C.T. Tindle, A.P. Stamp, K.M. Guthrie: J. Sound. Vib. *49*, 231-240 (1976)
A.O. Williams, Jr.: "Discrete, Continuous and Virtual Modes in Underwater Sound Propagation"; Tech. Rpt. ARL-TR-76-40, Applied Research Laboratory, Austin, Texas (1976)
2.26 R.G. Newton: *Scattering Theory of Waves and Particles* (McGraw-Hill, New York 1966)
2.27 C.W. Spofford (ed.): " A Synopsis of the AESD Workshop on Acoustic-Propagation Modelling by Non-Ray-Tracing Techniques"; TN-73-05, Acoustic Environment Support Detachment, Office of Naval Research (1973)
2.28 R.G. Newton: *The Complex J-Plane* (Benjamin, New York 1964)
R. Omnes, M. Froissart: *Mandelstam Theory and Regge Poles* (Benjamin, New York 1963)
2.29 N.N. Voitovich, A.D. Shatrov: Akusticheskii Zhurnal *18*, 516-523 (1972) [English transl.: Sov. Phys.-Acoust. *18*, 434-438 (1973)]
2.30 R. Vasudevan, K. Venkatesan, G. Jagannathan: Nuovo Cimento, Suppl. *5*, 621-641 (1967)
2.31 J.R. Wait: *Electromagnetic Waves in Stratified Media* (Pergamon Press, New York 1970) Chaps.3,4
B.S. Westcott: Quart. J. Appl. Math. *23*, 431-440 (1970)
2.32 L. Colin (ed.): "Mathematics of Profile Inversion"; NASA TMX-62, 150, Moffett Field, California (1971)
2.33 J.A. DeSanto: J. Math. Phys. *17*, 1752-1753 (1976)
2.34 R.D. Graves, A. Nagl, H. Überall, G.L. Zarur: J. Acoust. Soc. Am. *58*, 1171-1177 (1975)
2.35 M. Born, E. Wolf: *Principles of Optics* (Pergamon Press, New York 1965)
2.36 L.P. Soloman, L. Armijo: J. Acoust. Soc. Am. *50*, 960-963 (1971)
2.37 D. Ludwig: Comm. Pure Appl. Math. *19*, 215-250 (1966)
2.38 N. Bleistein: "A Brief Introduction to Ray Methods for Dispersive Hyperbolic Equations" [Ref.2.8, pp.133-142]
2.39 I.M. Blatstein: J. Acoust. Soc. Am. *52*, 1060-1061 (1972)
2.40 J.L. Synge: *Geometrical Mechanics and deBroglie Waves* (Cambridge University Press, Cambridge 1937)
2.41 C.T. Tindle, K.M. Guthrie: J. Sound. Vib. *34*, 291-295 (1974)
2.42 V.I. Chuprynin: Akust. Zh. *20*, 930-932 (1974) [English transl.: Sov. Phys. Acoust. *20*, 574-575 (1975)]

2.43 H.R. Krol: J. Acoust. Soc. Am. *54*, 96-98 (1973)
A. Barnes, L. Solomon: J. Acoust. Soc. Am. *53*, 147-155 (1973)
2.44 S.M. Flatte: J. Acoust. Soc. Am. *60*, 1020-1023 (1976)
H. Cox: J. Acoust. Soc. Am. *61*, 353-359 (1977)
2.45 M.A. Pederson, D. White, D.W. Johnson: J. Acoust. Soc. Am. *58*, 78-96 (1975)
2.46 R.F. Mackinnon, J.S. Partridge, S.H. Toole: J. Acoust. Soc. Am. *52*, 1471-1480 (1972)
2.47 E.L. Murphy: J. Acoust. Soc. Am. *47*, 899-908 (1970)
2.48 C.H. Harrison: J. Acoust. Soc. Am. *62*, 1382-1388 (1977)
2.49 H. Weinberg: J. Acoust. Soc. Am. *58*, 97-109 (1975)
2.50 E.S. Eby: J. Acoust. Soc. Am. *47*, 273-275 (1969)
2.51 W.D. Yang, G.S. Deschamps: Proc. IEEE *62*, 1541-1551 (1974)
2.52 "Geometrical Acoustics", Proc. SACLANTCEN Conf., La Spezia, Italy (1971)
2.53 F.D. Tappert: "The Parabolic Approximation Method", in *Wave Propagation and Underwater Acoustics*, ed. by J.B. Keller, J.S. Papadakis, Lecture Notes in Physics, Vol.70 (Springer, Berlin, Heidelberg, New York 1977)
2.54 S.T. McDaniel: J. Acoust. Soc. Am. *57*, 307-311 (1975); *58*, 1178-1185 (1975)
A.O. Williams, Jr.: J. Acoust. Soc. Am. *58*, 1320-1321 (1975)
2.55 R.M. Fitzgerald: J. Acoust. Soc. Am. *57*, 839-842 (1975)
D.R. Palmer: J. Acoust. Soc. Am. *60*, 343-354 (1976)
2.56 J. Corones: J. Math. Anal. Appl. *50*, 361-372 (1975)
F. Jensen, H. Krol: "The Use of the Parabolic Equation Method in Sound Propagation Modelling"; Tech. Rpt. Sm-72, NATO SACLANTCEN, La Spezia, Italy (1975)
2.57 J.A. DeSanto: J. Acoust. Soc. Am. *62*, 295-297 (1977)
2.58 E.A. Polyanskii: Akust. Zh. *20*, 142-143 (1974) [English transl.: Sov. Phys. Acoust. *20*, 90 (1974)]
2.59 J.A. DeSanto, J.S. Perkins, R.N. Baer: J. Acoust. Soc. Am. (1978)
2.60 H.K. Brock, R.N. Buchal, C.W. Spofford: J. Acoust. Soc. Am. *62*, 543-552 (1977)
2.61 J.A. DeSanto: "Connection between the solutions of the Helmholtz and parabolic equations for sound propagation"; Proc. NATO Conf. on Oceanic Acoustic Modelling, ed. by W. Bachmann, R.B. Williams (SACLANTCEN, La Spezia, Italy 1975) pp.43-1 to -17
2.62 D.H. Wood: "Elliptic partial differential equations may be related by a change of independent variables", in Lecture Notes in Mathematics, Vol.280, ed. by W.N. Everitt, B.D. Sleeman (Springer, Berlin, Heidelberg, New York 1972)
V.I. Ivanov: Zh. vych. mat. *1*, 246-254 (1961) [English transl.: USSR Comp. Math. and Math. Phys. *1*, 277-288 (1962)]
2.63 H. Weinberg, R. Burridge: J. Acoust. Soc. Am. *55*, 63-79 (1974)
2.64 N.N. Bojarski: "k-Space Formulation of the n-Dimensional Scattering Problem" [Ref.2.32, pp.8-13 to -25], and other reports available from the author (16 Pine Valley Lane, Newport Beach, Calif., USA)
2.65 R.N. Baer, W.S. Moseley: J. Acoust. Soc. Am. *61*, S11 (1977)
2.66 U. Frisch: "Wave Propagation in Random Media", in *Probabilistic Methods in Applied Mathematics*, Vol.1, ed. by A.T. Bharucha-Reid (Academic Press, New York 1968) pp.75-198
2.67 Yu.N. Barabanenkov, Yu.A. Kravtsov, S.M. Rytov, V.I. Tatarskii: Usp. Fiz. Nauk *102*, 3-42 (1970) [English transl.: Sov. Phys.-Usp. *13*, 551-575 (1971)]
2.68 G. Birkhoff, R. Bellman, C.C. Lin (eds.): *Hydrodynamic Instability*, Proc. Symp. Appl. Math., Vol.13 (American Mathematical Society, Providence, RI 1964)
2.69 R. Bellman (ed.): *Stochastic Processes in Mathematical Physics and Engineering*, Proc. Symp. Appl. Math., Vol.16 (American Mathematical Society, Providence, RI 1964)
2.70 V.I. Tatarskii: *The Effects of the Turbulent Atmosphere on Wave Propagation*, available as TT68-50464 from U.S. Dept. of Commerce, NTIS, Springfield, Va. (Translation of the Russian original)
2.71 J.B. Keller: "Wave Propagation in Random Media", in [Ref.2.68, pp.227-246]
2.72 J.M. Richardson: "Application of Truncated Hierarchy Techniques", in [Ref.2.69, pp.290-302]
2.73 P.-L. Chow: SIAM Rev. *17*, 57-81 (1975)
2.74 A.D. Varvatsis, M.I. Sancer: Radio Sci. *6*, 87-97 (1971)
D.A. DeWolf: J. Opt. Soc. Am. *58*, 461-466 (1968)

2.75 R.C. Bourret: Can. J. Phys. *44*, 2519-2524 (1966)
2.76 J.B. Keller: J. Opt. Soc. Am. *59*, 1003 (1969)
2.77 D.A. DeWolf: J. Acoust. Soc. Am. *54*, 1109-1110 (1973)
2.78 G.A. Baraff: J. Math. Phys. *11*, 1938-1941 (1970)
2.79 I. Lerche, E.N. Parker: J. Math. Phys. *14*, 1949-1954 (1973); *16*, 1838-1839 (1975)
2.80 J.L. Doob: Ann. of Math. *43*, 351-369 (1942)
2.81 M.J. Beran, G. Parrent, Jr.: *Theory of Partial Coherence* (Prentice-Hall, Englewood Cliffs, NJ 1964)
2.82 I.M. Besieris, F.D. Tappert: J. Math. Phys. *14*, 1829-1836 (1973)
2.83 K. Huang: *Statistical Mechanics* (Wiley and Sons, New York 1963)
I.M. Besieris, F.D. Tappert: J. Math. Phys. *17*, 734-743 (1976)
2.84 M.J. Beran, A.M. Whitman: J. Math. Phys. *16*, 214-217 (1975)
2.85 E. Collett, J.T. Foley, E. Wolf: J. Opt. Soc. Am. *67*, 465-467 (1977)
2.86 J.J. McCoy, M.J. Beran: J. Acoust. Soc. Am. *61*, 596-599 (1977)
2.87 R. Dashen: "Path Integrals for Waves in Random Media"; Tech. Rpt. JSR-76-1 (Stanford Research Institute, Menlo Park, Calif. 1977)
2.88 J.A. DeSanto: J. Acoust. Soc. Am. *62*, S18 (1977); and in preparation
2.89 W.C. Hoffman: "Wave Propagation in a General Random Continuous Medium", in [Ref.2.69, pp.117-144]
2.90 L.B. Dozier, F.D. Tappert: J. Acoust. Soc. Am. *63*, 353-365 (1978)
2.91 J.W.S. Rayleigh: *The Theory of Sound*, Vol.2 (Dover, New York 1945) pp.89-96
2.92 R. Petit, M. Cadilhac: C.R. Acad. Sci. Paris *262B*, 468-471 (1966)
2.93 R.F. Millar: Proc. Camb. Phil. Soc. *65*, 773-791 (1969); *69*, 217-225 (1971)
2.94 B.A. Lippmann: J. Opt. Soc. Am. *43*, 408 (1953)
2.95 J.L. Uretsky: Annals of Physics *33*, 400-427 (1965)
2.96 R.F. Millar: Radio Sci. *8*, 785-796 (1973)
2.97 R.L. Holford: "Scattering of Sound Waves at a Periodic, Pressure Release Surface: An Exact Solution"; Tech. Rpt., Bell Telephone Laboratories (1971)
2.98 V.I. Kantorovich, V.I. Krylov: *Approximate Methods of Higher Analysis* (Interscience, New York 1958)
2.99 R.K. Rosich, J.R. Wait: Radio Sci. *12*, 719-729 (1977)
2.100 J.A. DeSanto: J. Math. Phys. *12*, 1913-1923 (1971)
2.101 J.A. DeSanto: J. Math. Phys. *13*, 336-341 (1972)
2.102 J.A. DeSanto: J. Acoust. Soc. Am. *53*, 719-734 (1973)
2.103 J.A. DeSanto: J. Acoust. Soc. Am. *56*, 1336-1341 (1974)
2.104 R. Mittra, S.W. Lee: *Analytical Techniques in the Theory of Guided Waves* (Macmillan, New York 1971)
2.105 C.J. Bouwkamp: Physica *12*, 467-474 (1946)
2.106 A. Hessel, A.A. Oliner: Appl. Opt. *4*, 1275-1297 (1965)
2.107 E.V. Jull, J.W. Heath, G.R. Ebbeson: J. Opt. Soc. Am. *67*, 557-560 (1977)
2.108 J.A. DeSanto: J. Acoust. Soc. Am. *57*, 1195-1197 (1975)
2.109 G. Whitman, F. Schwering: IEEE Trans. AP-*25*, 869-876 (1977)
2.110 P.C. Waterman: J. Acoust. Soc. Am. *57*, 791-802 (1975)
2.111 P. Beckmann, A. Spizzichino: *The Scattering of Electromagnetic Waves from Rough Surfaces* (Pergamon Press, New York 1963)
2.112 F.G. Bass, I.M. Fuks: *Scattering of Waves by Statistically Irregular Surfaces I*; Rpt. JPRS 66061-1 and *II*; Rpt. JPRS 66061-2, National Technical Information Service, Springfield, Va. (1975). (Translation of RASSEYANIYE VOLN NA STATISTICHESKI NEROVNOY POVERKHNOSTI, Moscow 1972)
2.113 A.B. Shmelev: Usp. Fiz. Nauk *106*, 459-480 (1972) [English transl.: Sov. Phys.-Usp. *15*, 173-183 (1972)]
2.114 L. Fortuin: J. Acoust. Soc. Am. *47*, 1209-1228 (1970)
2.115 C.W. Horton, Jr.: J. Acoust. Soc. Am. *51*, 1049-1061 (1972)
2.116 B.G. Hurdle, K.D. Flowers, J.A. DeSanto: In Proc. NATO Conf. on Oceanic Acoustic Modelling, ed. by W. Bachmann, R.B. Williams (SACLANTCEN, La Spezia, Italy 1975) pp.20-1 to -28
2.117 H. Trinkhaus: In Proc. NATO Conf. on Oceanic Acoustic Modelling, ed. by W. Bachmann, R.B. Williams (SACLANTCEN, La Spezia, Italy 1975) pp.12-1 to -36
2.118 K.M. Mitzner: J. Math. Phys. *7*, 2053-2060 (1966)
2.119 R.E. Kleinman: "Low Frequency Methods in Classical Scattering Theory"; Lecture Notes, The Technical University of Denmark, Lyngby (1966)

2.120 D.N. Pattanayak, E. Wolf: Phys. Rev. *D13*, 2287-2290 (1976)
2.121 G.G. Zipfel, J.A. DeSanto: J. Math. Phys. *13*, 1903-1911 (1972)
2.122 J.A. DeSanto: J. Acoust. Soc. Am. *58*, S66 (1975)
2.123 J.A. DeSanto: In Proc. NATO Conf. on Oceanic Acoustic Modelling, ed. by W. Bachmann, R.B. Williams (SACLANTCEN, La Spezia, Italy 1975) pp.17-1 to -13
2.124 L.M. Brekhovskikh: *Waves in Layered Media* (Academic Press, New York 1960)
2.125 J.A. DeSanto: J. Math. Phys. *14*, 1566-1573 (1973)
2.126 J.A. DeSanto: J. Math. Phys. *15*, 283-288 (1974)
2.127 H.H. Essen, K. Hasselmann: Z. Geophysik *36*, 655-678 (1970)
2.128 J.A. DeSanto, O. Shisha: J. Comp. Phys. *15*, 286-292 (1974)
2.129 R.M. Brown, A.R. Miller: "Geometric Optics Theory for Coherent Scattering of Microwaves from the Ocean"; NRL Rpt. 7705, Naval Research Laboratory (1974)
2.130 C.S. Clay, H. Medwin, W.M. Wright: J. Acoust. Soc. Am. *53*, 1677-1682 (1973)
2.131 P.J. Lynch, R.J. Wagner: J. Math. Phys. *11*, 3032-3042 (1970)
2.132 D.E. Barrick: Radio Sci. *5*, 647-654 (1970)
2.133 W.S. Ament: Trans. IRE AP-*4*, 369-373 (1956)
E. Jakeman, P.N. Pusey: J. Phys. A: Math. Gen. *8*, 369-391 (1975)
2.134 M.L. Boyd, R.L. Deavenport: J. Acoust. Soc. Am. *53*, 791-801 (1973)
2.135 J.C. Novarini, H. Medwin: J. Acoust. Soc. Am. *64*, 260-268 (1978)
2.136 F.M. Labianca, E.Y. Harper: J. Acoust. Soc. Am. *62*, 1144-1157 (1977)
2.137 G.R. Valenzuela, J.W. Wright: IEEE Trans. AP-*21*, 536-537 (1972)
2.138 W.A. Kuperman, F. Ingenito: J. Acoust. Soc. Am. *61*, 1178-1187 (1977)
2.139 H.-G. Schneider: Acustica *35*, 18-25 (1976)
2.140 W. Bachmann: J. Acoust. Soc. Am. *54*, 712-716 (1973)
2.141 R.A. Brockelman, T. Hagfors: IEEE Trans. AP-*14*, 621-629 (1966)
2.142 J.C. Huang: J. Acoust. Soc. Am. *49*, 1600-1608 (1971)
2.143 E. Wolf, E.W. Marchand: J. Opt. Soc. Am. *54*, 587-594 (1964)
2.144 C. Eckart: J. Acoust. Soc. Am. *25*, 566-570 (1953)
2.145 S.K. Numrich: "Scattering of Acoustic Waves from Rough Surfaces"; Ph.D. dissertation, American University, Washington, DC (1978)
2.146 R.K. Rosich: "Electromagnetic Scattering from Periodic Rough Surfaces"; Ph.D. dissertation, University of Colorado, Boulder, Colo. (1977)

# 3. Numerical Models of Underwater Acoustic Propagation

F. R. DiNapoli and R. L. Deavenport

With 21 Figures

In this chapter an attempt is made to summarize those models of propagation loss in the field of underwater acoustics which have been converted into an automated computer code capable of being executed by someone other than the originator for a wide variety of problems. No single model currently exists which is adequate for all applications. This is perhaps not surprising considering the diversity of the ocean environment and its boundaries, and the concomitant fact that the acoustic frequencies of interest span the regime from less than 10 Hz to greater than 100 kHz. As a result a large number of models, each with its own domain of validity which in many cases is difficult to precisely define, have been developed. Their sheer number makes an exhaustive summary impossible within these limited pages. Thus it was decided to limit consideration to those models which purport to be a solution of the wave equation found in Sect.2.2.1. Fundamentally these models consider the ocean to be a deterministic environment for which the speed of sound is only a function of the spatial coordinates. Non-deterministic effects, if accounted for at all, are included in an ad hoc fashion following the determination of the deterministic propagation loss result. Model development work for the more general problem is required and is in progress. However, this effort has not reached the point where "hands off" computer codes are available. This is due in part to a lack of available experimental/environmental data and the need for larger and faster computers.

The models to be discussed can then be further segregated into range independent and range dependent categories. The former assume that the ocean is cylindrically symmetrical, the speed of sound an arbitrary function of only the depth (z) coordinate, and that all boundaries are parallel with the range (r) coordinate. These models discussed, in Sect.3.1, are in a fairly complete state of development as evidenced by the concern with reducing computer execution time and memory without significantly sacrificing accuracy. The range-dependent models, taken up in Sect.3.2, allow for the speed of sound to be an arbitrary function of either 2 or 3 spatial coordinates, and the boundaries need not be parallel. Their state of development is not as complete.

In general, the models which fall within these two subdivisions consider the ocean surface to be a pressure release boundary due to the large mismatch in characteristic impedance between water and air. The water column itself is treated as an ideal fluid incapable of supporting shear and having a uniform or at most piecewise constant density variation. For some models it is necessary to specify sound speed, attenuation and density values within the bottom. In these instances, the bottom is treated in a manner analogous to that given the water column, i.e., as an ideal fluid. Otherwise the effect of the ocean bottom is accounted for by ascribing to it a reflection loss versus grazing angle. The treatment given to both boundaries is, of course, approximate and in the case of the ocean bottom delineates a clear point of departure between underwater acoustics and seismic propagation. The approximations nonetheless have been found to be generally adequate when comparing theoretical and experimental propagation loss results. This is perhaps related to the second point of departure; the concern in underwater acoustics with the steady state, single frequency propagation loss versus range as opposed to a detailed analysis of waveform structure versus time in the seismic field.

## 3.1 Range-Independent Models

Of particular interest in this section are the phenomena associated with wave propagation in a plane stratified medium whose characteristic properties are arbitrary functions of only one spatial coordinate. A point monochromatic source, of angular frequency $\omega$, is assumed to be located at $(0,z_s)$ in an azimuthally symmetric cylindrical coordinate system $(r,\theta,z)$. It has been shown in Chap.2 that the time-independent pressure field $\mathscr{P}(r,z,\omega)$ at range $(r)$ and depth $(z)$ can be expressed by the Fourier-Bessel transform

$$\mathscr{P}(r,z,\omega) = \frac{S_\omega}{2\pi} \int_0^\infty G(z,z_s;\xi,\omega) J_0(\xi r)\xi \, d\xi \quad , \tag{3.1}$$

where $S_\omega$ represents the source strength, $J_0(\xi r)$ the zero order Bessel function, and $G(z,z_s)$ is a depth dependent Green's function.

The available models for computing the propagation loss versus range in such an environment can be grouped into five major categories (shown schematically in Fig.3.1):

1) Direct numerical integration,
2) residue (normal-mode) theory,
3) multipath expansion,
4) ray theory with corrections, and
5) classical ray theory

according to the method used for solving (3.1). Within each group one or two particular models have been identified as being representative.

These models are discussed in some detail to provide a feeling for the overall features or shortcomings of all models falling within that category versus those in

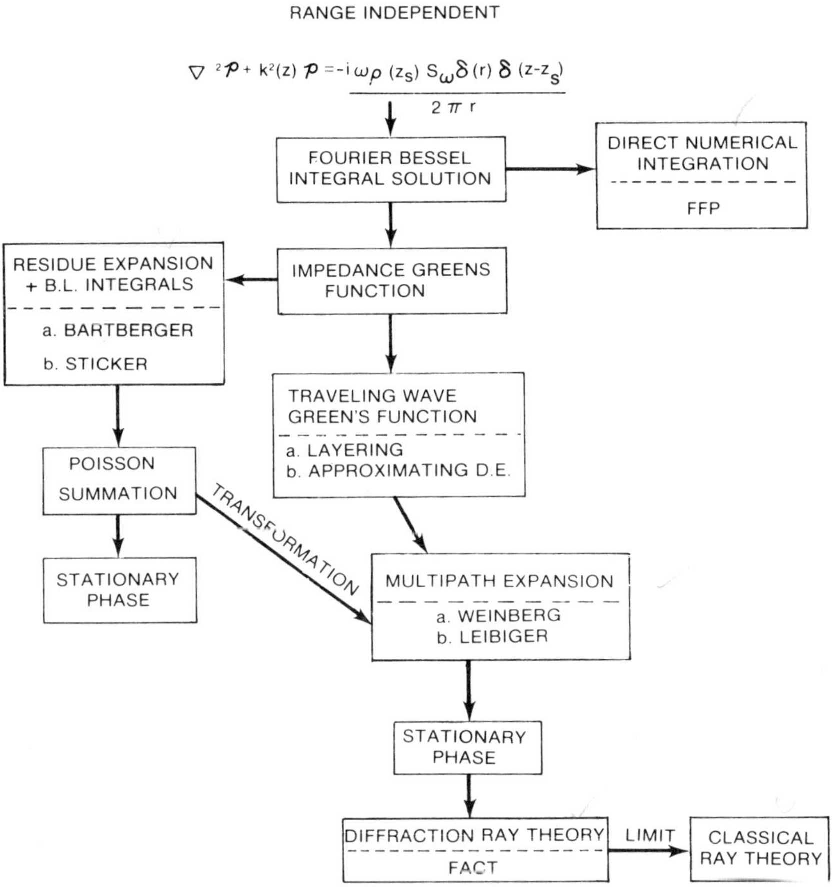


Fig. 3.1. Range-independent flow diagram of theoretical solutions

another. It is also possible to quantitatively assess the accuracy of the models of this section. This is provided along with a description of the methodology used to obtain the results. The same convention and notation for mathematical functions have been used throughout, rather than adopting those utilized by the originators. Thus, the depth (z) is always taken to be measured positively downwards from the ocean surface, and the harmonic function exp(-iωt) has been uniformly suppressed.

Many of the features/shortcomings of proceeding down one path in Fig.3.1 versus another are intimately related to the particular manner in which the depth-dependent Green's function G appearing in the integral solution, is obtained and ultimately evaluated numerically. For that reason its discussion is taken up for the case of a piecewise continuous layered media.

### 3.1.1 Depth-Dependent Green's Function (Impedance Formulation)

The integral solution for the acoustic pressure $\mathscr{P}(r,z,\omega)$, contains a product of a range (r) dependent Bessel function, and depth (z) dependent Green's function. The derivation of the former is straightforward since the index of refraction is assumed to be independent of range. The result for the depth-dependent Green's function is rather complicated due to the piecewise nature assumed for the depth dependence of the index of refraction. A derivation of G for this geometry is summarized below and given in more detail in [3.1].

In cylindrical coordinates with azimuthal symmetry the depth-dependent Green's function $G(z,z_s)$ must satisfy

$$\mathscr{L}(G) = \frac{d}{dz}\left(\frac{1}{i\omega\rho(z)}\frac{dG}{dz}\right) + q(z)G = -\delta(z-z_s) \quad , \tag{3.2}$$

where $q(z) = [k^2(z)-\xi^2][i\omega\rho(z)]^{-1}$, $k(z)$ the wavenumber [=2π/wave length=ω/sound speed variation $c(z)$] and $\rho(z)$ the density. All the models of this section assume that the density varies in a piecewise constant fashion within any layer. The differential operator $\mathscr{L}$ is defined as

$$\mathscr{L} = \frac{d}{dz}\left(\frac{1}{i\omega\rho(z)}\frac{d}{dz}\right) + q(z) \quad .$$

The solution of (3.2) is found by using the matrizant method [3.2-5] to obtain two linearly independent functions, $\tilde{F}$ and F, which satisfy (3.2) for $z < z_s$ and $z > z_s$ and the respective continuity conditions on $G(z)$ and $G'(z)/[i\omega\rho(z)]$ in those regions, where $G'(z) = dG/dz$. These functions represent the depth-dependent pressure and are unique, up to an arbitrary constant, which may be chosen such that (3.2) is also satisfied at the source depth $z_s$.

The notation used in this chapter was instituted to conserve space. Although compact, it is unorthodox, and will be explained at points where it is felt that confusion might arise. The symbol U without a subscript or superscript is an abbreviation for the depth-dependent particle velocity $U = (dF/dz)[i\omega\rho(z)]^{-1}$, for $z > z_s$. When it appears with a subscript and superscript, e.g., $U_k^{k+1}$, the subscript (k) refers to the layer in which F is defined. The superscript (k+1) refers to the depth ($z_{k+1}$, Fig.3.2) at which the operation is evaluated. Thus

$$U_k^{k+1} = \frac{1}{i\omega\rho_k(z)} \left.\frac{dF_k}{dz}\right|_{z=z_{k+1}} \quad ,$$

and as such $U_k^{k+1}$ is the depth dependent particle velocity in the $k^{th}$ layer evaluated at the depth $z_{k+1}$. If the subscript is absent and the superscript is z, then $U^z$ is the depth dependent particle velocity in any layer ($LS \leq k \leq N-1$) evaluated at any depth ($z_s \leq z \leq z_N$). The layer in which the source is located will be designated by LS and the receiver layer by LR. The depth dependent particle velocity

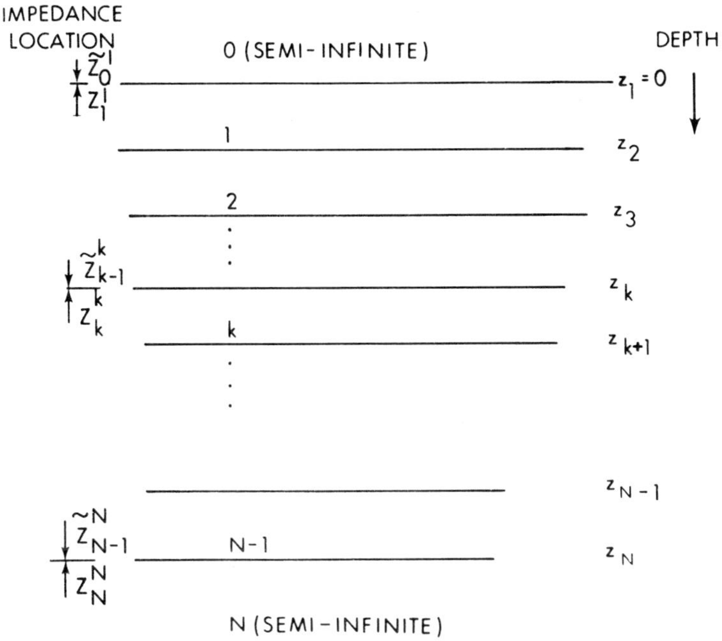


Fig. 3.2. Range-independent environmental description

$\tilde{U} = (d\tilde{F}/dz)[i\omega\rho(z)]^{-1}$ for $z < z_s$ is abbreviated similarly. Functions other than the depth-dependent particle velocity will appear with a subscript and superscript (e.g., $\gamma_k^{k+1}$). The meaning of the subscript and superscript when it occurs with these functions will be exactly the same as that explained above for U, unless otherwise stated.

Terms of the form $(df/dz)[i\omega\rho(z)]^{-1}$ and $(dg/dz)[i\omega\rho(z)]^{-1}$ appeared frequently throughout the analysis and have been abbreviated as (Df) and (Dg), respectively. Thus

$$(Df)_k^{k+1} = \left.\frac{1}{i\omega\rho_k(z)}\frac{df_k}{dz}\right|_{z=z_{k+1}} \quad .$$

Although closely related to the depth dependent particle velocity $U_k^{k+1}$, they differ from it because the function being operated on is not the total depth dependent pressure, but one of the linearly independent (fundamental) solutions used to form F.

*a) Matrizant Method*

The equation for the depth dependent pressure F, $z > z_s$, may be written as

$$\mathscr{L}(F) = \frac{d}{dz}\left(\frac{1}{i\omega\rho(z)}\frac{dF}{dz}\right) + q(z)F = 0 \quad , \tag{3.3}$$

where F is subject to the continuity conditions on G and $G'/i\omega\rho$ for $z > z_s$. In the matrizant method (3.3) is written as the two equations

$$U(z) = \frac{1}{i\omega\rho(z)}\frac{dF}{dz} \quad ; \quad \frac{dU}{dz} + q(z)F = 0 \quad ,$$

or equivalently as the matrix equation

$$\frac{d\psi(z)}{dz} = A(z)\psi(z) \quad , \tag{3.4}$$

where

$$\psi(z) = \begin{pmatrix} F(z) \\ U(z) \end{pmatrix} \quad ; \quad A(z) = \begin{pmatrix} 0 & i\omega\rho(z) \\ -q(z) & 0 \end{pmatrix} \quad .$$

The solution of (3.4) satisfying initial conditions at $z_i$ is given by

$$\psi(z) = M(z,z_i)\psi(z_i) \quad , \tag{3.5}$$

where $M(z,z_i)$ is the fundamental matrix (matrizant, propagator, etc.) which satisfies the matrix equation

$$\frac{dM(z,z_i)}{dz} = A(z)M(z,z_i) \quad ,$$

such that $M(z_i,z_i) = I$, where I is the identity matrix.

It is not possible to find a closed-form solution, other than the Peano-Baker expansion, when the entire interval $(z_1 \leq z \leq z_N)$ is considered as a single layer with k(z) arbitrary. An alternative is to subdivide the total interval into N-1 subintervals, as in Fig.3.2, such that known closed-form solutions to a second-order differential equation in $\gamma^z$ $(=\gamma(z))$ can be found associated with some $k_\ell(z)$ which is a good approximation to the given k(z) over the subinterval $(z_\ell \leq z \leq z_{\ell+1})$. If two fundamental solutions, $f(\gamma^z)$ and $g(\gamma^z)$ of $\mathscr{L}(\cdot) = 0$, can be expressed in terms of the known solutions, then the matrizant

$$M(z,z_i) = M(\gamma^z,\gamma^{z_i}) = \begin{pmatrix} Q(\gamma^z,\gamma^{z_i}) & P(\gamma^z,\gamma^{z_i}) \\ S(\gamma^z,\gamma^{z_i}) & R(\gamma^z,\gamma^{z_i}) \end{pmatrix} \quad , \tag{3.6}$$

over some arbitrary subinterval $[z,z_i]$ may be expressed as

$$M(\gamma^z,\gamma^{z_i}) = \begin{pmatrix} f(\gamma^z) & g(\gamma^z) \\ (Df)^z & (Dg)^z \end{pmatrix} \begin{pmatrix} f(\gamma^{z_i}) & g(\gamma^{z_i}) \\ (Df)^{z_i} & (Dg)^{z_i} \end{pmatrix}^{-1} \quad . \tag{3.7}$$

The use of the argument $\gamma^z$ is intended as a visual reminder that f and g are to be expressed in terms of known solutions of a differential equation in $\gamma^z$. In Sect.2.2.1 a procedure is given for relating the differential equation in z to the one in $\gamma^z$. The matrizant for the entire interval can then be obtained by making use of its group property

$$M\left(\gamma^{z},\gamma^{z_i}\right) = M\left(\gamma^{z},\gamma^{z_k}\right)M\left(\gamma^{z_k},\gamma^{z_i}\right) \quad , \quad z,z_i,z_k \in [z_1,z_N] \quad , \tag{3.8}$$

and $M^{-1}\left(\gamma^{z},\gamma^{z_i}\right)=M\left(\gamma^{z_i},\gamma^{z}\right)$. From (3.7) the elements of (3.6) are found to be

$$Q\left(\gamma^{z},\gamma^{z_i}\right) = \frac{\left[f(\gamma^{z})(Dg)^{z_i}-(Df)^{z_i}g(\gamma^{z})\right]}{\left[f\left(\gamma^{z_i}\right)(Dg)^{z_i}-(Df)^{z_i}g\left(\gamma^{z_i}\right)\right]} \quad , \quad Q\left(\gamma^{z_i},\gamma^{z_i}\right) = 1 \quad ,$$

$$R\left(\gamma^{z},\gamma^{z_i}\right) = \frac{\left[-(Df)^{z}g\left(\gamma^{z_i}\right)+(Dg)^{z}f\left(\gamma^{z_i}\right)\right]}{\left[f\left(\gamma^{z_i}\right)(Dg)^{z_i}-(Df)^{z_i}g\left(\gamma^{z_i}\right)\right]} \quad , \quad R\left(\gamma^{z_i},\gamma^{z_i}\right) = 1 \quad ,$$

$$S\left(\gamma^{z},\gamma^{z_i}\right) = \frac{\left[(Df)^{z}(Dg)^{z_i}-(Df)^{z_i}(Dg)^{z}\right]}{\left[f\left(\gamma^{z_i}\right)(Dg)^{z_i}-(Df)^{z_i}g\left(\gamma^{z_i}\right)\right]} \quad , \quad S\left(\gamma^{z_i},\gamma^{z_i}\right) = 0 \quad ,$$

$$P\left(\gamma^{z},\gamma^{z_i}\right) = \frac{\left[-f(\gamma^{z})g\left(\gamma^{z_i}\right)+g(\gamma^{z})f\left(\gamma^{z_i}\right)\right]}{\left[f\left(\gamma^{z_i}\right)(Dg)^{z_i}-(Df)^{z_i}g\left(\gamma^{z_i}\right)\right]} \quad , \quad P\left(\gamma^{z_i},\gamma^{z_i}\right) = 0 \quad . \tag{3.9}$$

The Green's function solution to (3.2) can now be constructed from the following two matrizant equations

$$\begin{pmatrix} G(z) \\ G'(z)[i\omega\rho(z)]^{-1}\end{pmatrix} = \tilde{\psi}(z) = \begin{pmatrix}\tilde{F}(z)\\ \tilde{U}(z)\end{pmatrix} = M\left(\gamma^{z},\gamma^{z_1}\right)\tilde{\psi}(z_1) \quad , \quad (z_1 \leqq z < z_s)$$

$$\begin{pmatrix} G(z) \\ G'(z)[i\omega\rho(z)]^{-1}\end{pmatrix} = \psi(z) = \begin{pmatrix}F(z)\\ U(z)\end{pmatrix} = M\left(\gamma^{z},\gamma^{z_N}\right)\psi(z_N) \quad , \quad (z_s < z \leqq z_N) \quad , \tag{3.10}$$

where $z_1$ and $z_N$ represent the lower and upper boundaries (Fig.3.2) respectively of two homogeneous half spaces between which the index of refraction is allowed to vary piecewise continuously. The initial values are thus known up to some multiplicative constants, $a_0$ and $a_N$, respectively, since

$$\tilde{F}(z) = a_0 \exp[-i\beta_0(z-z_1)] \quad , \quad \tilde{U}(z) = \frac{-\beta_0}{\omega\rho_0(z)}\tilde{F}(z) \quad , \quad z < z_1 \quad ,$$

$$F(z) = a_N \exp[i\beta_N(z-z_N)] \quad , \quad U(z) = \frac{\beta_N}{\omega\rho_N(z)}F(z) \quad , \quad z > z_N \quad ,$$

where $\beta_0$ and $\beta_N$ are the z components of the wave numbers $k_0$ and $k_N$, respectively, and $\text{Im}\{\beta_0\}$ and $\text{Im}\{\beta_N\} > 0$ are required to satisfy the radiation condition.

In what follows it is useful to rewrite $\tilde{\psi}(z)$ and $\psi(z)$ in terms of the impedance $\tilde{Z}$ above the source and Z below the source defined as

$$\tilde{Z}(z) = \frac{\tilde{F}(z)}{\tilde{U}(z)} \quad \text{and} \quad Z(z) = \frac{F(z)}{U(z)} \quad , \tag{3.11}$$

since the terminal impedances

$$Z_N^N = \frac{\omega\rho_N}{\beta_N} \quad \text{and} \quad \tilde{Z}_0^1 = \frac{-\omega\rho_0}{\beta_0} \quad , \tag{3.12}$$

are known uniquely. The convention adopted is that the superscript always refers to the depth at which Z is evaluated and the subscript to the layer where the evaluation takes place. Then the impedance anywhere above the source can be written uniquely in terms of its terminal impedance $\tilde{Z}_0^1$ by using the group property (3.8) to obtain

$$\begin{pmatrix} \tilde{Z}_k^z \\ 1 \end{pmatrix} = \frac{M_k\left(\gamma_k^z,\gamma_k^k\right) \prod\limits_{p=k-1}^{1} M_p(b,a) \begin{pmatrix} \tilde{Z}_0^1 \\ 1 \end{pmatrix}}{\left[\tilde{Z}_{k-1}^k S_k\left(\gamma_k^z,\gamma_k^k\right)+R_k\left(\gamma_k^z,\gamma_k^k\right)\right] \prod\limits_{p=k-1}^{1} \left[\tilde{Z}_{p-1}^p S_p(b,a)+R_p(b,a)\right]} \quad , \tag{3.13}$$

and similarly below the source

$$\begin{pmatrix} Z_k^z \\ 1 \end{pmatrix} = \frac{M_k\left(\gamma_k^z,\gamma_k^{k+1}\right) \prod\limits_{p=k+1}^{N-1} M_p(a,b) \begin{pmatrix} Z_N^N \\ 1 \end{pmatrix}}{\left[Z_{k+1}^{k+1} S_k\left(\gamma_k^z,\gamma_k^{k+1}\right)+R_k\left(\gamma_k^z,\gamma_k^{k+1}\right)\right] \prod\limits_{p=k+1}^{N-1} \left[Z_{p+1}^{p+1} S_p(a,b)+R_p(a,b)\right]} \quad , \tag{3.14}$$

where $a = \gamma_p^p$, $b = \gamma_p^{p+1}$ are evaluated at the upper and lower boundaries of the $p^{th}$ layer, respectively. The symbols a, b will be used as abbreviations for the arguments throughout except where explicitly noted to the contrary.

If the total interval $(z_1 \leq z \leq z_N)$ is partitioned into N-1 subintervals, then the matrix $\tilde{\psi}_k(z)$ in the $k^{th}$ layer may be expressed as

$$\tilde{\psi}_k(z) = M_k\left(\gamma_k^z,\gamma_k^{k+1}\right) \begin{pmatrix} \tilde{Z}_k^{k+1} \\ 1 \end{pmatrix} \tilde{U}_k^{k+1} \quad , \tag{3.15}$$

where $z_k \leq z \leq z_{k+1}$ for $1 \leq k \leq LS-1$ and $z_{LS} \leq z < z_s$ for $k = LS$, the source layer. By applying (3.8) over $[z_{k-1}, z_{k+1}]$ yields

$$\tilde{U}_k^{k+1} = \tilde{U}_{k-1}^k\left(\tilde{Z}_{k-1}^k S_k(b,a)+R_k(b,a)\right) \quad , \quad a = \gamma_k^k \quad \text{and} \quad b = \gamma_k^{k+1} \quad ,$$

thus (3.15) may be written as

$$\tilde{\psi}_k(z) = M_k\left(\gamma_k^z,\gamma_k^{k+1}\right)\begin{pmatrix}\tilde{Z}_k^{k+1}\\ 1\end{pmatrix}\left\{\tilde{U}_{k-1}^k\left[\tilde{Z}_{k-1}^k S_k(b,a)+R_k(b,a)\right]\right\} \quad .$$

Likewise the matrix $\psi_k(z)$ may be expressed as

$$\psi_k(z) = M_k\left(\gamma_k^z,\gamma_k^{k+1}\right)\begin{pmatrix}Z_{k+1}^{k+1}\\ 1\end{pmatrix}U_{k+1}^{k+1} \quad ,$$

where $z_k \leq z \leq z_{k+1}$ for $LS+1 \leq k \leq N-1$ and $z_{LS+1} \geq z > z_s$ for $k = LS$. The impedances $\tilde{Z}_k^{k+1}$ and $Z_{k+1}^{k+1}$ are the terminal impedances $\tilde{Z}_0^1$ and $Z_N^N$ referenced to the bottom of the $k^{th}$ layer and may be easily found from (3.13) and (3.14). The Green's function in the $k^{th}$ layer for $z < z_s$ may now be written as

$$G_k(z) = \left[\tilde{Z}_k^{k+1}Q_k\left(\gamma_k^z,\gamma_k^{k+1}\right) + P_k\left(\gamma_k^z,\gamma_k^{k+1}\right)\right]\left\{\tilde{U}_{k-1}^k\left[\tilde{Z}_{k-1}^k S_k(b,a)+R_k(b,a)\right]\right\} \quad . \qquad (3.16)$$

Similarly the Green's function in the $k^{th}$ layer for $z > z_s$

$$G_k(z) = \left[Z_{k+1}^{k+1}Q_k\left(\gamma_k^z,\gamma_k^{k+1}\right) + P_k\left(\gamma_k^z,\gamma_k^{k+1}\right)\right]U_{k+1}^{k+1} \quad . \qquad (3.17)$$

The subscript on $G_k(z)$ will generally be omitted since the receiver is always located at $(r,z)$ in the $LR^{th}$ layer.

Thus the Green's function $G(z)$, $z < z_s$, may be written in terms of $\tilde{Z}_k^{k+1}$, and the multiplicative constant $\tilde{U}_{k-1}^k\left[\tilde{Z}_{k-1}^k S_k(b,a)+R_k(b,a)\right]$. Similarly $G(z)$, $z > z_s$, is given in terms of $Z_{k+1}^{k+1}$ and the constant $U_{k+1}^{k+1}$. Since the local terminal impedances $\tilde{Z}_k^{k+1}$ and $Z_{k+1}^{k+1}$ are uniquely determined from (3.13) and (3.14) in terms of the known half space impedances $\tilde{Z}_0^1$ and $Z_N^N$, respectively, the Green's function $G(z)$ is completely specified upon determination of the multiplicative constants.

If continuity of $\tilde{F}_k$ and $\tilde{U}_k$ is to hold everywhere above the source then $\tilde{\psi}_k(z_k) = \tilde{\psi}_{k-1}(z_k)$ or equivalently $\tilde{U}_{k-1}^k\left[\tilde{Z}_{k-1}^k S_k(b,a)+R_k(b,a)\right] = \tilde{U}_k^{k+1}$. The relation to the constant in the source layer is for $1 \leq k \leq LS-1$

$$\tilde{U}_{k-1}^k\left[\tilde{Z}_{k-1}^k S_k(b,a)+R_k(b,a)\right] = \frac{\tilde{U}_{LS-1}^{LS}\left[\tilde{Z}_{LS-1}^{LS}S_{LS}\left(\gamma_{LS}^{LS+1},\gamma_{LS}^{LS}\right)+R_{LS}\left(\gamma_{LS}^{LS+1},\gamma_{LS}^{LS}\right)\right]}{\prod_{p=k+1}^{LS}\left[\tilde{Z}_{p-1}^p S_p(b,a)+R_p(b,a)\right]}$$

Similarly if the continuity of $F_k$ and $U_k$ is to hold everywhere below the source then $\psi_k(z_{k+1}) = \psi_{k+1}(z_{k+1})$ or $U_{k+1}^{k+1} = U_k^k / \left[Z_{k+1}^{k+1}S_k(a,b)+R_k(a,b)\right]$. This may be written in terms of the source constant as

$$U_{k+1}^{k+1} = \frac{U_{LS+1}^{LS+1}}{\prod_{p=LS+1}^{k+1} \left[Z_{p+1}^{p+1} S_p(a,b) + R_p(a,b)\right]} \quad , \quad LS+1 \leq k \leq N-1 \quad .$$

The two constants $\tilde{U}_{LS}^{LS-1}$ and $U_{LS+1}^{LS+1}$ are determined by integrating (3.2) between the limits $z_s - \varepsilon$ and $z_s + \varepsilon$ and taking the limit as $\varepsilon$ goes to zero. This is equivalent to satisfying the source conditions

$$F_{LS}^{S} - \tilde{F}_{LS}^{S} = 0 \quad , \quad U_{LS}^{S} - \tilde{U}_{LS}^{S} = -1 \quad .$$

The superscript S denotes that the evaluation occurs at the source depth $z_s$. One finds that the above conditions will be satisfied if the constants are defined according to

$$U_{LS+1}^{LS+1} = a_{12}/W \quad , \quad \tilde{U}_{LS-1}^{LS}\left[\tilde{Z}_{LS-1}^{LS} S_{LS}(b,a) + R_{LS}(b,a)\right] = a_{11}/W \quad ,$$

where with $e = \gamma_{LS}^{S}$, $a = \gamma_{LS}^{LS}$ and $b = \gamma_{LS}^{LS+1}$

$$a_{11} = \begin{pmatrix} Z_{LS+1}^{LS+1} & 1 \end{pmatrix} \begin{pmatrix} Q_{LS}(e,b) \\ P_{LS}(e,b) \end{pmatrix} \quad , \quad a_{12} = \begin{pmatrix} Q_{LS}(e,b) & P_{LS}(e,b) \end{pmatrix} \begin{pmatrix} \tilde{Z}_{LS}^{LS+1} \\ 1 \end{pmatrix} \quad ,$$

and the Wronskian is found to be given by

$$W = Z_{LS+1}^{LS+1} - \tilde{Z}_{LS}^{LS+1} \quad , \tag{3.18}$$

the difference between the terminal impedance $Z_N^N$ referenced up to the level $z = z_{LS+1}$ and the terminal impedance $\tilde{Z}_0^1$ referenced down to that same level. The constants in each layer are then found to be

$$\left[\tilde{Z}_{k-1}^{k} S_k(b,a) + R_k(b,a)\right]\tilde{U}_k^{k-1} = \left(\frac{a_{11}}{W}\right) \Big/ \prod_{p=k+1}^{LS} \left[\tilde{Z}_{p-1}^{p} S_p(b,a) + R_p(b,a)\right] \quad , \quad 1 \leq k \leq LS-1, \tag{3.19}$$

and

$$U_{k+1}^{k+1} = \left(\frac{a_{12}}{W}\right) \Big/ \prod_{p=LS+1}^{k+1} \left[Z_{p+1}^{p+1} S_p(a,b) + R_p(a,b)\right] \quad , \quad LS+1 \leq k \leq N-1 \quad . \tag{3.20}$$

Thus $G(z,z_s)$ in any layer can be found by using these constants in (3.16) and (3.17). If both the source and receiver are in the same layer (i.e., LS=LR), the Green's function for $z_{LS} \leq z \leq z_s$ and $1 \leq LS \leq N-1$ is

$$G(z,z_s) = \frac{\left[\tilde{Z}_{LR}^{LR+1} Q_{LR}\left(\gamma_{LR}^{z},\gamma_{LR}^{LR+1}\right)+P_{LR}\left(\gamma_{LR}^{z},\gamma_{LR}^{LR+1}\right)\right]\left[Z_{LS+1}^{LS+1} Q_{LS}\left(\gamma_{LS}^{S},\gamma_{LS}^{LS+1}\right)+P_{LS}\left(\gamma_{LS}^{S},\gamma_{LS}^{LS+1}\right)\right]}{Z_{LS+1}^{LS+1}-\tilde{Z}_{LS}^{LS+1}} . \quad (3.21)$$

The solution valid for $z_s \leqq z \leqq z_{LS+1}$ is obtained by interchanging z and $z_s$ in the above equation.

If the receiver is below the source and in a different layer (i.e., LR>LS), the Green's function for $z_{LR} \leqq z \leqq z_{LR+1}$, $LS+1 \leq LR \leqq N-1$ is

$$G(z,z_s) = \frac{\left[Z_{LR+1}^{LR+1} Q_{LR}\left(\gamma_{LR}^{z},\gamma_{LR}^{LR+1}\right)+P_{LR}\left(\gamma_{LR}^{z},\gamma_{LR}^{LR+1}\right)\right]\left[\tilde{Z}_{LS}^{LS+1} Q_{LS}\left(\gamma_{LS}^{S},\gamma_{LS}^{LS+1}\right)+P_{LS}\left(\gamma_{LS}^{S},\gamma_{LS}^{LS+1}\right)\right]\Delta_{LS+1}^{LR}}{Z_{LS+1}^{LS+1}-\tilde{Z}_{LS}^{LS+1}} , \quad (3.22)$$

where

$$\Delta_{LS+1}^{LR} = \left\{\prod_{k=LS+1}^{LR} \left[Z_{k+1}^{k+1} S_k(a,b)+R_k(a,b)\right]\right\}^{-1} .$$

The form of the solution for G provided above is the familiar way of expressing the Green's function solution. FELSEN and MARCUVITZ [3.6] and also CODDINGTON and LEVINSON [3.7] provided solutions which are identical in form to (3.22) except for the absence of the factor $\Delta_{LS+1}^{LR}$. This factor is absent if both the source and receiver are in the same layer (LS=LR) as can be seen by examining (3.21). Thus, the above result represents an extension of this earlier work in that the source and receiver may be in different layers in a piecewise layered media. The results for either LS or LR equal to zero or N are special cases. The Green's function for these cases is provided in [3.1].

The Green's function in the LR$^{th}$ layer may be expressed in various alternative forms using the relations provided in this section. One such form which displays the dependence on the known terminal impedances $Z_N^N$ and $\tilde{Z}_0^1$ explicitly is found to be

$$G(z,z_s) = \frac{\left(1-\tilde{Z}_0^1\right)K_1^{LS}\Phi_{LS}\Phi_{LR}K_{LR+1}^{N-1}\begin{pmatrix} Z_N^N \\ 1 \end{pmatrix}}{\left(1-\tilde{Z}_0^1\right)K_1^{N-1}\begin{pmatrix} Z_N^N \\ 1 \end{pmatrix}} , \quad (3.23)$$

where $z_{LR} \leqq z \leqq z_{LR+1}$ for $LS+1 \leq LR \leq N-1$ and $z_s \leqq z \leqq z_{LS+1}$. All of the terms in the above expression are matrices. In particular

$$\Phi_{LS} = \begin{pmatrix} P_{LS}\left(\gamma_{LS}^{S},\gamma_{LS}^{LS+1}\right) \\ -Q_{LS}\left(\gamma_{LS}^{S},\gamma_{LS}^{LS+1}\right) \end{pmatrix} \qquad \Phi_{LR} = \left[Q_{LR}\left(\gamma_{LR}^{z},\gamma_{LR}^{LR+1}\right)P_{LR}\left(\gamma_{LR}^{z},\gamma_{LR}^{LR+1}\right)\right]$$

$$K_\ell^p = \prod_{k=\ell}^{p} M_k\left(\gamma_k^k,\gamma_k^{k+1}\right) \quad . \tag{3.24}$$

The expression for $G(z,z_s)$ valid above the source is obtained from (3.23) by interchanging LS and LR and also $z_s$ and $z$.

### 3.1.2 Direct Numerical Integration: Fast Field Program (FFP)

The observation that the Fast Fourier Transform (FFT) algorithm could be used to evaluate the Bessel transform was first made by MARSH and ELAM [3.8] in 1967 and subsequently developed into a general purpose model by DINAPOLI [3.9].

The first step in this method consists of replacing the Bessel function in the integral solution (3.1) with the Hankel function associated with outward propagation. If then the first term in the asymptotic expansion

$$H_0^{(1)}(\xi r) \simeq (2/\pi i)^{\frac{1}{2}}\, e^{i\xi r}/\sqrt{\xi r} \quad , \tag{3.25}$$

is substituted, the integral may be written as

$$\mathscr{P}(r,z,z_s) \simeq (2/\pi i)^{\frac{1}{2}}\, \frac{S_\omega}{4\pi}\int_0^\infty \sqrt{\xi/r}\; G(z,z_s;\xi,\omega)\, e^{i\xi r}\, d\xi \quad . \tag{3.26}$$

Next let the horizontal wave number and the horizontal range be evaluated at the discrete values

$$\xi_m = \xi_0 + m\Delta\xi \quad , \quad r_n = r_0 + n\Delta r \quad , \quad (m,n) = 0,1,2,\ldots,L-1 \quad ,$$

with the added restriction that $\Delta r\Delta\xi = 2\pi/L$ where $L$ is equal to 2 raised to some integer power. Then (3.26) is given by the discrete Fourier transform

$$\mathscr{P}(r_n,z,z_s) \simeq \Delta\xi(2/\pi i)^{\frac{1}{2}}\, \frac{S_\omega}{4\pi}\, \frac{\exp(i\xi_0 r_n)}{r_n^{\frac{1}{2}}} \sum_{m=0}^{L-1} E_m\, e^{i2\pi mn/L} \quad , \tag{3.27}$$

where the input values are obtained from

$$E_m = G(z,z_s;\xi_m,\omega)\,\xi_m^{\frac{1}{2}}\,\exp(imr_0\Delta\xi) \quad . \tag{3.28}$$

The evaluation of (3.27) via the FFT yields the value of the field at each of the unaliased L/2 ranges essentially simultaneously. This format is thus ideally suited for the rapid calculation of the field as a function of range for a fixed source and receiver depth.

The only restriction imposed upon the matrizant solution for G was that the sound-speed variation $c_k(z)$ associated with the known solutions be a good approximation to the given sound-speed variation within the subinterval $z_k \leq z \leq z_{k+1}$. The solution for the pressure field via the FFT necessitates the evaluation of (3.28) for each of the L equi-spaced discrete values $\xi_m$ (m=0,1,2,...,L-1) of the horizontal component of the wavenumber. This calculation represents the major portion of the required execution time. Thus, it is evident that the particular manner in which $f_k$, $g_k$ depend upon $\xi$ will have a strong bearing upon the ease with which G can be found as a function of $\xi_m$.

Let the sound speed vary exponentially within each layer according to

$$c_k(z) = a_k \exp[(z-z_k)/H_k] \quad , \quad z_k \leq z \leq z_{k+1} \quad , \quad 1 \leq k \leq N-1 \quad ,$$

where $a_k$ is the sound speed at the top of the $k^{th}$ layer, which need not equal $c_{k-1}(z_k)$, and $H_k$ is an arbitrary scale factor. The fundamental solutions are then found to be

$$f_k\left(\gamma_k^z,\xi_m\right) = J_{(\nu_k)_m}\left(\gamma_k^z\right) \quad , \quad g_k\left(\gamma_k^z,\xi_m\right) = Y_{(\nu_k)_m}\left(\gamma_k^z\right) \quad , \tag{3.29}$$

that is, cylindrical functions of complex order

$$(\nu_k)_m \simeq (\xi_m H_k - i\alpha H_k) \quad \text{where} \quad k(z) \sim [\omega/c_k(z) - i\alpha] \quad , \tag{3.30}$$

with $\alpha$ some positive constant representing attenuation. The real argument is

$$\gamma_k^z = \frac{\omega H_k}{C_k(z)} \quad ; \quad \frac{d\gamma_k^z}{dz} = \gamma_k^z/H_k \quad . \tag{3.31}$$

The dependence upon $\xi_m$ can then be efficiently found by evaluating (3.29) once for a pair of starting values say $(\nu_k)_0$, $(\nu_k)_1$ and utilizing recurrence relations [3.10].

The wave number domain sampling distance $\Delta\xi$ is found from (3.30) to be

$$\Delta\xi = 1/H_k \quad , \tag{3.32}$$

upon noting that the recurrence relations require that the change in $\nu$ be unity. The fact that the FFT requires equi-spaced values of $\xi_m$ is in conflict with (3.32) since $H_k$ is different for each of the (N-1) possible layers. Then let $\Delta\xi$ equal the reciprocal of the largest scale factor $H_{max}$, namely

$$\Delta\xi = 1/H_{max} \quad , \tag{3.33}$$

and restrict the remaining scale factors to be some arbitrary integer multiple $p_k$ of this value, i.e., $H_{max}/H_k = p_k$ with $k = 1,2,\ldots,N-1$. This restriction limits the ability to approximate the given sound speed variation arbitrarily close but leaves enough flexibility to model most cases of interest in underwater acoustics. Then the discrete values of $\xi_m$ ($m=0,1,2,\ldots,L-1$) yield the discrete values of the order of the cylindrical functions in each layer ($1 \leqq k \leqq N-1$) according to $(\nu_k)_m = (\xi_0 H_k - i\alpha H_k) + m/p_k$. The condition that the change in $\nu_k$ be unity can be satisfied if $p_k$ pairs of starting values are calculated and used in the recurrence relations.

Since the minimum value of $\xi$ should be zero and $G(z,z_s)$ decays when $\xi > \omega/C_{min}$, where $C_{min}$ is the minimum sound speed within the region $z_1 \leqq z \leqq z_N$, one finds that the number of sample points, L, needed is approximately given by

$$L \sim \frac{\omega}{C_{min}} H_{max} \quad , \tag{3.34}$$

Typically $H_{max} \sim 10^5$ and thus at high frequencies not only does the number of calculations increase but available core storage may be exceeded. A technique has been implemented [3.11] for rigorously circumventing the core storage problem when it arises. It is noted that although the execution time and storage requirements increase with increasing frequency, the solution remains perfectly valid and accurate.

The only remaining point involves the accuracy of replacing the Hankel function with the first term in its asymptotic expansion (3.25) which was introduced without justification. A discussion of this approximation is provided in [3.1].

### 3.1.3 Normal Mode (and Branch Line Integral) Models

#### *a) Stickler (EJP Cuts) Bartberger (Pekeris Cuts)*

Normal mode models may be viewed as the result obtained when the integral solution to the wave equation is solved by the utilization of Cauchy's residue theory. In that case the pressure field is given by a sum (possibly infinite) of residues associated with discrete eigenvalues plus a branch cut integral arising from the lower half space (Nth layer in Fig.3.2) representing the contribution from the continuous portion of the eigenvalue spectrum. The particular manner in which the branch cuts are chosen determines the extent of the discrete portion of the eigenvalue spectrum and thus the physical interpretation of the modes (eigenfunctions).

The primary mathematical difference between the two models to be discussed is the manner in which the branch cuts are chosen. STICKLER [3.12] uses the EJP (Ewing, Jardetzky, Press) cuts [3.13], BARTBERGER [3.14] prefers the PEKERIS [3.15] cuts (Fig.3.3).

The sound speed within each ocean layer for both models is assumed to be well approximated by $c_k^{-2}(z) = a_k + b_k(z_k - z)$, $z_k \leqq z \leqq z_{k+1}$ where $a_k$ is the sound speed at the top of the $k^{th}$ layer and need not equal $c_{k-1}^{-2}(z_k)$, and $b_k$ is proportional to the

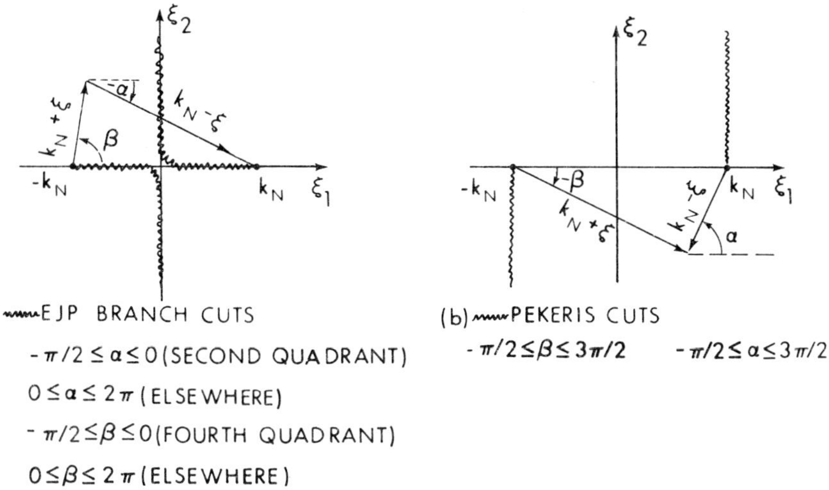


Fig. 3.3a and b. Definition of α and β for EJP and Pekeris branch cuts

sound-speed gradient. In this case the solutions f and g are given in terms of Airy functions according to $f_k\left(\gamma_k^z,\xi\right) = \mathrm{Ai}\left(\gamma_k^z\right)$, $g_k\left(\gamma_k^z,\xi\right) = \mathrm{Bi}\left(\gamma_k^z\right)$, where the argument is found to be

$$\gamma_k^z = -L_k^2\left[\omega^2 c_k^{-2}(z) - \xi^2\right] \quad , \quad L_k^2 = \omega^{-4/3} b_k^{-2/3} \quad .$$

The Green's function formulation previously given as (3.22) could be used to obtain a solution which would be equivalent to those obtained by BARTBERGER and STICKLER but not of the same form. In order to obtain their results, (3.22) must first be rewritten. This is accomplished by substituting for the impedance (3.11) which yields

$$G(z,z_s) = \frac{\left[F_{LR+1}^{LR+1} Q_{LR}(d,b) + U_{LR+1}^{LR+1} P_{LR}(d,b)\right]\left[\tilde{F}_{LS}^{LS+1} Q_{LS}(e,b) + \tilde{U}_{LS}^{LS+1} P_{LS}(e,b)\right]\Delta_{LS+1}^{LR}}{U_{LR+1}^{LR+1}\tilde{U}_{LS}^{LS+1}\left(Z_{LS+1}^{LS+1} - \tilde{Z}_{LS}^{LS+1}\right)} \quad , \qquad (3.35)$$

where $d = \gamma_{LR}^z$, $e = \gamma_{LS}^S$ and b is the function γ evaluated at the lower boundary of the layer (in this case either LS or LR) in which it is defined.

Next the difference in impedances evaluated at LS+1 occurring in the denominator is transferred to the level $z_N$. In doing this use is made of (3.13) and (3.14) which immediately yield

$$\begin{pmatrix} \tilde{Z}^{LS+1}_{LS} \\ 1 \end{pmatrix} = \prod_{p=LS+1}^{N-1} M_p(a,b) \begin{pmatrix} \tilde{Z}^{N}_{N-1} \\ 1 \end{pmatrix} \prod_{p=LS+1}^{N-1} \left[\tilde{Z}^{p}_{p-1} S_p(b,a) + R_p(b,a)\right]$$

$$\begin{pmatrix} Z^{LS+1}_{LS+1} \\ 1 \end{pmatrix} = \frac{\prod_{p=LS+1}^{N-1} M_p(a,b) \begin{pmatrix} Z^{N}_{N} \\ 1 \end{pmatrix}}{\prod_{p=LS+1}^{LR} \left[Z^{p+1}_{p+1} S_p(a,b) + R_p(a,b)\right] \prod_{p=LR+1}^{N-1} \left[Z^{p+1}_{p+1} S_p(a,b) + R_p(a,b)\right]} \quad .$$

The constant in the denominator of the last equation has been written as a product of two constants in anticipation of cancellation of terms arising from the definition of $\Delta^{LR}_{LS+1}$ (3.22). Then

$$\begin{pmatrix} Z^{LS+1}_{LS+1} - \tilde{Z}^{LS+1}_{LS} \\ 0 \end{pmatrix} = \prod_{p=LS+1}^{N-1} M_p(a,b) \begin{pmatrix} \alpha Z^{N}_{N} - \beta \tilde{Z}^{N}_{N-1} \\ \alpha - \beta \end{pmatrix}$$

where

$$\beta = \prod_{p=LS+1}^{N-1} \left[\tilde{Z}^{p}_{p-1} S_p(b,a) + R_p(b,a)\right] \quad ,$$

$$\alpha = \left\{\prod_{p=LS+1}^{N-1} \left[Z^{p+1}_{p+1} S_p(a,b) + R_p(a,b)\right]\right\}^{-1} \quad .$$

Since the determinant of a product of matrizants is unity when the trace of A(z) is zero, one obtains the desired relationship

$$Z^{LS+1}_{LS+1} - \tilde{Z}^{LS+1}_{LS} = \left(Z^{N}_{N} - \tilde{Z}^{N}_{N-1}\right)\alpha\beta \quad . \tag{3.36}$$

Substitution of the above into (3.35) gives for the right side

$$\frac{\left[F^{LR+1}_{LR+1} Q_{LR}(d,b) + U^{LR+1}_{LR+1} P_{LR}(d,b)\right]\left[\tilde{F}^{LS+1}_{LS} Q_{LS}(e,b) + \tilde{U}^{LS+1}_{LS} P_{LS}(e,b)\right]}{U^{LR+1}_{LR+1} \tilde{U}^{LS+1}_{LS} \left(Z^{N}_{N} - \tilde{Z}^{N}_{N-1}\right) \dfrac{\prod_{p=LS+1}^{N-1} \left[\tilde{Z}^{p}_{p-1} S_p(b,a) + R_p(b,a)\right]}{\prod_{p=LR+1}^{N-1} \left[Z^{p+1}_{p+1} S_p(a,b) + R_p(a,b)\right]}} \quad ,$$

which allows for the source and receiver to be in any layer but the evaluation of the Wronskian determinant occurs at $z_N$. The denominator may be simplified considerably more by first noting that if the identity

$$Z_N^N - \tilde{Z}_{N-1}^N = \left(Z_N^N/\tilde{U}_{N-1}^N\right)\left(\tilde{U}_{N-1}^N - \tilde{F}_{N-1}^N/Z_N^N\right) \quad ,$$

is substituted into the denominator and use is made of (3.20), one finds that

$$\frac{U_{LR+1}^{LR+1}}{U_N^N \prod_{p=LR+1}^{N-1} \left[Z_{p+1}^{p+1} S_p(a,b) + R_p(a,b)\right]} = 1 \quad . \tag{3.37}$$

Similarly it can be shown that

$$\frac{\tilde{U}_{LS}^{LS+1}}{\tilde{U}_{N-1}^N} \prod_{p=LS+1}^{N-1} \left[\tilde{Z}_{p-1}^p S_p(b,a) + R_p(b,a)\right] = 1 \quad . \tag{3.38}$$

The required result for the Green's function is then

$$G(z,z_s) = \frac{\left[F_{LR+1}^{LR+1} Q_{LR}(d,b) + U_{LR+1}^{LR+1} P_{LR}(d,b)\right]\left[\tilde{F}_{LS}^{LS+1} Q_{LS}(e,b) + \tilde{U}_{LS}^{LS+1} P_{LS}(e,b)\right]}{F_N^N\left(\tilde{U}_{N-1}^N - \tilde{F}_{N-1}^N/Z_N^N\right)} \quad . \tag{3.39}$$

The residue contribution to the pressure field occurs for some value of $\xi$ say $\xi_m$ called the eigenvalue for which $Z_N^N - \tilde{Z}_{N-1}^N = 0$, that is the impedance of the last half space $Z_N^N$ exactly equals the impedance of the entire medium above it $\tilde{Z}_{N-1}^N$ evaluated at $z_N$. Assuming that these eigenvalues give rise to simple poles, the residue contribution is found to be

$$\frac{iS_\omega}{2} \sum_{m=0} \left. \frac{\left[F_{LR+1}^{LR+1} Q_{LR}(d,b) + U_{LR+1}^{LR+1} P_{LR}(d,b)\right]\left[\tilde{F}_{LS}^{LS+1} Q_{LS}(e,b) + \tilde{U}_{LS}^{LS+1} P_{LS}(e,b)\right]}{F_N^N \frac{\partial}{\partial\xi}\left(\tilde{U}_{N-1}^N - \tilde{F}_{N-1}^N/Z_N^N\right)} \right|_{\xi=\xi_m}$$

$$\cdot\, H_0^{(1)}(\xi_m r)\xi_m \quad , \tag{3.40}$$

where the terminal impedance $Z_N^N$ is known to be $Z_N^N = \omega\rho_N/\sqrt{k_N^2-\xi^2}$.

The residue contribution obtained by both STICKLER and BARTBERGER can be obtained from (3.40) with some important exceptions which are related to the choice of branch cuts.

### *b) Stickler's Residue Contribution*

By choosing the EJP cuts (discussed in Sect.2.2.1) and invoking results from the spectral theory of ordinary differential equations [3.7], STICKLER was able to conclude that all of the eigenvalues which occur on the sheet of integration in the complex $\xi$ plane are real (it should be recalled that the entire medium is assumed to be a perfect fluid with no energy loss due to absorption). Furthermore, only a finite number of such values exist and they give rise to simple pole type singularities. These modes are thus trapped within the medium and the only loss of energy with range, assumed to be sufficiently far from the source, goes like $r^{-\frac{1}{2}}$, i.e., cylindrical spreading. The equivalent mode-ray (a plot of $\xi_m = k(z)\sin\theta(z)$ versus depth and range with $\theta$ measured from the vertical) should either have a turning point between $z_1$ and $z_N$ or be incident upon the boundary at $z_N$ at an angle for which total reflection occurs.

To obtain STICKLER's result note from (3.36) that at the eigenvalues the functions F and U are equal to $\tilde{F}$ and $\tilde{U}$ at all levels. The Green's function portion of the residue may then be written as

$$\left.\frac{\left[\tilde{F}_{LR+1}^{LR+1}Q_{LR}(d,b)+\tilde{U}_{LR+1}^{LR+1}P_{LR}(d,b)\right]\left[\tilde{F}_{LS}^{LS+1}Q_{LS}(e,b)+\tilde{U}_{LS}^{LS+1}P_{LS}(e,b)\right]}{\tilde{F}_{N-1}^{N}\frac{\partial}{\partial\xi}\left(\tilde{U}_{N-1}^{N}-\tilde{F}_{N-1}^{N}/Z_N^N\right)}\right|_{\xi=\xi_m} \quad . \qquad (3.41)$$

The entire denominator of (3.41) may then be written as

$$\tilde{F}_{N-1}^{N}\frac{\partial}{\partial\xi}\left(\tilde{U}_{N-1}^{N}-\frac{\tilde{F}_{N-1}^{N}}{Z_N^N}\right)=\frac{C_0^{-1}(\xi_m)}{i\omega} \quad ,$$

where

$$C_0^{-1}(\xi_m)=\left\{\left[\frac{2L_{N-1}\xi\gamma_{N-1}^{N}}{\rho_{N-1}}+\frac{i\omega\xi Z_N^N}{(\omega\rho_N)^2}-\frac{2(i\omega)^2\xi L_{N-1}^3\rho_{N-1}}{\left(Z_N^N\right)^2}\right]\left(\tilde{F}_{N-1}^{N}\right)^2 + \frac{\dot{B}_{N-1}A_{N-1}-\dot{A}_{N-1}B_{N-1}}{\rho_{N-1}L_{N-1}\pi}\right\}\Bigg|_{\xi=\xi_m} \quad , \quad \left(\dot{A}\equiv\frac{\partial A}{\partial\xi}\right) \qquad (3.42)$$

which is identical to [Ref.3.12,Eq.(12)] with the exception of a multiplicative factor of $k_m$ ($\xi_m$ in this notation) which arises from the original Bessel transform. Also a typographical error occurred by omitting the factor equivalent to $\gamma_{N-1}^{N}$ in the first term inside the bracketed term of (3.42). The residue contribution may then be written as

$$-\frac{\omega S_\omega}{2}\sum_{m=0} C_0(\xi_m)\left[\tilde{F}_{LR+1}^{LR+1}Q_{LR}(d,b)+\tilde{U}_{LR+1}^{LR+1}P_{LR}(d,b)\right]\left[\tilde{F}_{LS}^{LS+1}Q_{LS}(e,b)+\tilde{U}_{LS}^{LS+1}P_{LS}(e,b)\right] \cdot \xi_m H_0^{(1)}(\xi_m r) \quad , \qquad (3.43)$$

which is identical to STICKLER's corresponding residue contribution.

*c) Bartberger's Residue Contribution (Pekeris Cuts)*

Although BARTBERGER assumes an exp(iωt) time dependence, his work is summarized as if the harmonic factor exp(-iωt) had been used in keeping with the conventions adopted earlier.

The formulation used by BARTBERGER is based upon the Green's function given by (3.39). Thus the only difference in the form of his residue contribution versus STICKLER's is that the functions $F_{LR+1}^{LR+1}$, $U_{LR+1}^{LR+1}$ defined below the source are not replaced with their counterparts $\tilde{F}_{LR+1}^{LR+1}$, $\tilde{U}_{LR+1}^{LR+1}$ defined above the source.

In the evaluation of (3.40) it is required to evaluate the derivatives with respect to the wavenumber of the functions $\tilde{F}_{N-1}^{N}$ and $\tilde{U}_{N-1}^{N}$. It can be shown that [3.1]

$$\begin{pmatrix} \tilde{F}_1^1 \\ \tilde{U}_1^1 \end{pmatrix} = \prod_{p=1}^{N-1} M_p(a,b) \begin{pmatrix} \tilde{F}_{N-1}^{N} \\ \tilde{U}_{N-1}^{N} \end{pmatrix} \quad . \tag{3.44}$$

Thus an essential ingredient in the calculation of the above mentioned derivatives is $\partial M_p(a,b)/\partial\xi$ for which it is found by differentiating (3.9) that

$$\begin{aligned}
\frac{\partial Q_p(a,b)}{\partial\xi} &= 2L_p^3\xi i\omega\rho_p(z_p)S_p(a,b) - \frac{2L_p\xi bP_p(a,b)}{i\omega\rho_p(z_{p+1})} \\
\frac{\partial P_p(a,b)}{\partial\xi} &= -2L_p^3\xi i\omega\rho_p(z_p)R_p(a,b) - i\omega\rho_p(z_{p+1})2L_p^3\xi Q_p(a,b) \\
\frac{\partial S_p(a,b)}{\partial\xi} &= \frac{-2L_p\xi bR_p(a,b)}{i\omega\rho_p(z_{p+1})} + \frac{2L_p\xi aQ_p(a,b)}{i\omega\rho_p(z_p)} \\
\frac{\partial R_p(a,b)}{\partial\xi} &= \frac{2L_p aP_p(a,b)}{i\omega\rho_p(z_p)} - 2L_p^3\xi i\omega\rho_p(z_{p+1})S_p(a,b) \quad .
\end{aligned} \tag{3.45}$$

The quantitative difference between the two residue contributions arises from the type and number of eigenvalues which satisfy the dispersion equation $Z_{LS+1}^{LS+1} = \tilde{Z}_{LS}^{LS+1}$ or its equivalent form derived by setting (3.36) to zero. When the Pekeris branch cuts are chosen the number of discrete eigenvalues will be infinite. Furthermore, the discrete eigenvalue spectrum can be thought of as containing two subsets spanning the regions on either side of the Pekeris branch cut. The first contains a finite number of eigenvalues with real parts larger than $k_N$, that is lying to the right of the Pekeris branch cut. The imaginary part of these eigenvalues would be zero if the attenuation in each layer was set to zero. Then, in this instance, STICKLER's total residue contribution would be identical to that portion of BARTBERGER's residue series arising from this subset of his eigenvalues.

BARTBERGER's second subset of eigenvalues lie to the left of the Pekeris cut, are infinite in number and complex even with the attenuation set to zero. Then, for these eigenvalues $\xi_m = A_m + iB_m$ where

$$0 < A_m < k_N \quad , \quad A_{m+1} < A_m \quad ; \quad \lim_{m \to \infty} B_m = \infty \quad , \quad B_{m+1} < B_m \quad .$$

Examination of the range dependent terms in (3.43) under the assumption that the Hankel function is well approximated by the first term in its asymptotic expansion yields $\exp[(iA_m - B_m)r]/\sqrt{r}$. These modes, sometimes called "leaky modes", suffer the same cylindrical spreading loss as the trapped modes from the first set, but in addition are attenuated by an amount proportional to the imaginary part of the eigenvalue. It is common to physically interpret these modes in terms of bottom bounce energy since the equivalent mode-ray for $A_m$ would be incident upon the boundary $z_N$ at an angle between critical and normal incidence. The angle of incidence approaching $90^\circ$ as m increases. The magnitude of the reflection coefficient for these angles would be less than unity resulting in a loss of energy into the "bottom". On this basis it might be expected that as the mode number m increased the range at which these leaky modes would significantly contribute to the total field would diminish. BARTBERGER has found this to be precisely the case. On the other hand, more and more higher-order modes would have to be included in the sum as the range point approached the source. The maximum number of modes calculated in BARTBERGER's AP2 program is 500. Thus, for some combination of frequency, water depth, source/receiver depth, it is possible that some range interval exist close to the source but not necessarily starting at $r = 0$ for which his result for the total field will be in error due to the truncation of the infinite series. The end point of this interval is associated with that range at which the mode-ray equivalent for the first omitted mode would have reached the receiver. The beginning of the interval, usually marked by a sudden drop in level of the propagation loss versus range curve, depends on the source/receiver geometry and $c(z)$.

### *d) Branch Cuts and Branch Cut Integrals*

The two choices of branch cuts discussed in this section are shown schematically in Fig.3.3a and b. The behavior of the square root $\beta_N = \pm\sqrt{k_N^2 - \xi^2}$ in the complex plane $(\xi = \xi_1 + i\xi_2)$ is determined by first defining

$$k_N - \xi = \rho_+ e^{i\alpha} \quad , \quad k_N + \xi = \rho_- e^{i\beta} \quad ,$$

so that

$$\beta_N = (\rho_+ \rho_-)^{\frac{1}{2}} e^{i(\alpha+\beta)/2} = a_N + ib_N \quad .$$

The definitions of $\alpha$ and $\beta$ as well as values at critical points are exhibited in Fig.3.4 for the EJP cuts. With this information it is possible to show (Chap.2) that $b_N > 0$ everywhere on the top sheet and that $a_N > 0$ in quadrants 2 and 4 while $a_N < 0$ in quadrants 1 and 3.

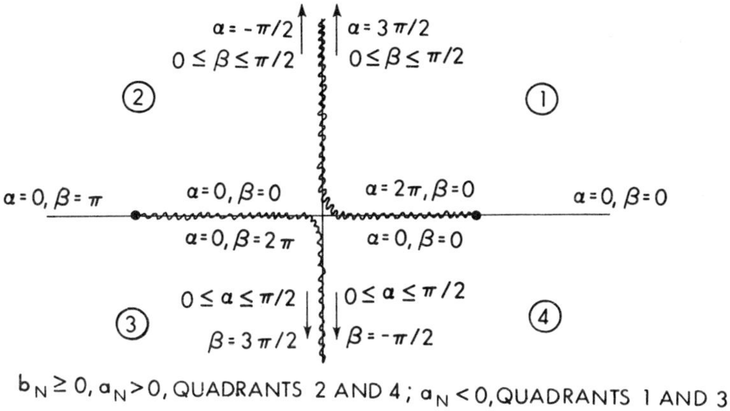


Fig. 3.4

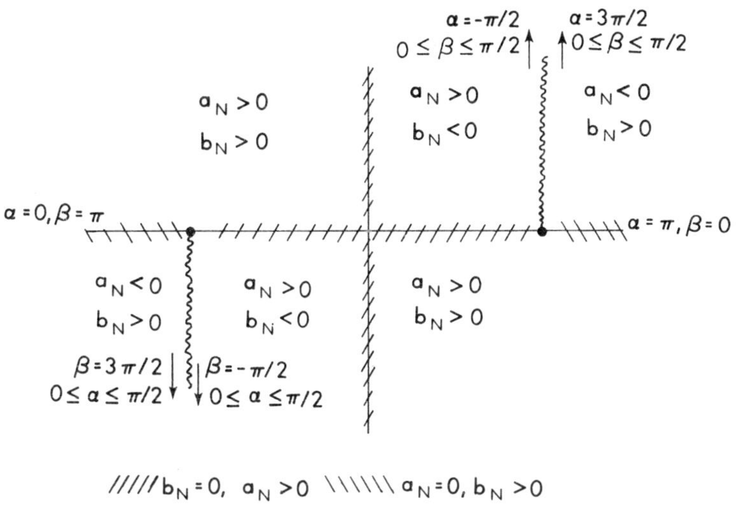


Fig. 3.5

Fig. 3.4. Definition of $\beta_N$ for EJP branch cut
Fig. 3.5. Definition of $\beta_N$ for Pekeris branch cut

The analogous information for the Pekeris cuts is given in Fig.3.5, where it can be seen that the imaginary part of $\beta_N$ is positive everywhere in the upper half plane except for that sector of the first quadrant between the Im{$\xi$} axis and the branch cut. It is in this sector that the complex eigenvalues associated with the leaky wave modes are located.

It can also be seen that the value of $\beta_N$ approaches a purely real positive value on the left-hand side of the branch cut in the first quadrant and a purely real negative value on the right as $\xi_2 \to \infty$. Thus, $\beta_N$ assumes purely real values at the extremities of the branch cut and complex values elsewhere. In both cases the contour is closed in the upper half plane after encircling the branch cut. Jordan's lemma will be satisfied for any depth with the EJP cuts and for depths between $z_1 \leq z \leq z_N$ for the Pekeris cuts. The satisfaction of Jordan's lemma with the Pekeris cuts and $z\to\infty$ is a moot point of some mathematical interest but of limited physical value, since in this case the environmental model itself has limited physical significance. A more detailed discussion of this point is contained in [3.1].

Having completely specified the behavior of the square root, $\beta_N$, on the top Riemann sheet it is now possible to obtain explicit results for the branch cut integrals. It becomes necessary to evaluate

$$(S\omega/4\pi)\int G(z,z_s;\xi,\omega)H_0^{(1)}(\xi r)\xi d\xi \quad ,$$

along paths on either side of the branch cut in the upper half plane. For this purpose the formulation of the Green's function (3.23) which explicitly displays the dependence upon the terminal impedance $Z_N^N$ will be used. Introduce the abbreviations

$$\begin{pmatrix} A & B \\ C & D \end{pmatrix} = K_1^{N-1} = \prod_{p=1}^{N-1} M_p(a,b) \quad , \quad \begin{pmatrix} E & X \\ Y & \Gamma \end{pmatrix} = K_1^{LS}\Phi_{LS}\Phi_{LR}K_{LR+1}^{N-1} \quad .$$

The Green's function may then be expressed as

$$G(z,z_s;\xi,\omega) = \left[\left(E-Y\tilde{Z}_0^1\right)+\left(X-\Gamma\tilde{Z}_0^1\right)Z_N^N\right]\left[\left(A-C\tilde{Z}_0^1\right)+\left(B-D\tilde{Z}_0^1\right)Z_N^N\right]^{-1} ,$$

and it is noted that for a pressure release surface $\tilde{Z}_0^1 = 0$.

For the Pekeris cut [$\xi=k_N+i\tau$, $0<\tau<\infty$ and on the right hand side of the cut where $\sqrt{k_N-\xi}=\tau^{\frac{1}{2}}\exp(-i\pi/4)$] one obtains

$$(iS\omega/4\pi)\int_\tau^0 \frac{[A_1+A_2]}{[A_3+A_4]}\xi H_0^{(1)}(\xi r)d\tau \quad ,$$

where

$$A_1 = E - Y\tilde{Z}_0^1 \quad , \quad A_2 = \left(X-\Gamma\tilde{Z}_0^1\right)\sqrt{k_N+\xi}/(\omega\rho_N)\tau^{\frac{1}{2}}e^{-i\pi/4} \quad ,$$
$$A_3 = A - C\tilde{Z}_0^1 \quad , \quad A_4 = \left(B-D\tilde{Z}_0^1\right)\sqrt{k_N+\xi}/(\omega\rho_N)\tau^{\frac{1}{2}}e^{-i\pi/4} \quad .$$

To the left of the cut, $\sqrt{k_N-\xi} = -\tau^{\frac{1}{2}}e^{-i\pi/4}$, which results in

$$(iS\omega/4\pi)\int_0^\tau \frac{[A_1-A_2]}{[A_3-A_4]}\xi H_0^{(1)}(\xi r)d\tau \quad .$$

Reversing the limits of integration of the first integral and combining yields

$$iS\omega/(2\pi)\int_0^\tau \frac{(EB-XA)}{Z_N^N\left[A^2-\left(B/Z_N^N\right)^2\right]}H_0^{(1)}(\xi r)\xi\, d\tau \quad .$$

This may be expressed in the same notation as that used for the residue series with some straightforward but lengthy manipulations omitted here but contained in [3.1], where it is shown that the Pekeris branch cut integral is given by

$$\lim_{\tau\to\infty} (-iS\omega/2\pi)\int_0^{\tau} \hat{G}(e,d;\xi)\Big/\left\{\left[\left(\tilde{U}^N_{N-1}\right)^2-\left(\tilde{F}^N_{N-1}/Z^N_N\right)^2\right]Z^N_N\right\}\xi H_0^{(1)}(\xi r)d\tau$$

$$\hat{G}(e,d;\xi) = \left[\tilde{F}^{LS+1}_{LS}Q_{LS}(e,b)+\tilde{U}^{LS+1}_{LS}P_{LS}(e,b)\right]\left[\tilde{F}^{LR+1}_{LR}Q_{LR}(d,b)+\tilde{U}^{LR+1}_{LR}P_{LR}(d,b)\right] \quad ,$$

where

$$\xi = k_N + i\tau \quad , \quad Z^N_N = \frac{\omega\rho_N}{\sqrt{k_N+\xi}\;\tau^{\frac{1}{2}}\, e^{-i\pi/4}} \quad , \quad \left(Z^N_N\right)^2 = \frac{(\omega\rho_N)^2}{k_N^2-\xi^2} \quad .$$

The form of the Green's function given by (3.23) may also be used to arrive at the EJP branch cut integral obtained by STICKLER. In this case explicit expressions are found for the integrals along the real and imaginary axis. Above the real axis let $\xi=\tau$, $0<\tau<k_N$ and $\sqrt{k_N-\xi}=\rho_1$ while below $\sqrt{k_N-\xi}=-\rho_1$. Following the same procedure as used with the Pekeris cut one obtains for the branch cut integral along the real axis

$$\frac{S_\omega}{2\pi}\int_0^{k_N}\frac{\hat{G}(e,d;\xi)\sqrt{k_N^2-\tau^2}\;(\omega\rho_N)^{-1}\;\tau H_0^{(1)}(\tau r)d\tau}{\left(\tilde{U}^N_{N-1}\right)^2-\left(\tilde{F}^N_{N-1}\right)^2\left(k_N^2-\tau^2\right)/(\omega\rho_N)^2} \quad ,$$

which is identical to STICKLER's integral containing $C_1(k)$ upon multiplying that answer by $i\omega\rho_0 S_\omega$, substituting for the difference in notation and noting that an extra multiplicative factor of k appears in STICKLER's result.

The result for the (branch cut) integral along the positive imaginary axis $\xi=i\tau$ is similar in form to the Pekeris branch cut integral, thus

$$\lim_{\tau\to\infty}\frac{-(S\omega)}{\pi^2}\int_0^{\tau}\hat{G}(e,d;\xi)\Big/\left[\left(\tilde{U}^N_{N-1}\right)^2-\left(\tilde{F}^N_{N-1}\right)^2\left(k_N^2+\tau^2\right)/(\omega\rho_N)^2\right]\left(k_N^2+\tau^2\right)^{\frac{1}{2}}(\omega\rho_N)^{-1}K_0(\tau r)d\tau$$

*e) Numerical Considerations*

As one might expect, the implementation problems are grossly similar for all normal mode programs. Thus a summary of one, namely BARTBERGER's AP2 program, should provide the reader with an overall feeling for what might be expected in general.

The program is broadly divided into a section for the computation of the eigenvalues and one for the calculation of propagation loss. The execution time for each section is comparable. The maximum number of modes allowed is 500 which is sufficient for typical deep ocean problems at frequencies below approximately 100 Hz. The modes excluded are the higher order modes ($m>500$ in the residue summation) which introduces an error at the short bottom bounce ranges but not beyond. The attenuation $\alpha_\omega$

[dB/distance] within the water column which is small at these frequencies is initially set to zero and accounted for later by adding the term $\alpha_\omega r$ to the propagation loss calculation. Almost all of the limitations imposed upon the applicability of normal mode programs stem from the difficulty encountered in numerically locating the eigenvalues which are the zeros of the Wronskian appearing in the denominator of (3.39). This is done iteratively. An initial estimate say $(\tilde{\xi}_m)_i$ is made and $\tilde{F}^N_{N-1}[(\tilde{\xi}_m)_i]$ and $\tilde{U}^N_{N-1}[(\tilde{\xi}_m)_i]$ are found from (3.44) where $\tilde{F}^1_1 = 0$ (pressure release surface). $\tilde{U}^1_1$ is an arbitrary normalization constant which is properly accounted for in the Green's function formulation. In addition, the derivatives of these functions with respect to $\xi$ are also calculated with the aid of (3.45). The initial conditions used by BARTBERGER in this case are $\partial\tilde{F}^1_1/\partial\xi = \partial\tilde{U}^1_1/\partial\xi = 0$. Since $W_{N-1}[(\tilde{\xi}_m)_i]$ will in general not be zero, a new estimate $(\tilde{\xi}_m)_{i+1}$ is obtained from

$$(\tilde{\xi}_m)_{i+1} = \left\{(\tilde{\xi}_m)_i - W_{N-1}/[\partial W_{N-1}/\partial\xi]\right\}\Big|_{(\tilde{\xi}_m)_i} \quad ,$$

where the derivative of the Wronskian determinant with respect to $\xi$ is given by (3.42). The iterations are terminated when the modulus of the last term in the equation above falls below a preset minimum value.

If the initial estimates are not sufficiently accurate the above procedure may fail to converge and the mode will be missed, or it may converge to the wrong mode resulting in a duplicate eigenvalue. If the eigenvalues are evenly spaced it usually suffices to obtain initial estimates by extrapolating from previously found eigenvalues. This procedure is used for higher-order modes (phase velocity) exceeding the maximum sound speed (including that of the first bottom layer). The spacing of the eigenvalues for lower order modes may be sufficiently irregular to render this procedure unreliable. Thus, BARTBERGER has devised a scheme for obtaining initial estimates of these modes based upon WKB approximations which displays the mode number explicitly. The total phase change of the mode depth function is calculated for a two-way transit from ocean surface to bottom. Included are the phase changes in the oscillatory region between and at mode turning points and discrete phase shifts at ocean boundaries. The value of $\xi$ is varied iteratively until the round trip phase change equals $2(m-1)\pi$. If multiple sound channels exist, which give rise to different families of trapped modes, the AP2 scheme computes the phase changes in all channels. Except for a few isolated instances of complete failure this process provides the initial estimates of the eigenvalues rapidly. The failures have been observed to occur in instances when $\xi$ is very close to the wavenumber associated with one of the layer boundaries. This case must also be treated with the care in the multipath expansion programs and has been examined by TOLSTOY [3.16]. In this case successive values for the initial estimate oscillate around this wavenumber and BARTBERGER then switches to an extrapolated value based upon previously computed eigenvalues.

The WKB iterations for the first mode begin with the initial estimate $\xi_0 = \omega/(V_{MIN}+\varepsilon)$, where $V_{MIN}$ is normally equal to the minimum sound speed. However, deep water problems are frequently encountered for which a number of low order modes are weakly excited since their phase velocities are appreciably less than the sound speed at the source and receiver depths. In such instances an algorithm exist which excludes the calculation of the associated eigenvalues.

### 3.1.4 Depth Dependent Green's Function (Traveling-Wave Formulation)

In this section results for the depth dependent Green's function analogous in form to (3.22) but expressed in terms of "generalized reflection coefficients" instead of impedances are summarized. A more detailed discussion may be found in [3.1].

The matrix $\tilde{\psi}_k(z)$ may be expressed directly in terms of the fundamental solutions by using (3.7,8,10)

$$\tilde{\psi}_k(z) = \begin{pmatrix} f_k(\gamma_k^z) & g_k(\gamma_k^z) \\ (Df)_k^z & (Dg)_k^z \end{pmatrix} \begin{pmatrix} f_k(\gamma_k^{z_k}) & g_k(\gamma_k^{z_k}) \\ (Df)_k^{z_k} & (Dg)_k^{z_k} \end{pmatrix}^{-1} \begin{pmatrix} \tilde{F}_{k-1}^k \\ \tilde{U}_{k-1}^k \end{pmatrix} \quad , \quad z_k \leqq z \leqq z_{k+1} \quad .$$

The above expression may then be put into the form

$$\tilde{\psi}_k(z) = \begin{pmatrix} f_k(\gamma_k^z) & g_k(\gamma_k^z) \\ (Df)_k^z & (Dg)_k^z \end{pmatrix} \begin{pmatrix} \tilde{R}_k^k \\ 1 \end{pmatrix} \tilde{\Delta}_k \quad , \quad \tilde{\Delta}_k = \frac{-(Df)_k^k \tilde{F}_{k-1}^k + f_k(a)\tilde{U}_{k-1}^k}{(Dg)_k^k f_k(a) - g_k(a)(Df)_k^k} \quad . \tag{3.46}$$

The "generalized reflection coefficient" in the $k^{th}$ layer evaluated at $z_k$ is

$$\tilde{R}_k^k = - \frac{(Dg)_k^k \left[\tilde{Z}_{k-1}^k - g_k(a)/(Dg)_k^k\right]}{(Df)_k^k \left[\tilde{Z}_{k-1}^k - f_k(a)/(Df)_k^k\right]} \quad , \quad a = \gamma_k^k \quad , \tag{3.47}$$

where $f_k$ and $g_k$ are chosen to be independent traveling wave solutions [3.6] to $\mathscr{L}(\cdot) = 0$. The ratios $g_k(a)/(Dg)_k^k$ and $f_k(a)/(Df)_k^k$ are partial impedances whose importance will be discussed shortly. The physical interpretation of $\tilde{R}_k^k$ (and its counterpart $R_k^{k+1}$ soon to be introduced) continues to be the subject of considerable discussion. For constant sound speed and density layers the partial impedances are independent of depth and it is possible to uniquely separate upgoing and downgoing waves. Thus the angle between the normal to the surface of constant phase and the vertical remains the same throughout the layer and a wave which initially starts out say in a downward direction never changes its direction of propagation within a layer. The reflection coefficient is thus independent of depth and can be thought of as originating at the boundary where a mismatch occurs between the local partial impedances and the total impedance $\tilde{Z}_{k-1}^k$ of the medium above. When the product of sound speed and density is a function of depth, the "generalized reflection coefficient" is also a function of depth through the partial impedances. Thus reflection occurs at all depths within the layer as well as at the boundary and the functions $f_k$ and $g_k$ can both be written as having upgoing and downgoing components. The ratio $f_k/g_k$ or vice versa then clearly cannot be interpreted in the same manner as was done for constant sound speed and density layers.

In underwater acoustics, however, the "generalized reflection coefficients" have been and continue to be treated as if they were identical to the plane wave counterparts. The error of this approximation has recently become apparent as evidenced by determination of negative experimental values for bottom loss [3.17].

With this short dialogue hopefully firmly entrenched, the quotation marks and the word generalized will be abandoned for the remainder of the section.

Proceeding in exactly the same manner as was done for $\tilde{\psi}_k$, the solution below the source may be re-expressed in the form

$$\psi_k(z) = \begin{pmatrix} f_k\left(\gamma_k^z\right) & g_k\left(\gamma_k^z\right) \\ (Df)_k^z & (Dg)_k^z \end{pmatrix} \begin{pmatrix} 1 \\ R_k^{k+1} \end{pmatrix} \Delta_k \quad , \quad b = \gamma_k^{k+1} \quad , \tag{3.48}$$

where

$$\Delta_k = \frac{(Dg)_k^{k+1} F_{k+1}^{k+1} - g_k(b) U_{k+1}^{k+1}}{(Dg)_k^{k+1} f_k(b) - g_k(b)(Df)_k^{k+1}} \quad , \quad R_k^{k+1} = \frac{-(Df)_k^{k+1}}{(Dg)_k^{k+1}} \frac{\left[Z_{k+1}^{k+1} - f_k(b)/(Df)_k^{k+1}\right]}{\left[Z_{k+1}^{k+1} - g_k(b)/(Dg)_k^{k+1}\right]} \quad . \tag{3.49}$$

The reflection coefficient in any layer may be written in terms of the reflection coefficients in any other layer and ultimately in terms of the terminal values according to

$$\begin{pmatrix} \tilde{R}_{k+1}^{k+1} \\ 1 \end{pmatrix} = \begin{pmatrix} -(Dg)_{k+1}^{k+1} & g_{k+1}(a) \\ (Df)_{k+1}^{k+1} & -f_{k+1}(a) \end{pmatrix} \prod_{p=k}^{1} M_p(b,a) \begin{pmatrix} f_1(a) & g_1(a) \\ (Df)_1^1 & (Dg)_1^1 \end{pmatrix} \begin{pmatrix} \tilde{R}_1^1 \\ 1 \end{pmatrix} \cdot \frac{\prod_{p=k}^{2} W[g_p, f_p; z_p]}{\prod_{p=k}^{1} [\tilde{R}_p^p \tilde{C}_p + \tilde{D}_p]} \quad . \tag{3.50}$$

$$\begin{pmatrix} R_{k-2}^{k-1} \\ 1 \end{pmatrix} = \begin{pmatrix} -(Df)_{k-2}^{k-1} & f_{k-2}(b) \\ (Dg)_{k-2}^{k-1} & -g_{k-2}(b) \end{pmatrix} \prod_{p=k-1}^{N-1} M_p(a,b) \begin{pmatrix} g_{N-1}(b) & f_{N-1}(b) \\ (Dg)_{N-1}^{N} & (Df)_{N-1}^{N} \end{pmatrix} \begin{pmatrix} R_{N-1}^{N} \\ 1 \end{pmatrix} \cdot \frac{\prod_{p=k-1}^{N-2} W[f_p, g_p; z_{p+1}]}{\prod_{p=k-1}^{N-1} \left[R_p^{p+1} C_p + D_p\right]} \quad , \tag{3.51}$$

where $W[g_p, f_p; z_p]$ is the Wronskian of $g_p$ and $f_p$, evaluated at $z_p$. The quantities $\tilde{A}_p$, $\tilde{B}_p$, $\tilde{C}_p$ and $\tilde{D}_p$ are defined by the matrix equation

$$\begin{pmatrix} \tilde{A}_p & \tilde{B}_p \\ \tilde{C}_p & \tilde{D}_p \end{pmatrix} = \begin{pmatrix} -(Dg)_{p+1}^{p+1} & g_{p+1}\left(\gamma_{p+1}^{p+1}\right) \\ (Df)_{p+1}^{p+1} & -f_{p+1}\left(\gamma_{p+1}^{p+1}\right) \end{pmatrix} \begin{pmatrix} f_p\left(\gamma_p^{p+1}\right) & g_p\left(\gamma_p^{p+1}\right) \\ (Df)_p^{p+1} & (Dg)_p^{p+1} \end{pmatrix}$$

and the quantities $A_p$, $B_p$, $C_p$ and $D_p$ are defined by the matrix equation

$$\begin{pmatrix} A_p & B_p \\ C_p & D_p \end{pmatrix} = \begin{pmatrix} -(Df)_{p-1}^{p} & f_{p-1}\left(\gamma_{p-1}^{p}\right) \\ (Dg)_{p-1}^{p} & -g_{p-1}\left(\gamma_{p-1}^{p}\right) \end{pmatrix} \begin{pmatrix} g_p\left(\gamma_p^{p}\right) & f_p\left(\gamma_p^{p}\right) \\ (Dg)_p^{p} & (Df)_p^{p} \end{pmatrix}$$

Following the procedure used in Sect.3.1.1 to satisfy the source conditions the Green's function with both the source and receiver in the same layer (LS=LR) is for $z_{LS} \leq z \leq z_s$

$$G(z,z_s) = \frac{\left[f_{LS}\left(\gamma_{LS}^{z}\right)\tilde{R}_{LS}^{LS} + g_{LS}\left(\gamma_{LS}^{z}\right)\right]\left[f_{LS}\left(\gamma_{LS}^{S}\right) + R_{LS}^{LS+1} g_{LS}\left(\gamma_{LS}^{S}\right)\right]}{W\left[f_{LS}, g_{LS}; z_S\right]\left[1 - \tilde{R}_{LS}^{LS} R_{LS}^{LS+1}\right]} \quad . \tag{3.52}$$

With LR > LS the result is found to be

$$G(z,z_s) = \frac{\left[f_{LR}\left(\gamma_{LR}^{z}\right) + R_{LR}^{LR+1} g_{LR}\left(\gamma_{LR}^{z}\right)\right]\left[f_{LS}\left(\gamma_{LS}^{S}\right)\tilde{R}_{LS}^{LS} + g_{LS}\left(\gamma_{LS}^{S}\right)\right]}{W\left[f_{LS}, g_{LS}; z_S\right]\left[1 - \tilde{R}_{LS}^{LS} R_{LS}^{LS+1}\right]} \Delta_{LS+1}^{LR} \quad , \tag{3.53}$$

where

$$\Delta_{LS+1}^{LR} = \prod_{p=LS+1}^{LR} \left\{ \frac{\left[f_{p-1}(b) + R_{p-1}^{p} g_{p-1}(b)\right]}{\left[f_p(a) + R_p^{p+1} g_p(a)\right]} \right\} \quad , \quad a = \gamma_p^p, \; b = \gamma_{p-1}^p \quad .$$

The above results are equivalent to those obtained (3.21), (3.22) with the impedance formulation. In what follows it will be required to have a form of the Green's function similar to (3.23) which displays the dependence upon the terminal reflection coefficients $\tilde{R}_1^1$ and $R_{N-1}^N$ explicitly. Substitution from the matrix form of (3.47) and (3.49) yields

$$\frac{\left(1-\tilde{R}_1^1\right)\begin{pmatrix} (Dg)_1^2 & -g_1(b) \\ -(Df)_1^2 & f_1(b) \end{pmatrix} K_2^{LS} \Phi_{LS} \Phi_{LR} K_{LR+1}^{N-1} \begin{pmatrix} g_{N-1}(b) & f_{N-1}(b) \\ (Dg)_{N-1}^N & (Df)_{N-1}^N \end{pmatrix}\begin{pmatrix} R_{N-1}^N \\ 1 \end{pmatrix}}{\left(1-\tilde{R}_1^1\right)\begin{pmatrix} (Dg)_1^2 & -g_1(b) \\ -(Df)_1^2 & f_1(b) \end{pmatrix} K_2^{N-1} \begin{pmatrix} g_{N-1}(b) & f_{N-1}(b) \\ (Dg)_{N-1}^N & (Df)_{N-1}^N \end{pmatrix}\begin{pmatrix} R_{N-1}^N \\ 1 \end{pmatrix}} \quad . \tag{3.54}$$

In order to provide some insight into the role played by the partial impedances, examine the inner product of matrices within $M_k(a,b)M_{k+1}(a,b)$. Then

$$\begin{pmatrix} (Dg)_k^{k+1} & -g_k(b) \\ -(Df)_k^{k+1} & f_k(b) \end{pmatrix} \begin{pmatrix} f_{k+1}(a) & g_{k+1}(a) \\ (Df)_{k+1}^{k+1} & (Dg)_{k+1}^{k+1} \end{pmatrix}$$

$$= \begin{pmatrix} f_{k+1}(a)g_k(b)\left[\dfrac{(Dg)_k^{k+1}}{g_k(b)} - \dfrac{(Df)_{k+1}^{k+1}}{f_{k+1}(a)}\right] & g_{k+1}(a)g_k(b)\left[\dfrac{(Dg)_k^{k+1}}{g_k(b)} - \dfrac{(Dg)_{k+1}^{k+1}}{g_{k+1}(a)}\right] \\ -f_{k+1}(a)f_k(b)\left[\dfrac{(Df)_k^{k+1}}{f_k(b)} - \dfrac{(Df)_{k+1}^{k+1}}{f_{k+1}(a)}\right] & g_{k+1}(a)f_k(b)\left[-\dfrac{(Df)_k^{k+1}}{f_k(b)} + \dfrac{(Dg)_{k+1}^{k+1}}{g_{k+1}(a)}\right] \end{pmatrix}$$

If the assumption is now made that the partial impedances are equal that is

$$\frac{(Df)_k^{k+1}}{f_k(b)} = \frac{(Df)_{k+1}^{k+1}}{f_{k+1}(a)} \quad , \quad \frac{(Dg)_k^{k+1}}{g_k(b)} = \frac{(Dg)_{k+1}^{k+1}}{g_{k+1}(a)} \quad , \tag{3.55}$$

one has that

$$\begin{pmatrix} (Dg)_k^{k+1} & -g_k(b) \\ -(Df)_k^{k+1} & f_k(b) \end{pmatrix} \prod_{p=k+1}^{L} M_p(a,b) = \begin{pmatrix} B_k^L & 0 \\ 0 & A_k^L \end{pmatrix} \begin{pmatrix} (Dg)_L^{L+1} & -g_L(b) \\ -(Df)_L^{L+1} & f_L(b) \end{pmatrix} \quad , \tag{3.56}$$

where

$$B_k^L = \frac{g_k(b)}{g_L(a)} \prod_{p=k+1}^{L-1} \left[\frac{g_p(b)}{g_p(a)}\right] \quad , \quad A_k^L = \frac{f_k(b)}{f_L(a)} \prod_{p=k+1}^{L-1} \left[\frac{f_p(b)}{f_p(a)}\right] \quad .$$

If the above result is substituted in (3.54), it can be shown [3.1] that the Green's function $G(z,z_S;\xi,\omega)$, written explicitly in terms of the terminal reflection coefficients in non-matrix form, is equal to

$$\frac{\left[A_1^{LS} f_{LS}\left(\gamma_{LS}^{S}\right)\tilde{R}_1^1 + B_1^{LS} g_{LS}\left(\gamma_{LS}^{S}\right)\right]\left[B_{LR}^{N-1} f_{LR}\left(\gamma_{LR}^{z}\right) + A_{LR}^{N-1} R_{N-1}^{N} g_{LR}\left(\gamma_{LR}^{z}\right)\right]}{W[f_{LR}, g_{LR}; z_{LR+1}]\, B_1^{N-1}\left[1 - \dfrac{A_1^{N-1}}{B_1^{N-1}} \tilde{R}_1^1 R_{N-1}^N\right]} \quad . \tag{3.57}$$

Also under the assumption of continuity of partial impedances one has that

$$\tilde{R}_{k+1}^{k+1} = \frac{A_1^{k+1}}{B_1^{k+1}} \tilde{R}_1^1 \quad , \quad R_{k-2}^{k-1} = \frac{A_{k-2}^{N-1}}{B_{k-2}^{N-1}} R_{N-1}^N \quad .$$

In order to carry the analysis further, specific forms for the functions f and g are required. For the moment let these solutions be represented by the WKB approximations. In order to avoid the problem of their validity at turning points, attention can be limited to values of $\xi$ associated with bottom bounce energy for which no turning point exist between $z_1$ and $z_N$. Then

$$f_k(z) \simeq \rho^{\frac{1}{2}}(z)\hat{q}_k^{-\frac{1}{4}}(z)\,\exp\left[i\left(\phi_k^z+\pi/4\right)\right] \quad , \quad g_k(z) \simeq \rho^{\frac{1}{2}}(z)\hat{q}_k^{-\frac{1}{4}}(z)\,\exp\left[-i\left(\phi_k^z+\pi/4\right)\right]$$

$$\hat{q}_k(z) = \left(\omega^2/c_k^2(z)-\sqrt{\rho(z)}\,\frac{d^2}{dz^2}\left(\frac{1}{\sqrt{\rho(z)}}\right)-\xi^2\right) \quad , \quad \phi_k^z = \int_k^z \hat{q}_k^{\frac{1}{2}}(t)dt \quad . \qquad (3.58)$$

Continuity of the partial impedances is obtained if $\hat{q}_k^{\frac{1}{2}}(z_{k+1}) = \hat{q}_{k+1}^{\frac{1}{2}}(z_{k+1})$, and $d\hat{q}_k/dz = d\hat{q}_{k+1}/dz$ at $z = z_{k+1}$, that is the sound speed and its first derivative with respect to depth must be continuous in addition to the density. Assuming this to be the case it is a straightforward matter to verify that the Green's function reduces to

$$G(z,z_s) = \frac{\hat{q}_{LS}^{-\frac{1}{4}}(z_s)\left\{\exp\left[i\left(\phi_1^s+\pi/4\right)\right]\tilde{R}_1^1+\exp\left[-i\left(\phi_1^s+\pi/4\right)\right]\right\}\hat{q}_{LR}^{-\frac{1}{4}}(z)\left\{\exp\left[i\left(\phi_{N-1}^z+\pi/4\right)\right]+R_{N-1}^N\,\exp\left[-i\left(\phi_{N-1}^z+\pi/4\right)\right]\right\}}{W[f_{LR},g_{LR};z_{LR+1}]\,\exp\left(-i\phi_1^{N-1}\right)\left[1-\tilde{R}_1^1R_{N-1}^N\,\exp\left(2i\phi_1^{N-1}\right)\right]},$$

(3.59)

where

$$W[f_{LR},g_{LR};z_{LR+1}] = -2i \quad , \quad R_{k-2}^{k-1} = \exp\left(2i\phi_{k-2}^{N-1}\right)R_{N-1}^N \quad , \quad \tilde{R}_{k+1}^{k+1} = \exp\left(2i\phi_1^{k+1}\right)\tilde{R}_1^1 \;.$$

Thus the absolute magnitude of the reflection coefficient at any level (with no turning points) below the source equals that of the reflection coefficient at the bottom and similarly above the source.

Furthermore, this result is identical to that which would be obtained if instead of layering, the solutions f and g over the entire interval ($z_1 \leq z \leq z_N$) were approximated by WKB solutions. It can be seen that for this approach to be useful the sound speed and its first derivative plus the density must be continuous throughout the interval. Viewed in the opposite way, if a discontinuity in the local partial impedances exist at any level, it gives rise to a reflection at that level whose amplitude is frequency dependent.

To properly account for the effects caused by a sound-speed structure which is not slowly varying with the approach which approximates the solutions over the entire interval, either more terms would have to be introduced into the approximate solution or the sound speed profile would have to be layered over the appropriate depth intervals.

### 3.1.5 Multipath Expansion Models

The models to be discussed in this section use a combination of wave and ray theory. The mathematics involved is centered around expanding the kernel of the field integral in terms of an infinite series (sometimes referred to as the Rainbow or Debye expansion [3.18-20]) of integrals which correspond to various types of multiple reflections. WEINBERG [3.21] did not consider the sound speed profile to be layered except under some special instances to be taken up separately. Instead approximate solutions for f and g are found for the entire depth interval $z_1 \leqq z \leqq z_N$ which are then used to construct the depth dependent Green's function. WEINBERG's model of the ocean with respect to Fig.3.2 would then consist of the three layers: air, water and bottom with the water layer contained between $z_1$ and $z_2$. In this case LS is always equal to LR and from (3.52) the depth dependent Green's function above the source is

$$G(z,z_s) = \frac{\left[f_1(z)\tilde{R}_1^1+g_1(z)\right]\left[f_1(z_s)+R_1^2 g_1(z_s)\right]}{W[f_1,g_1;z_s]\left[1-\tilde{R}_1^1 R_1^2\right]} \quad , \quad z_1 \leqq z \leqq z_s \quad . \tag{3.60}$$

The solutions for the functions used in the program

$$f_1(z) \simeq CV_0(z)[Bi(\gamma^z)+iJAi(\gamma^z)] \quad , \tag{3.61}$$

arise from taking the first term of an asymptotic expansion. The constant C is defined as

$$C = \pi^{\frac{1}{2}}\omega^{1/6}\exp[i\omega\hat{Q}(z_0,z_t;\lambda)]\exp(-ij\pi/4) \quad \text{and} \quad \hat{Q}(a,b;\lambda) = \int_a^b [c^{-2}(z)-\lambda^2]^{\frac{1}{2}}dz \quad . \tag{3.62}$$

The argument of the Airy functions is $\gamma^z = \omega^{2/3}\rho(z,\lambda)$ and the functions $\rho$, and $V_0$ must satisfy the Eikonal and transport equations [3.21], respectively. The constant j is either ±1 depending upon the juxtaposition of $z_t$ (the turning point depth) and $z_0$, some arbitrary depth usually taken as that depth where the sound speed is a minimum. In particular $j = 1$ when $z_0 < z_t$ (lower turning point), $j = 1$ when $z_0 > z$ (upper turning point). Finally the parameter $\lambda$ is related to the separation constant $\xi$ according to $\lambda = \omega\xi$.

The solution for $g_1$ is obtained from (3.61) by systematically replacing i with -i. Whereas the full solution (if it converges) accommodates an arbitrary sound speed variation it can be shown that (3.61) exactly satisfies $\mathscr{L}(f_1) = 0$ when $c^{-2}(z)$ is linear in depth. This, of course, is the same sound speed variation assumed by BARTBERGER and STICKLER (Sect.3.1.3) within each of their layers. An indication then of the approximation made by using the truncated solution (3.61) can be obtained by comparing the given sound speed profile with

$$c^{-2}(z) = \alpha_1 + \beta(z_1-z) \qquad z_1 \leqq z \leqq z_2 \quad , \tag{3.63}$$

which can have at the most one turning point. For this variation

$$\bar{\rho}(z,\lambda) = -\beta^{1/3}(z-z_t) \quad , \quad V_0(z,\lambda) = |\partial\bar{\rho}/\partial z|^{-\frac{1}{2}} \quad , \tag{3.64}$$

thus the argument of the Airy functions is zero at the turning point. If for large $\gamma^z$ the Airy functions are replaced by their asymptotic expansions with $\zeta = \omega^{2/3}\bar{\rho}(z;\lambda)$

$$Ai(-\zeta) \simeq \pi^{-\frac{1}{2}}\zeta^{-\frac{1}{4}} \sin\left(\frac{2}{3}\zeta^{2/3} + \frac{\pi}{4}\right) \quad , \quad Bi(-\zeta) \simeq \pi^{-\frac{1}{2}}\zeta^{-\frac{1}{4}} \cos\left(\frac{2}{3}\zeta^{2/3} + \frac{\pi}{4}\right) \quad , \tag{3.65}$$

they then reduce to the WKB solutions given by (3.58) (multiplied by $\omega^{\frac{1}{2}}$) with $z_k = z_0$. Another indication of the approximation introduced by using the truncated solution (3.61) was provided at the end of the previous section. There the approaches of modeling the ocean with layers and that of approximating the solutions over the entire (in this case $z_1<z<z_{2=N}$) interval, were discussed. It was noted that the latter approach fails whenever the exact partial impedances become discontinuous, which is related to the continuity of sound speed and its first derivative. In such instances the approximate solution approach must be modified either by the insertion of layers or the inclusion of higher-order terms from the full solution. For the time being then limit consideration to cases where at most one turning point exists and (3.61) is exact.

Before proceeding to the multipath expansion it is convenient to redefine the reflection coefficients $\left(\tilde{R}_1^1, R_1^2\right)$ previously given as (3.47) and (3.49) as

$$\tilde{R}_1^1 = \frac{g_1(z_1)}{f_1(z_1)}\tilde{\mathscr{R}}_1^1 \quad , \quad \tilde{\mathscr{R}}_1^1 = \frac{\tilde{Y}_0^1 - (Dg)_1^1/g_1(z_1)}{(Df)_1^1/f_1'(z_1) - \tilde{Y}_0^1} \quad , \tag{3.66}$$

$$R_1^2 = \frac{f_1(z_2)}{g_1(z_2)}\mathscr{R}_1^2 \quad , \quad \mathscr{R}_1^2 = \frac{Y_2^2 - (Df)_1^2/f_1(z_2)}{(Dg)_1^2/g_1(z_2) - Y_2^2} \quad , \tag{3.67}$$

with $\tilde{Y}_0^1$, $Y_2^2$ representing the admittance at the surface and bottom, respectively. The behavior of the reflection coefficients when z is close to a turning point is, from (3.61),

$$\tilde{R}_1^1 = \exp[-2i\omega\hat{Q}(z_0,z;\lambda)]\ \exp(j\pi/2)\ \exp\left\{-2ij\ \tan^{-1}\left[Ai\left(\gamma_1^1\right)/Bi\left(\gamma_1^1\right)\right]\right\} \quad ,$$

$$R_1^2 \simeq -\exp[2i\omega\hat{Q}(z_0,z;\lambda)]\ \exp(-j\pi/2)\ \exp\left\{2ij\ \tan^{-1}\left[Ai'\left(\gamma_1^2\right)/Bi'\left(\gamma_1^2\right)\right]\right\} \quad . \tag{3.68}$$

When z is not close to a turning point the WKB solutions yield

$$\tilde{R}_1^1 = \exp\{-i[2\omega\hat{Q}(z_0,z_1;\lambda)+\pi]\} \quad , \quad R_1^2 \simeq \exp[2i\omega\hat{Q}(z_0,z_2;\lambda)] \quad .$$

The Green's function may be expanded in two ways. Multiplying the components of the numerator gives four terms. Each term represents a ray path type which includes all cycles of the ray.

The individual cycles of the rays may be obtained by expanding the term $\left(1-\tilde{R}_1^1R_1^2\right)$ appearing in the denominators of the four terms in a binomial series. Doing this and simultaneously substituting from (3.66) and (3.67) gives the multipath expansion for $z < z_s$

$$\sum_{n=0}^{\infty}\left(\tilde{\mathscr{R}}_1^1\mathscr{R}_1^2\ A_1^1/B_1^1\right)^n\left[\frac{g_1(z)f_1(z_s)}{-2i\omega}+\frac{f_1(z)f_1(z_s)g_1(z_1)\tilde{\mathscr{R}}_1^1}{-2i\ f_1(z_1)}\right.$$
$$\left.+\frac{f_1(z)g_1(z_s)g_1(z_1)f_1(z_2)}{-2i\omega f_1(z_1)g_1(z_2)}\tilde{\mathscr{R}}_1^1\mathscr{R}_1^2+\frac{g_1(z)g_1(z_s)f_1(z_2)}{-2i\omega g_1(z_2)}\mathscr{R}_1^2\right]\ , \qquad (3.69)$$

where

$$B_1^1 = g_1(z_2)/g_1(z_1)\quad ,\quad A_1^1 = f_1(z_2)/f_1(z_1)\quad .$$

For the case of either no turning points (e.g., bottom bounce paths) or at sufficiently large distance from a turning point, the WKB solutions are appropriate. Then $A_1^1/B_1^1 = \exp[2i\hat{Q}(z_1,z_2;\lambda)]$ and upon replacing the Hankel function in the field integral with the first term in its asymptotic expansion[1] one obtains the multipath integral representation for the pressure $\mathscr{P}(z,z_s,r)$ as

$$\mathscr{P}(z,z_s,r) \qquad (3.70)$$
$$=\sum_{n=0}^{\infty}\int_0^{\infty}\mathscr{G}(z,z_s;\lambda,f)\ \exp\{i\omega[\hat{Q}(z,z_s;\lambda)+2n\hat{Q}\left(z_1,z_{2_1};\lambda\right)-\pi/4]\}\ \exp[i\omega\lambda r_{d\lambda}+(\ldots)+(\ldots)+(\ldots)]\ ,$$

where the terms in (...) would be of a form similar to that shown explicitly and $\mathscr{G}$ is an expression involving Airy functions which reduces to

$$\mathscr{G}(z,z_s;\lambda,f) = \frac{\hat{q}^{-\frac{1}{4}}(z)\hat{q}^{-\frac{1}{4}}(z_s)\omega\lambda}{-2i\sqrt{2+2\rho\omega\lambda r}}\left(\tilde{\mathscr{R}}_1^1\mathscr{R}_1^2\right)^n\quad ,$$

far away from turning points. The integrals above are of the form

$$I = \int_{\lambda_1}^{\lambda_2}\mathscr{G}(\lambda)\ e^{i\psi(\lambda)}d\lambda\quad , \qquad (3.71)$$

and can be evaluated by the method of stationary phase.

---

[1]WEINBERG modified the usual large argument asymptotic expansion so that

$$H_0^{(1)}(\omega\lambda r)\simeq(2+2\pi\omega\lambda r)^{-\frac{1}{2}}\exp[i(\omega\lambda r-\pi/4)]$$

provides a closer approximation for small values of the argument.

The stationary phase points can be related to the eigenrays of classical ray theory (a discussion of this is taken up in Sect.3.1.6). In the event that the stationary phase evaluation breaks down, WEINBERG evaluated the phase integrals in (3.70) numerically. The breakdown occurs in the vicinity of caustics which can be located from classical ray theory. Having located these regions, the difference in phase between a ray on the caustic and a ray near the caustic is calculated. If this difference is small the integrals are evaluated numerically otherwise the stationary phase technique is used. WEINBERG has found that for the deep ocean and acoustic frequencies greater than 100 Hz the majority of the integrals are evaluated by the method of stationary phase.

Before discussing the numerical integration procedure, this is an appropriate time to note a significant difference in emphasis between the models previously discussed and those of this section and also the classical ray-theory models with corrections (Sect.3.1.6) such as FACT [3.22]. In the latter case, the primary emphasis was upon accuracy and a minimum number of approximations were introduced after the initial statement of the problem. The resulting execution speed was fundamentally that needed to evaluate the mathematics. STICKLER's branch cut integral for example might be routinely evaluated even though in many instances it would not significantly contribute to the total field. Similarly the limits of integration for the FFP are generally larger than required from a practical standpoint.

In the former case, the models were developed with the goal of minimizing execution speed without significantly effecting accuracy. To meet this end approximations are introduced, based in many instances upon the long experience of the developer. It is impossible to summarize all such approximations. Indeed when examined in isolation a false perception of their validity to the overall answer may result. In order, however, to provide an indication of what is involved the major steps in the program are outlined below for a typical deep water profile shown schematically in Fig.3.6.

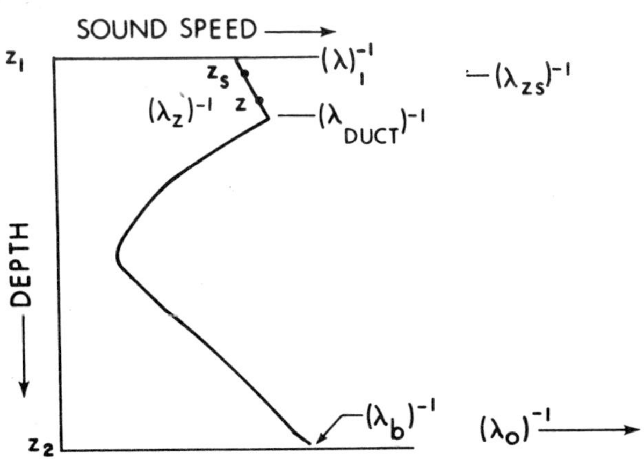


Fig. 3.6. Typical deep water profile

In order to perform the numerical integration, formulas for the amplitude of $\mathcal{G}$ must be obtained. To illustrate the process let it be assumed that both source and receiver are in the surface duct. In this case, the lower limit of (3.71) is set to zero (corresponding to infinite sound speed) and $\lambda_2$ is taken to be the sound speed at the surface ($z_1$) which is the local minimum sound speed. The $\lambda$ domain is then divided into the subintervals

$$0 \text{ to } \lambda_B \quad , \quad \lambda_B \text{ to } \lambda_{DUCT} \quad , \quad \lambda_{DUCT} \text{ to } \lambda_z \quad , \quad \lambda_z \text{ to } \lambda_{z_s} \quad , \text{ and } \lambda_{z_s} \text{ to } \lambda_{z_1} \quad ,$$

which are identified on Fig.3.6. Each of these subintervals is further divided by tracing 16 non-vertexing rays (0 to $\lambda_B$), 32 vertexing rays ($\lambda_B$ to $\lambda_{z_s}$) and 16 rays vertexing above the source ($\lambda_{z_s}$ to $\lambda_{z1}$). In addition, 30 additional rays are traced through (0 to $\lambda_{z1}$). In each instance the spacing between the rays is non-uniform in angle so as to improve the numerical evaluation of integrals to be discussed below.

The particular formulas used for the amplitudes of f and g within each subinterval are chosen based upon the size of $\zeta = \omega^{2/3}\bar{\rho}(z;\lambda)$. If $\zeta < -0.8477$ no turning points are encountered and the amplitudes are calculated from the WKB solutions. This might occur, for example, in the interval 0 to $\lambda_B$ where steep bottom bounce rays exist. On the other hand, if $\zeta > 20$, a turning point exists but it is far removed from either z or $z_s$ so that the amplitude would be exponentially decaying. Such a case would exist for ($\lambda_{z_S}, \lambda_{z_1}$) with rays vertexing far above either the source or receiver depth. In such cases, WEINBERG set the amplitude to zero on the basis that contributions from other subintervals would dominate the total answer. Finally, if $\zeta$ falls between these two limits turning points exist near either the source or receiver depths. In this instance the amplitudes are obtained from the Airy functions.

As mentioned earlier, the truncated solution approach used by WEINBERG must be modified when either the partial impedances are not continuous or if two or more turning points are in close proximity. Such a situation exists for the profile considered within the interval ($\lambda_{DUCT}, \lambda_z$). The Green's function for the multipath integral (3.70) corresponding to the direct path connecting source and receiver in the surface duct is modified by replacing $R_1^2$ with a reflection coefficient referenced to the bottom of the surface duct. This new reflection coefficient is obtained by replacing the profile below the duct with a semi-infinite layer within which the sound speed is given by (3.65) with $\gamma_1$ obtained from that portion of the original profile immediately below the surface duct. The reflection coefficient for this bilinear profile, obtained by satisfying the continuity conditions, is then approximated so that the final result is a ratio of the difference and sum of Airy functions.

The reflection coefficient $R_1^2$ is also modified for the interval ($\lambda_0, \lambda_{DUCT}$) which in this case corresponds to RSR ray paths. When the turning point is sufficiently above the bottom the magnitude of $R_1^2$ is set to unity and its phase to $-\pi/2$. In those cases where the turning point is close to the bottom $R_1^2$ is replaced with (3.68).

The interval of integration $(\lambda_1,\lambda_2)$ in (3.71) is initially approximated by $(0,\lambda_{z1})$. This is further divided into subintervals

$$(0<\lambda_1<\lambda_2 \ldots \lambda_i<\lambda_j \quad \ldots \lambda_{z_1}) \quad ,$$

where $\lambda_i$ and $\lambda_j$ correspond to caustic regions defined by either $\lambda_B$, $\min(\lambda_{z_s},\lambda_z)$ or $\lambda$'s for minimum and maximum range values. Thus a typical integral to be evaluated numerically is of the form

$$I = \int_{\lambda_i}^{\lambda_j} \mathscr{G}(\lambda) \; e^{i\psi(\lambda)} d\lambda \quad . \tag{3.72}$$

For this subinterval, the phase $\psi(\lambda)$ is fit with a cubic and the amplitude $\mathscr{G}(\lambda)$ is fit with a linear function. If $\lambda$ is sufficiently far from a point of stationary phase, $\psi(\lambda)$ is monotonic so that (3.72) can be transformed to

$$I = \int_{\lambda_i}^{\lambda_j} \left[\mathscr{G}(\psi)\, \frac{d\lambda}{d\psi}\right] e^{i\psi} d\psi \quad . \tag{3.73}$$

The quantity in brackets can then be approximated by a polynomial in $\psi$, and the expression can be evaluated in closed form. However, if $\lambda$ is near a point of stationary phase (but still close to a caustic), (3.72) is evaluated by the trapezoidal rule.

The philosophy adopted in LEIBIGER's Raymode X program [3.23] is to utilize, whenever possible, the simpler tools of ray theory while retaining the more exact formulation of mode theory. Two important benefits are thus realized: 1) It is possible to interpret mode theory expressions in a manner similar to ray theory[2] and, 2) the use of ray theory simplifies some of the computational aspects of the normal mode theory, allowing for considerable savings in computer execution time.

LEIBIGER initially assumed the medium to be layered with the sound speed given by (3.63) within each layer. A fundamental solution of $\mathscr{L}(\cdot)=0$ may then be written as $f(z;\xi)=V(z;\xi)\ \exp\left(i\phi_0^z\right)$, where

$$V(z;\xi) = K\left[\phi_0^z/\hat{q}^{\frac{1}{2}}(z;\xi)\right]^{\frac{1}{2}} H_{1/3}^{(1)}\left(\phi_0^z\right) \exp\left(-i\phi_0^z\right) \quad , \quad K = (\pi/2)^{\frac{1}{2}} \exp(+i5\pi/12) \quad .$$

In the definition of $\hat{q}$ and $\phi$ (3.58) $\rho(z)$ is assumed to be unity. The solution for $g(z;\xi)$ is taken to be the complex conjugate of f. Sufficiently far from a turning point $V(z;\xi)\sim\hat{q}^{-\frac{1}{4}}$, and the exact solutions reduce to the usual WKB approximations. It is also noted that LEIBIGER's fundamental solutions, f and g, are formally equivalent to WEINBERG's truncated solutions.

[2]The discussion of this aspect of Raymode X seemed more appropriately suited to Sect.3.1.6.

LEIBIGER made the approximation that the wave is totally transmitted across all interfaces until either a layer is reached where a turning point exists or the wave interacts with either the ocean surface or bottom. At a turning point the "reflection coefficient" R is assigned unit magnitude and a $\pi/2$ phase shift. The magnitude of R is obtained from a scattering model and the phase shift is assumed to be $-\pi$ at the ocean surface. At the ocean bottom the phase of R is zero, and the magnitude of R is obtained from an input table of values.

With these provisos, the approximate expression for the exact depth dependent Green's function (3.54) for $z \geqq z_s$ is

$$G(z,z_s) \sim \frac{\left[f(z_s;\xi)\tilde{R}_1^1+g(z_s;\xi)\right]\left[f(z;\xi)+R_{N-1}^N g(z;\xi)\right]}{2i\ \exp\left(-i\phi_1^N\right)\left[1-\tilde{\mathscr{R}}_1^1\mathscr{R}_{N-1}^N\ \exp\left(2i\phi_1^N\right)\right]} \quad .$$

The subscript of $\phi_1^N$ is a phase reference depth for the numerator term containing $\tilde{R}_1^1$ which is either the ocean surface or an upper turning point associated with a downward refracted wave. The superscript N is the phase reference depth for the remaining numerator term. It is either the bottom depth or the depth of a lower turning point corresponding to an upward refracted wave.

The conditions for which the above expression is a reasonable approximation for G are the same as those discussed in Sect.3.1.4, namely that the partial impedances be continuous. In the presence of a surface duct the formula is modified by replacing $R_{N-1}^N$ with the reflection coefficient at the bottom of the duct obtained from the continuity conditions for a bilinear profile.

Although LEIBIGER and WEINBERG initially approached the problem from different viewpoints (layering versus asymptotic solutions) both made fundamentally the same approximation (continuity of partial impedances) and arrived at basically the same Green's function. For the profile shown in Fig.3.6, LEIBIGER would replace the $(0,\infty)$ limits of integration with the subintervals $(\xi_N,\xi_{BOT})$, $(\xi_{BOT},\xi_{DUCT})$, $(\xi_{DUCT},\xi_{SURF})$, where $\xi_N$ is an input parameter related to the largest source angle to be considered. The integral expression for the pressure field is next rewritten in terms of the four integrals obtained when the numerator of the Green's function is expanded. These integrals over the previously defined limits of integration are evaluated by either normal mode theory, when the number of trapped modes (the default criterion is 10) within the $\xi$ partition is small; or via the multipath expansion.

In both instances then the integrals to be evaluated are of the form

$$\mathscr{P}(r,z,z_s) \sim \int_{\xi_A}^{\xi_B} \mathscr{G}\,\frac{(z,z_s,\xi)\ \exp\left[i\left(\phi_1^z-\phi_1^z s+\xi r\right)\right]}{\left[1-\tilde{\mathscr{R}}_1^1\mathscr{R}_{N-1}^N\ \exp\left(2i\phi_1^N\right)\right]}\,d\xi$$

$$\mathscr{G}(z,z_s;\xi) = i\left[e^{-i\pi/2}\ \xi/(2\pi r)\right]^{\frac{1}{2}}\ V^*(z_s;\xi)V(z;\xi) \quad , \qquad (3.74)$$

where for illustrative purposes only the integral resulting from the product of gf and substitution of the first term of the asymptotic expansion for $H_0^{(1)}$ is considered here.

When normal-mode theory is utilized the branch-cut contribution is not included. However, when (3.74) is evaluated by the multipath expansion for different limits of integration the branch-cut contribution to the total field is at least partially accounted for. The singularities $\xi_p$ associated with the modes are assumed to be simple poles obtained by solving $W(\xi_p) = 1-\tilde{\mathscr{R}}_1^1 \mathscr{R}_{N-1}^N \exp\left(2i\phi_1^N\right) = 0$. If the reflection coefficients are defined in terms of a magnitude and phase ($\theta_1$ and $\theta_N$) one has that

$$\left|\tilde{\mathscr{R}}_1^1\right| \left|\mathscr{R}_{N-1}^N\right| = \exp\left(+2\,\mathrm{Im}\left\{\phi_1^N\right\}\right) \quad ; \text{ and } \quad \theta_1 + \theta_N + 2\,\mathrm{Re}\left\{\phi_1^N\right\} = -2p\pi \quad , \quad p = 0,1,2,\ldots \quad . \tag{3.75}$$

The real part of the eigenvalue $\hat{\xi}_p = \mathrm{Re}\{\xi_p\}$ is found by solving (3.75) without recourse to iterative methods which results in a considerable savings in execution time. To accomplish this the extreme mode numbers $(N_A, N_B)$ are determined by substituting $\xi_A$ and $\xi_B$ into the second part of (3.75) and solving for p. In addition, differentiation yields

$$\frac{dp}{d\xi} = \frac{1}{2\pi}\left(R_c - \frac{\partial\theta_1}{\partial\xi} + \frac{\partial\theta_N}{\partial\xi}\right) \quad , \text{ where } \quad R_c = \frac{-\partial}{\partial\xi}\, 2\,\mathrm{Re}\left\{\phi_1^N\right\}\Big|_{\hat{\xi}_p} \quad ,$$

is the cycle range $R_c$ associated with the mode eigenvalue. In the next section it is shown that $R_c$ is also twice the horizontal distance traversed by an eigenray which connects the phase reference point 1 to the phase reference point N. The value for $R_c$ is available in closed form due to the assumed sound speed variation within each layer. If one assumes that the phase of the reflection coefficients is a slowly varying function of $\xi$ the above equation may then also be solved for the extreme values $\xi_A$, $\xi_B$. Thus a curve of $\hat{\xi}_p$ versus p passing through the end points $(\xi_A, N_A)$, $(\xi_B, N_B)$ and having the slopes prescribed above can be obtained. This curve is approximated by a cubic polynomial and the unknown coefficients are determined by interpolation. This expression is then evaluated for each integer p such that $N_A \leq p \leq N_B$ for a direct (non-iterative) evaluation of the real part of the eigenvalues.

The imaginary part of $\xi_p$ can be shown to be given approximately, with the use of WKB approximations, by

$$\mathrm{Im}\{\xi_p\} \sim \frac{-\ln\left(\left|\tilde{\mathscr{R}}_1^1\right| \left|\mathscr{R}_{N-1}^N\right|\right)}{R_c}\Bigg|_{\hat{\xi}_p} \quad .$$

The residue expansion associated with (3.74) then becomes

$$\mathscr{P}(r,z,z_s) \sim 2\pi i \sum_{p=N_A}^{N_B} \frac{\mathscr{G}(z,z_s;\xi_p)}{\left.\frac{\partial w}{\partial \xi}\right|_{\xi_p}} \exp\left[i\left(\phi_1^z-\phi_1^z s+\xi_p r\right)\right] \quad . \tag{3.76}$$

The normalization term in the denominator can be approximately written as

$$\left.\frac{\partial w}{\partial \xi}\right|_{\xi_p} \simeq -\frac{\partial \tilde{\mathscr{R}}_1^1}{\partial \xi}\left(\tilde{\mathscr{R}}_1^1\right)^{-1} + \frac{\partial\left(\mathscr{R}_{N-1}^N\right)}{\partial \xi}\left(\mathscr{R}_{N-1}^N\right)^{-1} - iR_c\Big|_{\hat{\xi}_p} \quad ,$$

and further approximated by retaining only the term involving $R_c$. The final form of the residue series is obtained by expanding the phase terms $\left(\text{e.g.}, \phi_1^z\right)$ in a Taylor's series about $\hat{\xi}_p$ and retaining only the first two terms since for trapped modes $\mathrm{Im}\{\xi_p\}$ is small. Then

$$\mathscr{P}(r,z,z_s) \sim 2\pi \sum_{p=N_A}^{N_B} \frac{\mathscr{G}(z,z_s;\hat{\xi}_p)}{(R_c)_p} \exp\left[i\left(\phi_1^z-\phi_1^z s+\hat{\xi}_p r\right)\right]$$

$$\cdot \exp(-\mathrm{Im}\{\xi_p\}[r(z_1,z)-r(z_1,z_s)-r]) \quad ,$$

where $r(z_1,z)$, $r(z_1,z_s)$ are the horizontal ranges obtained from ray theory from $z_1$ to z and $z_s$, respectively.

For the multipath approach the denominator of (3.74), for example, is expanded so that

$$\mathscr{P}(r,z,z_s) \sim \sum_{q=0}^{\infty} \int_{\xi_A}^{\xi_B} \mathscr{G}(z,z_s;\xi)\left(\tilde{\mathscr{R}}_1^1\mathscr{R}_{N-1}^N\right)^q \exp\left[i\left(\phi_1^z-\phi_1^z s+2q\phi_1^N+\xi r\right)\right]d\xi \quad .$$

The interval of integration is divided into a number of unequal sections based upon the number of rays traced which is an input parameter. The value of $\xi$ for some of these subsections is sufficiently far from a stationary phase point so that their contribution is excluded. Although stationary phase techniques are not used, the stationary phase point is available from ray calculations (see discussion at the end of Sect.3.1.6). The justification for neglecting such subsections is involved with the spiral-like nature of the cumulative result for the field taken up at the beginning of Sect.3.1.6. When the subsection does not meet this criteria the phase term is approximated by a quadratic expression in $\xi$. The amplitude of the kernel is assumed to be slowly varying over the subinterval, evaluated at an interior point and removed outside of the integral. The resulting integral can be expressed in terms of Fresnel integrals by a suitable transformation. Thus LEIBIGER evaluated the multipath integrals by quadrature whereas WEINBERG utilized a combination of quadrature and stationary phase. An indication of the relative accuracy of the two methods, for a single example, can be obtained upon examining the accuracy assessment results provided in Sect.3.1.7.

3.1.6 Connection Between Modes and Rays

PEDERSEN and GORDON [3.24] have compared normal-mode theory with ray theory and found many useful results regarding diffraction corrections to classical ray theory. BARTBERGER [3.25] has investigated the behavior of the mode (residue) summation at a fixed range in order to gain a better insight into normal-mode theory. LEIBIGER [3.23] has conducted a similar study as part of the development of Raymode X and utilized the results to considerably reduce the required execution time (Table 3.3, Sect.3.1.7) of that program. Before examining the underlying mathematics (the notation pertaining to Raymode X, Sect.3.1.5 will be used), it is instructive to consider the example of a cumulative mode sum [3.25] shown in Fig.3.7. A straight line is drawn from the origin in the complex plane to the point representing the first mode. From there another straight line is drawn to the point representing the sum of the first two modes and so on. The vector distance from the origin to the final point then represents the resultant acoustic pressure at the fixed horizontal range (12.6 kyd in this case) examined. At this range, which from ray theory is dominated by bottom bounce paths, a broad peak results in the plot of propagation loss versus range caused by multipath interference. The composite plot (Fig.3.8) is a useful tool for interpreting the behavior of the spiral like curve of Fig.3.7. A smoothed version of the propagation loss is shown in the upper right hand corner. Follow the dashed line, intersecting the peak at 12.6 kyds, to the lower right hand portion of the figure. The curve shown there is a plot of $\mathrm{Re}\{\xi_n - \xi_{n-1}\}r = 2\pi$, where the difference

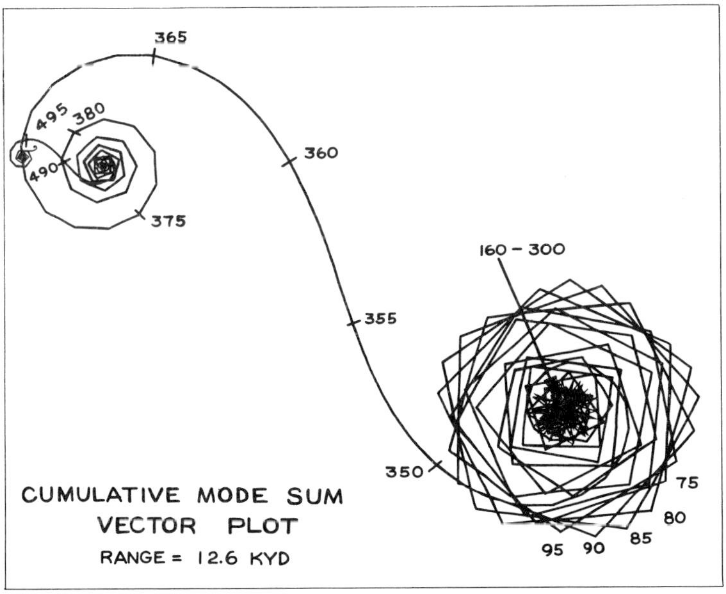


Fig. 3.7. Vector mode plot at 12.6 kyd

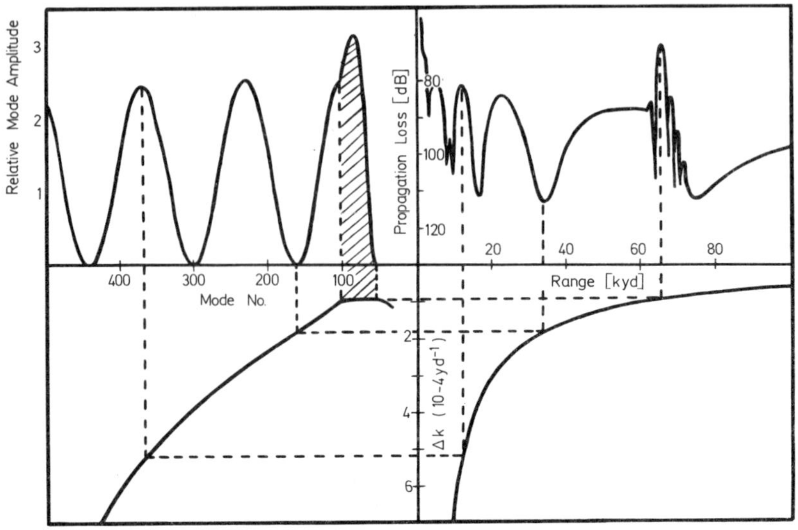


Fig. 3.8. Mode contribution to propagation loss

between the real parts of the eigenvalues is labeled Δk in Fig.3.8. Next follow the dashed line to the lower left hand portion where the locus of Δk's for successive modes is plotted versus mode number. Finally, the dashed line intersects the plot (upper left-hand corner of Fig.3.8) of relative mode amplitude (depth functions only) versus mode number at a peak value occurring between mode numbers 355 and 360.

Returning to Fig.3.7, it can be seen that the low order modes which have a large amplitude wrap around themselves to form large polygons and then return to the origin. These modes would be important at the convergence zone range, about 65.6 kyd, but contribute essentially nothing to the resultant intensity at the bottom bounce range considered here. As the mode number increases (160-300 which surrounds the second peak of the upper left-hand portion of Fig.3.8), the phase difference between successive modes gradually passes through 180 degrees, leading to the spiky appearance of the plot in Fig.3.7. This group of modes also contribute little to the final mode sum. As the mode number approaches the third peak (upper left portion of Fig.3.8) the phase difference passes through 360 degrees in the vicinity of mode 355 and the flat portion of the spiral is formed. Later, at about mode 480, the phase difference between successive modes approaches 720 degrees and the curve has another smaller flat portion. Finally, a third flat portion is barely visible at the end of the plot which corresponds to a phase difference of 1080 degrees.

Thus, the dominant contribution to the mode sum at this range is made by a relatively small group of modes (350-360) whose phase difference is approximately $2\pi$. These modes are associated with rays which reach the receiver after interacting with the bottom once. Their contribution to the propagation loss curve would be similar to the smooth result shown in Fig.3.8. Additional smaller contributions come from the second flat portion of the spiral associated with a group of higher order modes, whose successive phase difference is approximately $4\pi$, associated with rays reaching the receiver after interacting with the bottom twice. Their contribution to the propagation loss result would produce a higher frequency oscillation riding on top of the previously mentioned smooth curve. Less of a contribution is made by the group of modes (successive phase difference of approximately $6\pi$) associated with rays which have interacted with the bottom three times.

The explanation for the dominance of a small group of modes within the total mode sum can be obtained with the aid of the Poisson summation formula [3.26-32]

$$\sum_{p=N_A}^{N_B} I_p = \sum_{q=0}^{\infty} \int_{N_A}^{N_B} I_p \, e^{-i2\pi pq} \, dp \quad ,$$

where on the right-hand side $I_p$ is taken to be a continuous function of p. If this formula is applied to LEIBIGER's residue series (3.76) one of the four terms in the field expression is given by

$$\mathscr{P}(r,z,z_s) \sim \sum_{q=0}^{\infty} 2\pi i \int_{N_A}^{N_B} \frac{\mathscr{G}(z,z_s;\bar{\xi})}{\partial w/\partial\xi\big|_{\bar{\xi}}} \exp\left[i\left(\phi_1^z - \phi_1^z s + \bar{\xi}r - 2\pi pq\right)\right] dp \quad , \tag{3.77}$$

where $\bar{\xi} = \xi_p$.

Although the stationary phase technique will not be used to evaluate this expression it is informative to examine the points of stationary phase given by $\partial\left(\phi_1^z - \phi_1^z s + \bar{\xi}r\right)/\partial p = 2\pi q$.

This implies that for a fixed range only those successive modes whose phase change is an integer multiple of $2\pi$ will make a significant contribution to the total field. The phase difference of the remaining modes is such as to lead to cancellation and gives the non-flat regions of spirals similar to that discussed previously. The equation for locating points of stationary phase can also be written as

$$\frac{\partial}{\partial\bar{\xi}}\left(\phi_1^z - \phi_1^{z_s}\right) + r = 2\pi q \frac{\partial p}{\partial\bar{\xi}} \quad .$$

However, one has from the mode equation that

$$\frac{\partial\bar{\xi}}{\partial p} = 2\pi\left[\left(-2\frac{\partial}{\partial\bar{\xi}}\phi_1^N\right) + i\frac{\partial}{\partial\bar{\xi}}\ln\left(\tilde{\mathscr{R}}_1^1\mathscr{R}_{N-1}^N\right)\right]^{-1} \quad ,$$

and to a good approximation $\partial\bar{\xi}/\partial p \sim 2\pi/R_c$ where $R_c$ is the cycle range for the $p^{th}$ mode, i.e., the horizontal range needed for the angle associated with the eigenvalue to return to its initial value when plotted as a function of depth and range. Specifically

$$R_c = -2\frac{\partial}{\partial\hat{\xi}}\mathrm{Re}\left\{\phi_1^N\right\} \quad .$$

From classical ray theory one also has that the horizontal range between two points say $z_1$ and $z$ on a ray path is given by $r(z_1,z) = -\partial\mathrm{Re}\left\{\phi_1^z\right\}/\partial\hat{\xi}$.

Thus the stationary phase equation may also be written as

$$r \simeq r(z_1,z) - r(z_1,z_s) + qR_c(z_1,z_N) \quad ,$$

which is also the equation satisfied by an eigenray from classical ray theory.

Then although the number of modes in the residue summation is identical for all ranges, at a given range only a finite subset satisfying the equation $\partial\bar{\xi}/\partial p \simeq 2\pi q[-r(z_1,z)+r(z_1,z_s)+r]^{-1}$ will constitute the major portion of the field. Furthermore, this subset of modes can be associated with the eigenrays from classical ray theory which at that range connect the source and receiver. Looked at in another way, the acoustic pressure at a given range as given by mode theory is produced almost entirely by those modes whose skip distance $R_c$ times an integer multiple, corresponds to that range, since $[-r(z_1,z)+r(z_1,z_s)]$ is generally small.

Although (3.77) can be approximately evaluated by stationary phase it is more informative in showing the connection between modes and the multipath expansion to transform that result to

$$\mathscr{P}(r,z,z_s) \sim \sum_{q=0}^{\infty} \int_{\bar{\xi}_A}^{\bar{\xi}_B} \mathscr{G}(z,z_s;\bar{\xi})\left[\tilde{\mathscr{R}}_1^1(\bar{\xi})\mathscr{R}_{N-1}^N(\bar{\xi})\right]^q \exp\left[i\left(\phi_1^z-\phi_1^{z_s}+2q\phi_1^N+\bar{\xi}r\right)\right]d\bar{\xi} \quad ,$$

where use has been made of $\partial\left[-\tilde{\mathscr{R}}_1^1\mathscr{R}_{N-1}^N \exp\left(2i\phi_1^N\right)\right]/\partial\bar{\xi} = 2\pi i \; \partial p/\partial\bar{\xi}$ and $2q\phi_1^N - iq\ln\left(\tilde{\mathscr{R}}_1^1\mathscr{R}_{N-1}^N\right) = -2pq\pi$. The above integral is one of the four terms previously derived for the multipath expansion.

If this integral is evaluated by stationary phase the following is obtained

$$\mathscr{P}(r,z,z_s) \sim \sum_{q=0}^{\infty} [\pi/\psi''(\xi^*)]^{\frac{1}{2}} \mathscr{G}(\xi^*)\left[\tilde{\mathscr{R}}_1^1(\xi^*)\mathscr{R}_{N-1}^N(\xi^*)\right]^q e^{i\psi(\xi^*)}$$

$$\cdot \int_{V_a}^{V_b} \exp(iV^2\pi/2)dV \quad ; \quad V(\bar{\xi}) = [\psi''(\xi^*)/\pi]^{\frac{1}{2}}(\bar{\xi}-\xi^*) \quad ,$$

where the phase $\psi(\xi) = \phi_1^z - \phi_1^{z_s} + 2q\phi_1^N + \xi r$ has been expanded in a Taylor's series about the stationary phase point $\xi^*$ and only terms up to the second order retained. This result represents a diffraction correction to classical ray theory, since in the limit of increasing frequency it yields classical ray theory except for a factor $c(z_s)/c(z)$. For the details one should consult the illuminating paper by HASKELL [3.33]. HASKELL has also obtained corrections at caustics $[\psi''(\xi^*)=0]$ by retaining higher order terms in the Taylor expansion for $\psi(\xi)$. Similar corrections, sometimes called non-uniform, have also been derived by BREKHOVSKIKH [3.34]. Thus, it is possible to show the connection between the multipath expansion and classical ray theory as well as the non-uniform caustic corrections used to modify classical ray theory models such as the *F*ast *A*symptotic *C*oherent *T*ransmission (FACT) model [3.22].

### 3.1.7 Quantitative Model Assessment

The discussion to this point was tacitly concerned with the accuracy of those models of propagation loss in underwater acoustics which have been converted into an automated computer code capable of being executed by someone other than the originator for a wide variety of problems. An indication of accuracy can be obtained, in general terms, by understanding the approximations involved in going from the exact integral solution to the equations calculated in the computer code. However, a user must be able to translate this general understanding into a decision of suitability for a particular application. One recourse is to ask the opinion of experts, however, their subjective opinion is often found to be in conflict with that of other experts resulting in a fair amount of confusion.

Additionally, although accuracy is a concern it is not the only factor involved in the selection of a model as can be seen from examining Fig.3.9. The amount of required execution time is of obvious concern because of the cost involved. Similarly, if the model is accurate enough but requires more core storage than is available it is of limited viability. If the model is not up and running at your facility, time delays will be encountered with its implementation. Some models can only be properly run by the originator because of subjective choices for input parameters controlling the accuracy of approximations. If that person is not available, the user may decide to choose another, less accurate mode. Very often it may be possible to decrease the execution time of a model for the application at hand without seriously effecting

A. ASSESSMENT OF ACCURACY

B. RUNNING TIME

C. AMOUNT OF COMPUTER MEMORY REQUIRED

D. EASE OF IMPLEMENTATION

E. COMPLEXITY OF PROGRAM EXECUTION

F. EASE OF EFFECTING SLIGHT ALTERATIONS TO THE PROGRAM

G. AVAILABLE ANCILLARY INFORMATION

Fig. 3.9. Factors influencing model selection

accuracy. This would be difficult without extensive documentation. Finally, the model may provide only propagation loss between omni-directional sources and receivers when a beam-former output may be required.

Given a multitude of candidates the analyst must arrive at a decision based upon trade-offs between accuracy and the above mentioned factors in the context of the application at hand. The problem of merging the user's requirements with the proper model has received considerable attention recently. The effort of the *P*anel *O*n *S*onar *S*ystem *M*odels (POSSM) [3.35,36] is fairly typical of the activity in this area and will be summarized. That group decided that quantitative information should be provided about candidate models for the factors contained in Fig.3.9. With such information available the analyst could then arrive at objective decisions regarding trade-offs within the context of his application.

The assessment of accuracy is the most difficult portion of the matrix to complete. One reason being the desire that the methodology used be more objective than the old subjective technique of overplotting predictions and arriving at a value judgement. The methodology adopted is summarized in Fig.3.10.

The process begins by selecting a standard against which a model is to be evaluated for accuracy. The standard may be either an experimental data set or the output of a model. Since accuracy is assessed in terms of the mean level (which, of course, varies with the independent variable, be it range, azimuth, time, etc.) fluctuations about this mean must be eliminated before meaningful comparison can be made. In the discussion and examples to follow the data considered will be transmission loss versus range; the applicability of the comparison method, however, is not restricted to such data. Typically, data which is broadband or derived using incoherent phase addition shows little or no fluctuations and is suitable for comparison without modification. Conversely, narrowband data and/or CW coherent or semicoherent model outputs typically show significant rapid fluctuations about the mean. In such cases the range dependent mean is obtained by smoothing the data by applying a moving average. The width of the moving window is determined from considerations of sonar system integration times and own-ships and targets speed of advance. Given the range dependent means of standard and model, a difference of the two curves is obtained and divided into range intervals. These intervals correspond to ray path regimes such as direct path, first and second bottom bounce regions, first, second and third convergence zones. The next step is calculation of the mean and standard deviation of the differences within each interval. Finally, these means and standard deviations are appropriately weighted and averaged, resulting in *C*umulative *A*ccuracy *M*easures (CAM). In the example to follow, only a single scenario is examined, a fairly simple one environmentally, and hence the accuracy results should not be considered to be indicative of a model's expected performance in other environments.

The scenario used by the various models is described in Fig.3.11. The water depth is 3000 m, the source at 27 m, receiver at 117 m, the sound speed profile approximately bilinear. The bottom reflection loss, extracted in part from measured data,

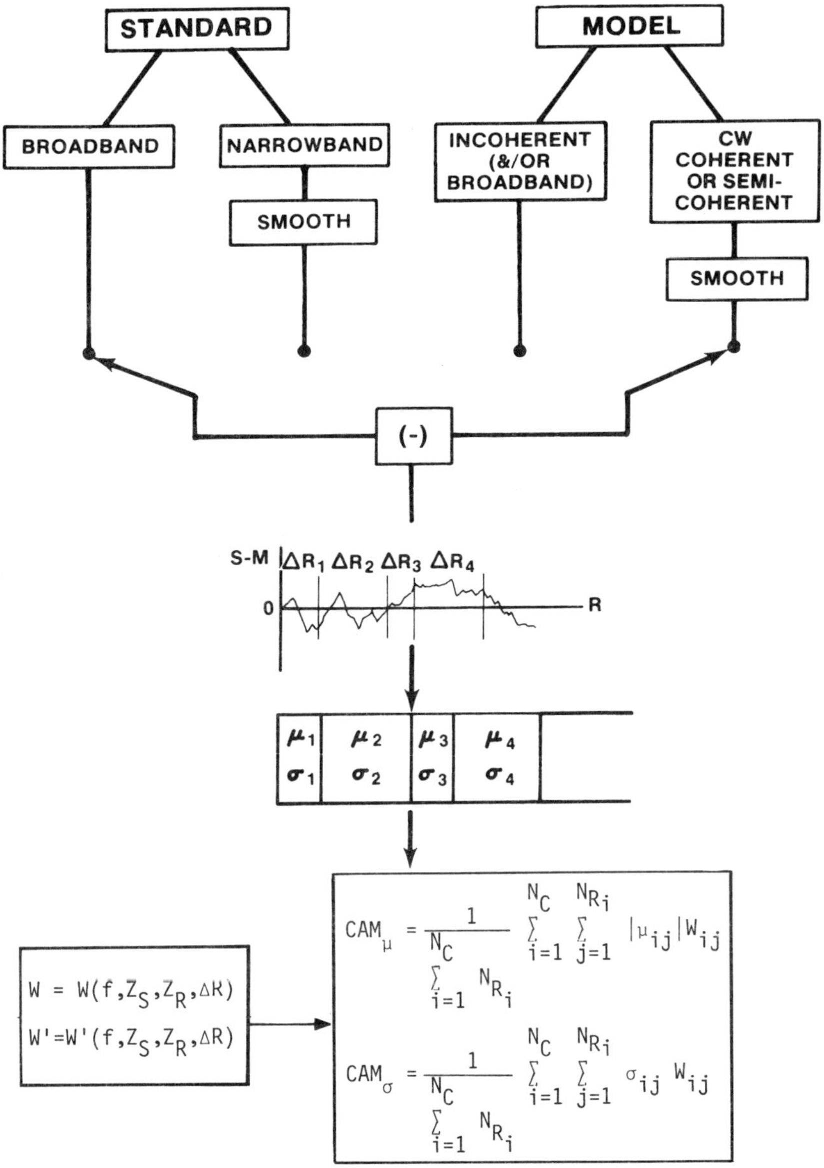


Fig. 3.10. Summary of model assessment methodology

**NAVYWIDE PROPAGATION LOSS MODEL COMPARISONS**

- MEDITERRANEAN SCENARIO
- FREQUENCIES
  35, 67.5, 100, 200 Hz
- STANDARDS
  HAYS-MURPHY DATA
  FAST FIELD PROGRAM
- MODELS
  SINGLE PROFILE
  FLAT BOTTOM
  INPUTS PROVIDED

Fig. 3.11. Scenario for accuracy assessment

**CONTRIBUTORS**

| LABORATORY | MODEL | PHASE ADDITION |
|---|---|---|
| FNWC | FACT | S |
| NADC | AP2<br>PLRAY | C<br>S,I |
| NOSC | GORDON N. M.<br>LORA<br>RAYWAVE II | C<br>C,S,I<br>I |
| NRL | RTRACE | C,I |
| NSWC | NSWC N. M. | C |
| NUSC | CONGRATS V<br>FACT<br>FFP<br>FFP (1/3-OCT)<br>RAYMODE X | C,I<br>C,S,I<br>C<br>C<br>C,I |

Fig. 3.12. Listing of models assessed. C = Coherent; I = Incoherent; S = Semi-Coherent

had a critical angle of 22°; at normal incidence the bottom loss is 6 dB. Four cases (I-IV) were examined, corresponding to the four frequencies shown. The standard chosen was the results of the *F*ast *F*ield *P*rogram (FFP). All models compared against the standard used a single sound speed profile and a flat bottom and were provided identical input information.

The models used in this example were supplied by various Navy laboratories[3] as shown in Fig.3.12. Note that two versions of the FACT model were used. The version resident at *F*leet *N*umerical *W*eather *C*entral (FNWC) is, of course, an operational model and cannot therefore be altered to generate results with the requested data density. Perhaps more importantly, as will be seen shortly, the running time of the FNWC version is an order of magnitude greater than what might be termed a "bare bones" version of FACT generally available at Navy laboratories. The bare bones version is free of the input/output requirements necessitated by a variety of fleet support needs. In a similar vein, it should be pointed out that the version of Raymode X used in this study is not the fleet operational version in that certain parameters which would be assigned frozen values for such usage were varied in this example. These parameters basically determine the trade-off between accuracy and running time. If the choice of phase addition is considered a total of 21 models were put through the accuracy assessment procedure.

In Fig.3.13 the results of the standard (FFP) before and after smoothing for 67.5 Hz are shown. The window used for the running average was 2 kilometers which is nearly equivalent to a 5 minute average on a 12 knot target. A similar plot is shown in Fig.3.14 for the Raymode X model.

---

[3]Information regarding documentation of these models is given in [3.9,14,21,22,37-43].

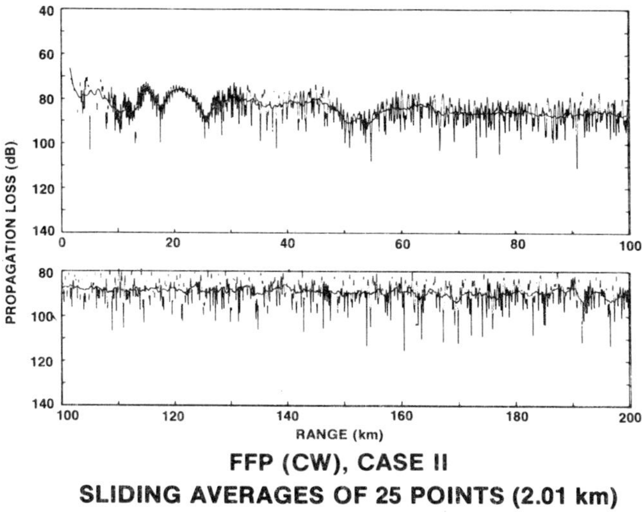


Fig. 3.13. Standard before and after smoothing

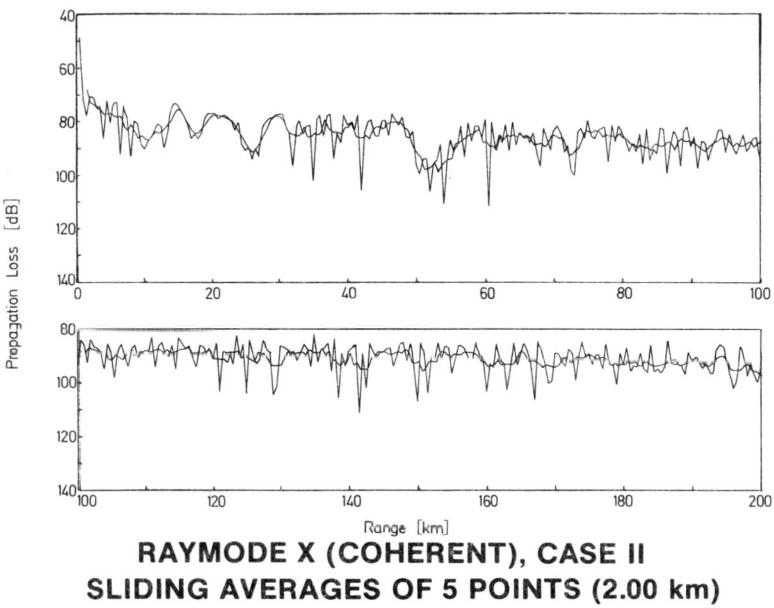


Fig. 3.14. Raymode X before and after smoothing

The difference between the smoothed versions of FFP and Raymode X is shown in Fig.3.15. Sets of curves such as these were generated for each model and type of phase addition. Although no convergence zone was obvious in the FFP and Raymode X results, the range intervals chosen for dividing the difference curves were direct path region, first bottom bounce region, CZ, second bottom bounce region, and 50 kilometer intervals thereafter. It is the large critical angle of the actual bottom loss that accounts for the "filling in" of the inter-CZ regions.

Table 3.1. Means and standard deviations of differences between the FFP (cw) and model results: Case II (in dB)

| Range interval | DP | | BB1 | | CZ | | BB2 | | to 100 km | | to 150 km | | to 200 km | |
|---|---|---|---|---|---|---|---|---|---|---|---|---|---|---|
| Model | μ | σ | μ | σ | μ | σ | μ | σ | μ | σ | μ | σ | μ | σ |
| AP2 normal mode | - | - | 0.0 | 2.0 | -0.5 | 1.9 | -1.0 | 1.8 | -1.6 | 2.7 | -1.3 | 2.7 | -1.9 | 2.6 |
| Congrats V coherent | - | - | -1.2 (-1.2) | 2.1 (1.5) | -2.1 (-0.7) | 2.7 (2.2) | -0.8 (-1.9) | 3.1 (2.7) | -2.2 (-0.5) | 3.9 (1.8) | -2.3 (-1.5) | 3.2 (2.5) | -3.3 (-1.4) | 3.0 (2.5) |
| Congrats V incoherent | - | - | 0.8 (1.0) | 3.7 (3.8) | -0.6 (-0.2) | 2.0 (2.1) | -0.5 (-0.4) | 2.8 (2.9) | -1.1 (-0.6) | 1.4 (1.0) | -3.0 (-0.6) | 1.6 (0.9) | -4.6 (-0.8) | 2.0 (1.5) |
| FACT/FNWC incoherent | - | - | 1.6 | 3.6 | 1.9 | 1.6 | 0.2 | 2.8 | -0.3 | 1.2 | 0.7 | 1.1 | 1.0 | 1.4 |
| FACT/NUSC coherent | - | - | -1.1 | 2.6 | 2.2 | 1.4 | -0.2 | 1.7 | -0.8 | 1.6 | -0.5 | 1.2 | -0.4 | 1.4 |
| FACT/NUSC incoherent | - | - | 1.4 | 3.6 | 0.9 | 2.2 | 0.1 | 3.0 | -2.2 | 2.0 | -2.0 | 1.3 | -1.9 | 1.5 |
| FFP cw | - | - | - | - | - | - | - | - | - | - | - | - | - | - |
| FFP 1/3 octave | - | - | | | | | | | | | | | | |
| Gordon normal mode | - | - | 0.6 | 2.0 | 0.4 | 2.2 | -0.8 | 1.9 | 0.7 | 1.9 | -0.3 | 2.6 | -1.0 | 2.5 |
| LORA coherent | - | - | - (0.1) | - (3.2) | - (0.1) | - (1.7) | - (-04) | - (3.1) | - (-1.6) | - (4.3) | - (-2.5) | - (3.6) | - (-4.0) | - (4.4) |

Table 3.1 (continued)

| Range interval | DP | | B31 | | CZ | | BB2 | | to 100 km | | to 150 km | | to 200 km | |
|---|---|---|---|---|---|---|---|---|---|---|---|---|---|---|
| Model | μ | σ | μ | σ | μ | σ | μ | σ | μ | σ | μ | σ | μ | σ |
| LORA semi-coherent | - | - | -<br>(0.2) | -<br>(2.9) | -<br>(0.8) | -<br>(2.0) | -<br>(0.2) | -<br>(2.4) | -<br>(0.0) | -<br>(1.0) | -<br>(-0.8) | -<br>(1.6) | -<br>(-1.8) | -<br>(1.9) |
| LORA incoherent | - | - | -<br>(0.9) | -<br>(3.7) | -<br>(0.9) | -<br>(2.0) | -<br>(0.4) | -<br>(2.6) | -<br>(0.0) | -<br>(1.2) | -<br>(-0.8) | -<br>(1.7) | -<br>(-1.8) | -<br>(2.0) |
| NSWC normal mode | - | - | 1.9 | 3.7 | 0.9 | 1.6 | -0.2 | 1.4 | -0.2 | 1.9 | -0.1 | 2.0 | -0.4 | 2.3 |
| PL RAY semi-coherent | - | - | 0.0<br>(0.5) | 4.8<br>(4.1) | 2.2<br>(2.1) | 2.9<br>(2.9) | 0.4<br>(0.4) | 5.0<br>(5.1) | 1.6<br>(1.6) | 1.2<br>(1.1) | 1.2<br>(1.4) | 0.9<br>(1.0) | 0.3<br>(0.2) | 1.6<br>(1.5) |
| PL RAY incoherent | - | - | 1.4<br>(1.3) | 3.7<br>(3.7) | 1.2<br>(1.1) | 2.3<br>(2.3) | 0.4<br>(0.4) | 2.6<br>(2.7) | 0.5<br>(0.6) | 1.1<br>(1.1) | 0.1<br>(0.3) | 0.9<br>(1.1) | -0.9<br>(-1.1) | 1.5<br>(1.5) |
| RAYMODE X coherent | - | - | -0.9<br>(-0.7) | 2.1<br>(2.1) | -2.0<br>(-1.5) | 2.5<br>(2.4) | -3.2<br>(-3.3) | 2.3<br>(2.6) | -2.0<br>(-1.6) | 2.0<br>(2.1) | -3.8<br>(-1.5) | 2.1<br>(2.2) | -6.7<br>(-2.4) | 2.9<br>(1.9) |
| RAYMODE X incoherent | - | - | 1.1<br>(1.3) | 3.7<br>(3.7) | 0.3<br>(0.5) | 2.1<br>(2.0) | -0.8<br>(-0.4) | 2.7<br>(2.7) | -0.8<br>(-0.3) | 1.2<br>(1.1) | -2.0<br>(-0.4) | 1.3<br>(1.0) | -3.5<br>(-0.8) | 1.8<br>(1.6) |
| RAYWAVE II incoherent | - | - | 2.9 | 3.0 | 1.9 | 2.9 | 2.2 | 3.5 | 1.4 | 2.1 | 0.9 | 1.7 | 0.7 | 1.8 |
| RTRACK coherent | - | - | -0.6 | 2.7 | -0.1 | 2.3 | -2.1 | 3.7 | -0.1 | 2.5 | -0.5 | 1.9 | -0.7 | 2.4 |
| RTRACE incoherent | - | - | - | - | - | - | - | - | - | - | - | - | - | - |

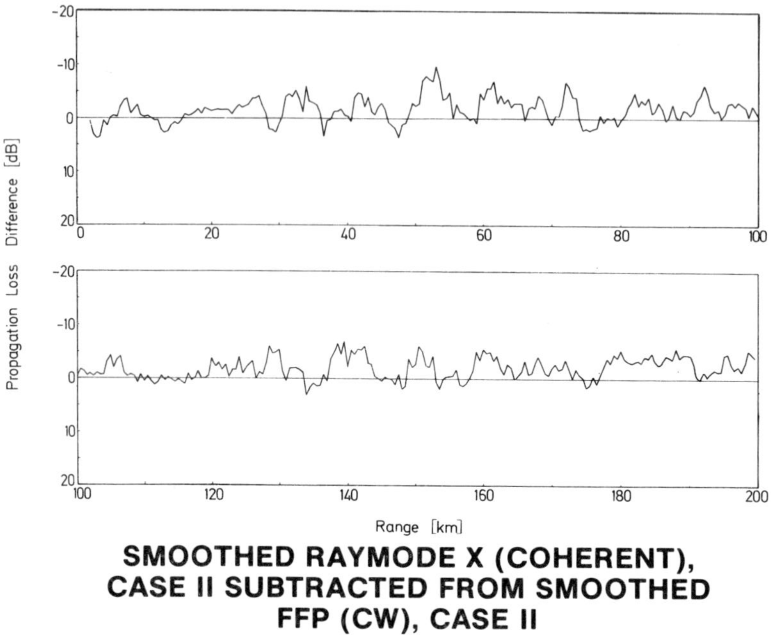


Fig. 3.15. Difference curves

Table 3.1 shows the means and standard deviations of the difference curve for each model in each range interval. Numbers in parentheses represent the results of a second model submission. These resulted from the follow-up contact with each modeler giving him the results of the assessment of his model and also how it compared with other models. Also provided were running time, core storage, a model description, references to his model, etc. The modeler was then given the option of rerunning his model possibly trading-off accuracy for running time. Four modelers availed themselves of the resubmission option.

Recall that the CAM for the means and standard deviations is obtained by taking weighted averages of these quantities over all cases and range intervals. Since this example is for illustrative purposes unit weights were assigned. The cumulative accuracy measures reduce to the grand mean and standard deviations shown in Table 3.2. Each model's accuracy is now condensed to two numbers, both in decibels, giving the mean level of accuracy and the spread in accuracy over all cases and ranges. Note that all models were able to achieve values for grand means between 0.9 to 2.2 dB (upon consideration of only second submissions) and 2.1 to 5 dB for the standard deviation. This happy state of affairs would not be expected to persist for more challenging environmental scenarios.

Two facts regarding this accuracy assessment are worthy of note. First, the methodology results in a hierarchy ranging from the basic data to a single number.

Table 3.2. CAM with FFP as standard. [Averages of means and standard deviations over all cases and range intervals (standard of comparison: FFP model results)]

| | $\overline{\lvert\mu\rvert}$ | $\bar{\sigma}$ |
|---|---|---|
| AP2 normal mode | 1.1 | 2.4 |
| CONGRATS V coherent | 2.4 (1.3) | 3.5 (2.2) |
| CONGRATS V incoherent | 1.8 (0.9) | 2.5 (2.1) |
| FACT/FNWC semi-coherent | 1.2 | 2.6 |
| FACT/NUSC coherent | 1.0 | 2.5 |
| FACT/NUSC incoherent | 1.2 | 2.4 |
| FFP cw | - | - |
| FFP 1/3-octave | | |
| Gordon normal mode [a] | 0.6 | 2.3 |
| LORA coherent[b] | 2.0 (1.6) | 5.0 (4.5) |
| LORA semi-coherent [b] | 1.4 (1.2) | 3.2 (2.9) |
| LORA incoherent [b] | 1.5 (1.0) | 3.0 (2.8) |
| NSWC normal mode | 1.0 | 2.8[c] |
| PLRAY semi-coherent | 1.0[c] (1.3) | 2.6[c] (2.5) |
| PLRAY incoherent | 1.3[c] (1.2) | 2.2 (2.3) |
| RAYMODE X coherent | 3.1 (2.2) | 2.7 (2.5) |
| RAYMODE X incoherent | 1.5 (0.9) | 2.4 (2.3) |
| RAYWAVE II incoherent | 2.0 | 2.5 |
| RTRACE coherent | 0.9 | 3.5 |
| RTRACE incoherent [d] | 1.9 | 2.5 |

[a] Cases I-III; [b] Cases III-IV;
[c] Cases II-IV; [d] Case III.

Thus a model's accuracy can be examined at many levels of detail. Further, this allows the results to be used in a diagnostic manner. The second point is the statement of a limitation: Accuracy results should not be extended to other environmental scenarios, particularly those which are more complex.

Table 3.3. Model running times. Average run time per case

| Model | Computer | Average run time per case [s] | No. of points per prediction [to 200 km] | Remarks |
|---|---|---|---|---|
| AP2 | CDC 6600 | 60.6 | 400 | Run time is frequency dependent. |
| CONGRATS V [a] | UNIVAC 1108 | 42.3 (70.0) | 200 | |
| FACT/FNWC | CDC 6500 | 25.0 | 216 | Includes calculations besides those needed for propagation loss. |
| FACT/NUSC | UNIVAC 1108 | 2.5 | 200 | |
| FFP | UNIVAC 1108 | 373.0 | 2371 | Run time is frequency dependent. |
| Gordon normal mode | UNIVAC 1108 | 84.8 [b] | 400 | Run time proportional to frequency f = 200 Hz not run. |
| LORA [a] | UNIVAC 1108 | 28.7 (21.2) | 219 | First submission: f = 37.5, 67 Hz not run. Run time is frequency dependent. |
| NSWC normal mode | CDC 6500 | 702.0 | 400 | Run time is strongly frequency dependent. |
| PLRAY | CDC 6600 | ? (24.0) | 400 (200) | First submission: f = 37.5 Hz not run. |
| RAYMODE [a] | UNIVAC 1108 | 19.2 (41.7) | 400 | Run time is frequency dependent. |
| RAYWAVE II | UNIVAC 1108 | 50.7 | 200 | |
| RTRACE | CDC 3800 | 45.0 | 400 | |

[a] For these models, it is assumed that the various types of phase addition used were part of the same computer run and that the run time as compared to a run with only coherent phase addition is negligibly different.

[b] Includes estimate for case IV of 162 s although this case was not run.

In most cases, a second submission improved accuracy by 1 dB or less. In the case of WEINBERG's multipath expansion model the larger improvement is due to two factors: 1) Alteration of logic involving bottom bounce paths, and 2) an error discovered in the sound speed profile input.

The cost of the modest accuracy improvements in terms of running time is provided in Table 3.3. Note that the execution times are not directly comparable due to the use of different computers. Examination of the first and second submissions reveal an approximate doubling of running time for Raymode X and WEINBERG's multipath model. In the second submission of HOFFMAN's [3.65] *L*ong-*R*ange *A*ctive *S*onar *S*ystems (LORA) model, cases I and II were added but the basic program parameters were left unchanged.

Table 3.4. Mode core storage. (Words of computer storage required)

| Model | Computer storage (decimal words) |
|---|---|
| Raymode X | 14,115 |
| NSWC Normal mode | 18,940 |
| AP2 | 20,000 |
| PLRAY | 20,480 |
| FACT(NUSC) | 21,855 |
| LORA | 35,000 |
| RAYWAVE II | 37,000 |
| FFP | 51,572 |
| CONGRATS V | 51,760 |
| Gordon normal mode | 53,000 |
| RTRACE | 58,100 |
| FACT(FNWC) | 60,000 |

Table 3.4 presents the core requirements for the models. Only one model, Raymode X, has less than 16K words. Four models require less than 32K and the remaining six models less than 64K words.

### 3.1.8 Waveform Prediction Models

The major emphasis in modeling underwater acoustics has been with the prediction of propagation loss for an infinite *c*ontinuous *w*ave (cw) point source. When information about the arrival structure or frequency dispersive characteristics of the waveform was desired it was either obtained in a gross fashion from classical ray-theory programs [3.44] or for low frequencies by the use of stationary phase techniques in conjunction with normal mode theory [3.45] following the approach used by PEKERIS [3.15].

Recently the alternative approach of direct numerical evaluation of the double-integral expression for the pressure field via the FFT has been successfully implemented by DINAPOLI [3.46]. In this approach the real part $\hat{P}_R$ of the pressure field is given by the Fourier synthesis

$$\hat{P}_R(r,z,t) = \int_{-\infty}^{\infty} \mathscr{P}(r,z,f)\Omega(f)\, \exp(-i2\pi ft)df \quad ,$$

where $\Omega(f)$ is the Fourier transform of the input source waveform and $\mathscr{P}(r,z,f)$ is the transfer function of the medium which can be expressed in terms of the Fourier-

Bessel transform previously examined for cw propagation. The impulse response of the medium is then given by

$$h(r,z,t) = \int_{-\infty}^{\infty} \mathscr{P}(r,z,f)\ \exp(-i2\pi ft)df \quad .$$

Let the continuous variables t and f be evaluated at the discrete points $t_k$ and $f_p$ where with $\Delta t\Delta f = 1/M$, $t_k = t_0+k\Delta t$, $f_p = f_0+p\Delta f$; $(k,p) = 0,1,2,\dots,M-1$. The constant $t_0$ is usually set equal to the approximate arrival time of the beginning of the waveform and $f_0$ would correspond to the lowest frequency present after filtering. Since $\Delta f$ is predetermined from sampling theory, information about the waveform will be obtained from $t_0$ to approximately $t_0 + (\Delta f)^{-1}$. Generally the time duration of the received waveform will increase with increasing range. Thus $\Delta f$ is chosen small enough so that the time duration of the received waveform is adequate at the largest range of interest. As might be anticipated the required computer storage and execution time increases as the frequency resolution $\Delta f$ decreases, resulting in a significant data management problem.

The pressure field waveform is then obtained upon evaluation of

$$\hat{P}_R(r,z,t_k) \simeq \Delta f\ \exp(-i2\pi f_0 t_k) \sum_{m=0}^{2M-1} E_p(r,z,f_p)\ \exp(-i2\pi pk/2M) \quad ,$$

where in order to obtain a real answer the input is arranged according to

$$E_p(r,z,f_p) = \Omega(f_p)\mathscr{P}(r,z,f_p)\ \exp(-i2\pi\rho\Delta ft_0)$$

$$E_{M+p}(r,z,f_p) = E^*_{M-p} \quad , \quad p = 1,2,3,\dots,M-1 \quad ,$$

where the * indicates the complex conjugate, and additionally

$$E_0 = \mathrm{Re}\{\Omega(f_0)\mathscr{P}(r,z,f_0)\} \qquad E_M = \mathrm{Re}\{\Omega(f_M)\mathscr{P}(r,z,f_M)\ \exp(-i2\pi M\Delta ft_0)\} \quad .$$

If the FFP (3.26) is used to obtain the transfer function the evaluation of the double FFT

$$\hat{P}_R(r_n,z,t_k) \simeq Ar_n^{-\frac{1}{2}} \sum_{p=0}^{2M-1} \left\{\Omega(f_p)\ \exp(-i2\pi\rho\Delta ft_0)\left[\sum_{m=0}^{L-1} E_{m,p}\exp(i2\pi mn/L)\right]\right\}\exp(-i2\pi\rho k/2M),$$

where

$$A = \Delta f\Delta\xi\left(\frac{2}{\pi i}\right)^{\frac{1}{2}} \exp[i(\xi_0 r_n - 2\pi f_0 t_k)] \quad , \quad E_{m,p} = G(z,z_s;\xi_m,f_p)\xi_m^{\frac{1}{2}}\ \exp(imr_0\Delta\xi) \quad ,$$

provides the received pressure waveform as a function of range and time.

As an illustration of the above procedure consider the typical Arctic profile shown in Fig.3.16. The rough under ice cover effectively filters out high frequencies

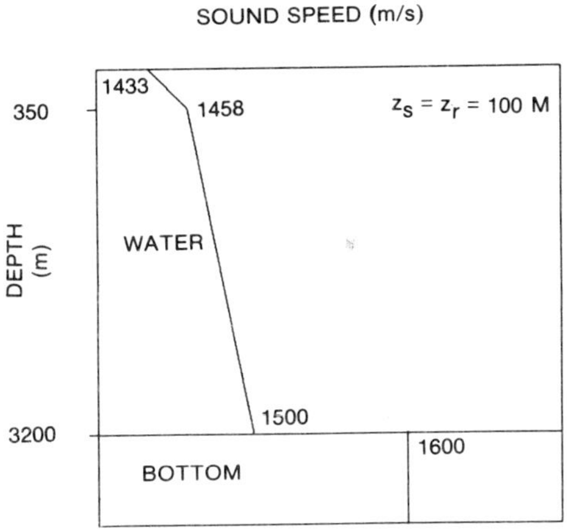


Fig. 3.16: Typical Arctic profile

at significant ranges from the source. Then limit the frequency range to $0 \leq f \leq 250$ Hz. With $\Delta f = 0.12207\ s^{-1}$, $\Delta t = 0.002$ s, $M = 2048$ the time duration of the predicted waveform will be 8.192 s. This is adequate for ranges less than a few hundred kilometers. The value of $t_0$ is set equal to $t_0 = r_n/(1475)$ which is roughly the average arrival time of the deep RSR paths which come in first. The bottom was assumed to be a semi-infinite half-space fluid layer with a constant sound speed of 1600 $ms^{-1}$ and its density and that of the water column was set to unity. The treatment of the bottom is of course unrealistic and justifiable only in the sense that the example is for illustrative purposes only.

The FFP must then be run at each of the above discrete frequencies in order to obtain the input $E_p$. The parameters used for this calculation were

$$\Delta\xi = 9.964727793 \times 10^{-6} \qquad \xi_0 = 0.02372488 \qquad L = 8192$$

$$\Delta r = 76.9705\ \text{m} \qquad r_0 = \Delta r$$

$$z_s = 100\ \text{m} \qquad z = 100\ \text{m}$$

which provides sampling of all significant wave numbers.

The impulse response at $r_n \simeq 100$ km is shown in Fig.3.17. The three groups of spikes near the end of the figure from right to left correspond to bottom bounce energy associated with increasing angles of incidence. Three spikes are evident within each group which correspond to the four rays which interact with the bottom. The middle spike representing two rays whose travel time difference is so small that they cannot be resolved with the frequency resolution used. Excellent agreement with ray-theory travel times is found to exist between the groups and also within each group.

The remainder of the figure, with the exception of the beginning, represents RSR rays which are trapped within the water column. The pattern of low amplitude arrival

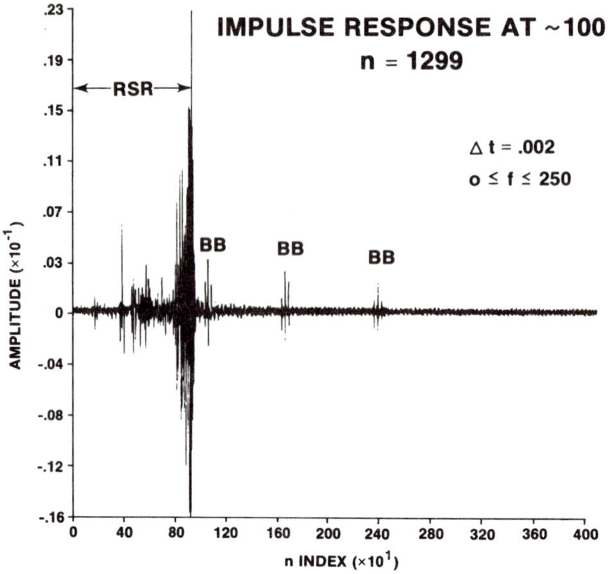


Fig. 3.17. Impulse response

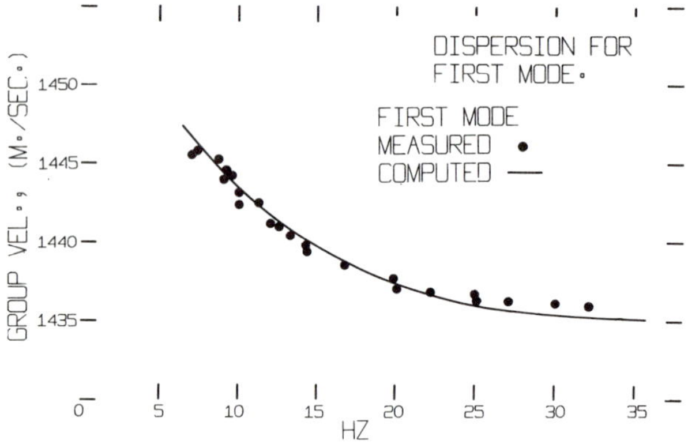


Fig. 3.18. Dispersion of first mode

followed by arrivals of successively higher amplitude and then abrupt termination is typical of that commonly observed in the deep SOFAR channel which in the case of the Arctic has its axis at the ocean surface. Finally the spikes at the beginning correspond to a combination of the deepest penetrating RSR arrivals and bottom bounce rays incident on the bottom between critical and grazing.

KUTSCHALE [3.47,48] has also examined this same problem with the FFP technique. In one case study he compared the predicted frequency dispersion in that portion of the waveform which would correspond to the first mode with the analogous experimental results. That comparison is reproduced in Fig.3.18 where it can be readily seen that excellent agreement was found. This agreement would not be expected to persist at higher frequencies where nondeterministic effects would be more significant.

## 3.2 Range-Dependent Models

Computer models for the range-dependent problem exist but are still in the developmental stage of evolution. In some instances the validity of the solution is understood but its numerical implementation for underwater acoustic scenarios poses severe practical problems with existing computer technology. Alternate solutions are available which by their nature obviate to some extent the implementation problems. It is difficult, however, to precisely translate the required mathematical approximations into conditions of applicability of the solution. A limited amount of comparisons against experimental data has occurred. However, this is often not a totally satisfying process due to the incompleteness of the associated environmental data. For these reasons a definitive appraisal of these models is best left for the future. It was felt that a brief survey of active areas of research would be more appropriate and is provided below.

### 3.2.1 Split Step Algorithm for Parabolic Equation

The conditions under which the solution to the range-dependent Helmholtz equation is well approximated by the parabolic approximation (2.111) is discussed in [3.49] and in Sect.2.2.2. Here attention is focused on the inherent approximations made in solving the parabolic equation (2.114) by the split-step algorithm introduced by HARDIN and TAPPERT [3.50] in 1973.

If in (2.114), with k equal to $k_0$, the change of variables $U(r,z) = 2ik_0 p(r,z)$ is made the parabolic equation may be written as

$$\frac{\partial U(r,z)}{\partial r} = i[A(r,z)+B(z)]U(r,z) \quad , \tag{3.78}$$

where the operators A and B are given by

$$A(r,z) = (k_0/2)[n^2(r,z)-1] \quad ; \quad B(z) = (1/2k_0)(\partial^2/\partial z^2) \quad . \tag{3.79}$$

To obtain the split-step solution at $r = r_0 + \Delta r$ assume that

$$U(r,z) = \exp\left\{i\int_{r_0}^{r_0+\Delta r} A(r,z)dr + i\int_{r_0}^{r_0+\Delta r} B(z)dr\right\}U(r_0,z) \quad , \tag{3.80}$$

which is only partially true, since the quantity [A(r,z)+B(z)] does not commute with its integral. Next assume that the index of refraction is a slowly varying function of range so that

$$U(r_0+\Delta r,z) \simeq \exp(i\Delta rA+i\Delta rB)U(r_0,z) \quad . \tag{3.81}$$

The exponential operator can be approximated (split) in various ways. Originally TAPPERT assumed

$$U(r_0+\Delta r,z) = [\exp(i\Delta rA)\ \exp(i\Delta rB)]U(r_0,z) \tag{3.82}$$

and later [3.51]

$$U(r_0+\Delta r,z) = [\exp(i\Delta rB/2)\ \exp(i\Delta rA)\ \exp(i\Delta rB/2)]U(r_0,z) \quad . \tag{3.83}$$

If the splitting given by (3.82) is assumed it can be shown that the error incurred by assuming commutativity of the operators A and B is

$$\left\{\exp(i\Delta rA)\ \exp(i\Delta rB) - \exp[i\Delta r(A+B)]\right\}U(r,z) = +\frac{(\Delta r)^2}{2}\left[n\frac{\partial U}{\partial z}\frac{\partial n}{\partial z} + \frac{nU}{2}\frac{\partial^2 n}{\partial z^2} + \frac{U}{2}\left(\frac{\partial n}{\partial z}\right)^2\right] + \dots$$

In order to obtain the split-step algorithm, let $V = [\exp(i\Delta rB)]U(r_0,z)$ and since $V = \mathscr{I}^{-1}[\mathscr{I}(V)]$ one has that

$$[\exp(i\Delta rB)]U(r_0,z) = \mathscr{I}^{-1}\left\{\exp\left(-i\Delta rs^2/2k_0\right)\hat{U}(r_0,s)\right\} \quad ,$$

where

$$\mathscr{I}\{U(r_0,z)\} = \hat{U}(r_0,s) = \int_{-\infty}^{\infty} U(r_0,z)\ e^{-izs}\ dz \quad . \tag{3.84}$$

Then from (3.82) the split-step solution is $U(r_0+\Delta r,z) = \exp\left[ik_0\Delta r(n^2-1)/2\right]\cdot \mathscr{I}^{-1}\left\{\exp\left(-i\Delta rs^2/2k_0\right)\mathscr{I}[U(r_0,z)]\right\}$, where the Fourier integral transform, indicated by $\mathscr{I}$, is numerically evaluated with the FFT at each range step. The solution can thus be marched out in range in a stable manner. In the evaluation of (3.84), attenuation is added to the medium below the depth of interest to insure that the field decays to zero for large depths.

A feeling for the composite effects of the three assumptions:

a) $[A+B, \int (A+B)dr] = 0$ [needed for (3.80) to be valid]

b) $(ik_0/2)\int_{r_0}^{r_0+\Delta r} [n^2(r,z)-1]dr \simeq (ik_0/2)[n^2(r,z)-1]\Delta r$ (3.85)

[needed for (3.81) to be valid]

c) $\exp[i\Delta r(A+B)] \simeq \exp(i\Delta rA)\ \exp(i\Delta rB)$ ,

can be obtained by determining $\partial U/\partial r$ from (3.82) and then comparing the result with (3.78). From (3.82) one obtains

$$\frac{\partial U(r,z)}{\partial r} = \left[iA(r,z)+i\Delta r\frac{\partial A}{\partial r}\right]U(r,z) + \left\{\exp[i\Delta rA(r,z)]iB(z)\ \exp[i\Delta rB(z)]U(r_0,z)\right\} \tag{3.86}$$

The second term may be simplified upon rewriting (3.82) to obtain

$$\exp(iA\Delta r)iB\ \exp(iB\Delta r)U(r_0,z) = [\exp(iA\Delta r)iB\ \exp(-iA\Delta r)]U(r,z)] \quad . \tag{3.87}$$

The term in brackets can be written by the Baker-Hausdorff expansion [3.52] with the commutator $[A,B] = AB - BA$ as

$$\exp(i\Delta rA)(iB)\ \exp(-i\Delta rA) = iB - \Delta r[A,B] - i(\Delta r)^2/2![A,[A,B]] + \dots$$

Then (3.86) may be written as

$$\begin{aligned}\frac{\partial U(r,z)}{\partial r} &= i[A(r,z)+B(z)]U(r,z) + i\Delta rk_0n(\partial n/\partial r)U(r,z)\\ &+ \left\{-\Delta r[A,B]-i[(\Delta r)^2/2!][A,[A,B]]+\dots\right\}U(r,z) \quad ,\end{aligned}$$

and the error (difference in differential equations) incurred by using the split-step solution

$$ik_0\Delta rn\frac{\partial n}{\partial r}U + \Delta r\left[n\frac{\partial n}{\partial z}\frac{\partial U}{\partial z}+\frac{nU}{2}\frac{\partial^2 n}{\partial z^2}+\frac{U}{2}\left(\frac{\partial n}{\partial z}\right)^2\right] - \frac{ik_0U}{2}\left(n\Delta r\frac{\partial n}{\partial z}\right)^2 + \dots \quad , \tag{3.88}$$

is seen to depend on the step size $\Delta r$, the frequency and the gradient of the index of refraction. Thus, whenever significant energy from the field interacts with the bottom where $(\partial n/\partial z)$ is large the $\Delta r$ needed to make the error acceptable may lead to prohibitive computer execution times.

The result (3.88) was also obtained by JENSEN and KROL [3.53] who assumed $n \simeq 1$ and ignored higher-order terms in the Baker-Hausdorff expansion. Their work also contains additional analysis regarding the applicability of the split-step algorithm and a comparison of execution times between the split-step solution and other models for several cases.

### 3.2.2 Parabolic Decomposition Method

PAPADAKIS and WOOD [3.54] sought an exact solution $\mathscr{P}(z,z_s;r)$ to the Helmholtz equation for range-dependent problems as an integral of the solutions of two parabolic equations. Specifically with cylindrical symmetry and a separable index of refraction given by $n^2(z,r) = n_2^2(z) + n_1^2(r)$ one has

$$\mathscr{P}(z,z_s;r) = \int_0^\infty G_1^*(t,r)G_2(t,z,z_s)dt \quad , \tag{3.89}$$

where the asterisk denotes the complex conjugate and $G_1$ and $G_2$ satisfy the parabolic equations

$$-2ik\,\frac{\partial G_1}{\partial t} + \frac{1}{r}\frac{\partial}{\partial r}\left(r\,\frac{\partial G_1}{\partial r}\right) + k^2\left[n_1^2(r)+1\right]G_1 = \frac{-2\delta(t)\delta(r)}{r} \tag{3.90}$$

$$2ik\,\frac{\partial G_2}{\partial t} + \frac{\partial^2 G_2}{\partial z^2} + k^2\left[n_2^2(z)-1\right]G_2 = \delta(t)\delta(z) \quad . \tag{3.91}$$

This approach reduces to the parabolic approximation usually obtained (Sect.2.2.2) which involves the solution of a single parabolic equation multiplied by a function depending upon range (2.111). To see this let $n_1(r)$ be zero. Then

$$G_1(r,t) = [H(t)/(4\pi t)]\,\exp\left[-i\,\frac{k}{2}\left(\frac{r^2}{t}+t\right)\right] \quad ,$$

where $H(t)$ is the unit step function. A stationary phase evaluation of (3.89), with the stationary phase point at $t=r$, yields

$$\mathscr{P}(z,z_s;r) \simeq \frac{\exp[i(kr+\pi/4)]}{2\sqrt{\pi kr}}\,G_2(r,z,z_s) \quad .$$

For arbitrary $n_1^2(r)$, $G_1$ is in general not known and would have to be approximated. The solution for $\mathscr{P}$ could then be obtained by stationary phase techniques. Alternatively, $G_1$ and $G_2$ could both be found numerically (see the discussion in Sect.3.2.3) and the solution for $\mathscr{P}$ obtained by quadrature. In this instance attention would not have to be limited to an index of refraction which is slowly varying and complex boundary value problems could be accommodated.

Finally, if the index of refraction is an arbitrary function of depth and range and non-separable the representation (3.89) remains valid but $G_1$ depends on r, z and t. This case is also discussed in Sect.2.2.2b. Research on evaluating (3.89) for this case is in progress.

### 3.2.3 Finite Differences

The application of finite difference schemes to range-dependent problems in underwater acoustics has received limited attention. SMITH [3.55] used finite difference to solve the coupled time-dependent fundamental equations. McDANIEL [3.56] has investigated a finite difference solution of the parabolic wave equation and compared it with the split-step algorithm solution. The appeal of finite difference schemes lies with the ability to solve complicated boundary value problems which would be intractable with other approaches. The shortcomings stem from the excessive demands made upon computer memory and execution time. It should be noted, however, that the exact nature of these demands within the context of underwater acoustic propagation has yet to be generally defined. Work along this line has recently been initiated by LEE and PAPADAKIS [3.57] and a brief summary is provided below.

The solution of the elliptic wave equation given in terms of the parabolic decomposition integral (3.89) is to be solved numerically. If the variation in both range and depth is rapid, the associated parabolic equations (3.90,91) would be solved by finite differences. On the other hand, if the range variation, for example, is slow an asymptotic solution for that parabolic equation may be used. In any event finite difference solutions of parabolic equations must be obtained. Then consider as an illustrative example the parabolic equation

$$U_r = a(r,z)U + b(r,z)U_{zz} = \left(a+b\frac{\partial^2}{\partial z^2}\right)U = LU \quad . \tag{3.92}$$

The conventional development of implicit schemes can be expressed by

$$e^{-kL/2}\, U_m^{n+1} = e^{kL/2}\, U_m^n \quad ,$$

and approximated by

$$(1-kL/2)U_m^{n+1} = (1+kL/2)U_m^n \quad , \tag{3.93}$$

where m is the depth index for a step size of h and n the range index associated with the range step size k.

Introducing the second-order central difference operator

$$D_z^2 = \frac{\delta_z^2}{h^2}\,[1-\rho(k,h)] \quad ,$$

where $\rho$ is a parameter chosen for optimum efficiency yields

$$-b_m^{n+1}\frac{(1-\rho)k}{2h^2}U_{m+1}^{n+1} + \left[1+b_m^{n+1}\frac{k}{h^2}(1-\rho) - \frac{ka_m^{n+1}}{2}\right]U_m^{n+1} - b_m^{n+1}\frac{(1-\rho)k}{2h^2}U_{m-1}^{n+1}$$

$$= b_m^n\frac{(1-\rho)k}{2h^2}U_{m+1}^n + \left[1-b_m^n\frac{k}{h^2}(1-\rho) + \frac{ka_m^n}{2}\right]U_m^n + b_m^n\frac{(1-\rho)k}{2h^2}U_{m-1}^n \quad .$$

The above may be written in matrix form, viz., $AU^{n+1} = BU^n + V_1^n + V_2^{n+1}$, where A and B are tridiagonal matrices and $V_J$ (J=1,2) contain information about the boundary conditions. Appropriate selections for $\rho$ will result in the familiar Crank-Nicolson and Douglas schemes.

The initial local truncation error is of order $\mathcal{O}(k^3+kh^2)$ for range-independent problems and of order $\mathcal{O}(k^2+kh^2)$ for range-dependent problems. The finite difference methods can be shown to be consistent and the stability condition is

$$\left| \frac{1-b(1-\rho)k/h^2[1-\cos(Jh)]-ka/2}{1+b(1-\rho)k/h^2[1-\cos(Jh)]+ka/2} \right| \leq 1 \quad , \quad J = 1,2 \quad .$$

Unconditional stability results if all $a,b,k,h > 0$. The convergence of the finite difference scheme can be shown by examining the norm inequalities $\| t.s.-n.s. \| \leq \| t.s.-f.s. \| + \| f.s.-n.s. \|$ . where

t.s. stands for the *theoretical solution* of (3.92),
n.s. stands for the *numerical solution* of (3.92),
f.s. stands for the *finite* difference *solution* of (3.92).

The convergence is established by applying the consistency criteria to the first norm on the right-hand side and applying the error control to the second norm.

A second approach involves the use of finite difference schemes developed to solve *O*rdinary *D*ifferential *E*quations (ODE). Let the $U_{zz}$ term in (3.92) be discretized by a second order central difference. Then

$$(U_m)_r = a_m U_m + \frac{b_m}{h^2}(U_{m+1} - 2U_m + U_{m-1}) \quad ,$$

which leads to a system of ordinary differential equations

$$\frac{dU_m}{dr} = \left(a_m - \frac{2b_m}{h^2}\right) U_m + \frac{b_m}{h^2}(U_{m+1} + U_{m-1}) \quad .$$

They may be written in matrix form as $U' = AU + g(r,z,U)$. A family of *N*on-*L*inear *M*ulti-*S*tep (NLMS) methods [3.58] as well as *L*inear *M*ulti-*S*tep methods (LMS) [3.59] have been developed to solve such systems. The conditions for consistency and stability for both the NLMS and LMS methods are well known and these methods are convergent. In addition to the existence of well developed numerical programs for such methods, powerful step size adjustment techniques based upon the use of predictor-corrector algorithms automatically adjust the step size to maintain a prescribed accuracy.

To provide a feeling for the execution time consider (3.92) with $b(r,z) = 1$, $a(r,z) = r^2 - z^2$ and the boundary conditions

$$U(0,z) = 1 \quad , \quad U(r,0) = 1 \quad , \quad U(r,1) = e^{-r} \quad ,$$

for which the exact solution is $U(r,z) = \exp(-rz)$. The answers at a range of 1609 meters and a depth of 804.5 meters as obtained on a PDP 11/70 computer are given in Table 3.5.

### 3.2.4 Range-Dependent Normal Mode Theory

KANABIS [3.60] has developed a computer model for the range-dependent problem which has its foundations in normal mode theory. The rate of change in the stratification

Table 3.5

| Method | k [m] | h [m] | Time [s] | Solution |
|---|---|---|---|---|
| Finite Difference | 1.609 | 160.9 | 77 | 0.6066E + 00 |
| O.D.E. | 1.609 | 160.9 | 92 | 0.6067E + 00 |
| Exact | | | | 0.6065E + 00 |

with range is not limited to be "slowly varying" since the effect of mode conversion is approximately included. Hence, large changes in the sound speed profile, water depth, and bottom composition with range may in principle be accommodated.

The total range interval is divided into segments wherein the sound speed profile and bottom composition are arbitrary functions of depth but do not change with range over the subinterval. The "trapped modes" within any range interval are calculated as if the problem were range-independent. These modes are then summed to give the total field at the junction of the next segment. Next it is assumed that this field may be adequately represented by a vertical distribution of appropriately weighted point sources. The trapped modes of the new range interval are calculated for each point source and summed to give the total field at the end of that section. In the early development [3.61] each range segment was rectangular which resulted in a bottom height with a staircase behavior. The backscattering of energy from this bottom structure as well as that which would result from the difference in impedance of the water column between segments was neglected. This approximation was subsequently [3.60] improved upon. This model has undergone limited comparison with both experimental data and analytical test cases. As a result it is difficult at this time to reach a conclusion regarding the adequacy of the approximations made. In addition, the number of required range segments seems best answered by running the model for an increasing number of range segments until convergence in the answer is observed.

### 3.2.5 Range-Dependent Ray Theory Models

Two methods for implementing range-dependent ray theory are in common use [3.62]. The first method [3.63] allows the sound-speed representation to be arbitrary in depth but linear, quadratic, or cubic in range at fixed depth. The second method [3.64] is based upon segmenting the region between the profiles into triangular sectors where two vertices of the triangle correspond to two points on one of the profiles and the third vertex corresponds to a point on the other profile. Along the connecting legs of the triangle the sound speed varies as $c(z,x) = c_0 + az + bx$. Both methods have their drawbacks. The first method is easy to automate but leads to ray-tracing difficulties since closed form expressions for the ray paths are not available. Additionally, this method can lead to totally specified profiles. The second method is quite difficult to automate in that it usually requires an oceanographer to determine the required connections. Aside from this problem, however, the linearity of the sound speed leads to a closed form expression for a ray's path within each triangular sector enabling a relatively rapid trace.

### 3.2.6 Finite Element Approach

KALINOWSKI [3.65] has recently examined the applicability of the *F*inite *E*lement *M*ethod (FEM) solution to acoustic propagation problems in a range-dependent environment. A synopsis of that survey has been provided by KALINOWSKI and is presented below.

The FEM solution (in its current form) was initially developed [3.66] in the structural mechanics field of the aircraft industry. A complete historical treatment may be found in [3.67]. The initial generalization to other fields [3.68] has been followed by papers [3.69-71] in nonstructural applications such as fluid mechanics, acoustics, electromagnetic field theory, heat transfer, etc. The FEM is well documented in introductory books [3.72-75] as well as more detailed theoretical developments [3.76,77]. Further [3.78] provides a collection of 7115 references on the general subject as a whole.

The propagation of acoustic energy in the ocean involves the interaction between the areas of acoustic wave propagation in fluids and stress wave propagation in solids. A good deal of finite element related work is found exclusively in the fluids area (either fluids alone or interacting with submerged structures) [3.69-71,79,80]; or, exclusively in the solids area (either solids alone or interacting with buried structures), see, for example, [3.81-89] and numerous publications in the Earthquake Engineering & Structural Dynamics Journal and the Bulletin of the Seismological Society of America. However, very little [3.65,90,91] finite element orientated work appears to be published specifically in the area of acoustic wave propagation in the ocean wherein the ocean bottom is treated as a coupled part of the solution (i.e., the bottom is modeled with more detail than implied by the usual approach of treating the fluid-bottom interaction with either a rigid or known impedance type boundary condition).

In [3.65,91] a displacement formulation FEM is suggested for acoustic propagation problems which discusses modeling both the fluid and irregularly shaped multilayered bottom with either rotationally symmetric ring elements or with planar elements. This technique allows for a sound-speed profile which can vary in both range and depth; a nonflat ocean surface or bottom; both dilatational and shear waves in the bottom; dissipative loss in the bottom; a simply connected fluid domain (e.g., voids to represent a school of fish); 3-d directional sources modeled via a Fourier expansion of the field solution in angular harmonics; and a methodology adaptable to existing general purpose [3.92] programs. A sample solution taken from [3.65] is provided later in this section.

In [3.90], a Galerkin type variational scheme, although not specifically stated as such, is employed which leads to an FEM type formulation. The fluid domain is treated as rotationally symmetric, bounded by a flat free surface and an arbitrarily shaped (but rigid) bottom. The main contribution of this work is an accurate procedure for handling the infinite domain (i.e., "transparent boundary") at the truncation end of the finite element mesh.

*a) The Finite Element Method.*

As previously pointed out, the finite element method has been firmly established over the past 20 years, consequently its development is given briefly for completeness[4]. Consider the case where the response of the continuum is expressed by the solution to the partial differential equation(s)

$$[A(\phi)] = 0 \quad , \tag{3.94}$$

which applies to a domain $\Omega$ where boundary conditions

$$[B(\phi)] = 0 \quad , \tag{3.95}$$

are satisfied on the boundary $\Gamma$; [A] and [B] are general partial differential operators, $\{\phi\}$ are function(s) representing the continuous field solution[5]. Often, $[B(\phi)]$ is defined as a mixed boundary value problem (i.e., $[B] = [B_1(\phi)]$ on $\Gamma_1$ and $[B] = [B_2(\phi)]$ on $\Gamma_2$ where $\Gamma = \Gamma_1 + \Gamma_2$). The specific meaning of $\{\phi\}$ depends on the formulation selected by the analyst (a specific example for a fluid domain is $\phi$ = pressure in the fluid and $\phi$ = field displacements in a solid ocean bottom).

I) The unknown response function(s) $\{\phi\}$ is approximated $\{\hat{\phi}\}$ by a series of prescribed shape functions $[N(\bar{x})]$ and associated unknown multiplying factors $\{a\}$, where

$$\{\phi(\bar{x})\} \approx \{\hat{\phi}(\bar{x})\} = [N(\bar{x})]\{a\} \quad . \tag{3.96}$$

The column vector $\{a\}$ has I components $[a_1, a_2, \ldots, a_I]^T$. The specific structure of (3.96) is displayed more clearly with a specific example; let $\{\phi\}$ denote, say, two dependent functions of a particular problem (like displaced components in a solid); thus the more detailed form of (3.96) becomes

---

[4]Paper [3.93] gives a concise description of the FEM, which is condensed here with some notational differences and with emphasis on the aspects of these methods relevant to the ocean-bottom interaction problem.

[5]The standard [ ], { } notation for matrices and column vectors, respectively, are employed throughout the finite element section.

$$\{\hat{\phi}(\bar{x})\} = \begin{Bmatrix} {}_1\hat{\phi}(\bar{x}) \\ {}_2\hat{\phi}(\bar{x}) \end{Bmatrix} = \begin{bmatrix} N_1(\bar{x}) & 0 & N_2(\bar{x}) & 0 & \ldots & N_J(\bar{x}) & 0 \\ 0 & N_1(\bar{x}) & 0 & N_2(\bar{x}) & \ldots & 0 & N_J(\bar{x}) \end{bmatrix} \begin{Bmatrix} {}_1a_1 \\ {}_2a_1 \\ {}_1a_2 \\ {}_2a_2 \\ \cdot \\ \cdot \\ \cdot \\ {}_1a_J \\ {}_2a_J \end{Bmatrix}$$

where $I = 2J$ for this example.

II) The problem unknows in the FEM are represented by discrete $\{a\}$ values rather than the continuous $\{\hat{\phi}(\bar{x})\}$ function. The $\{a\}$ values are determined as solutions to a set of "approximating" equations which have integral forms like

$$F_j(a_i) \equiv \int_\Omega G_j(\hat{\phi})d\Omega + \int_\Gamma g_j(\hat{\phi})d\Gamma = 0 \quad , \quad j = 1,2,\ldots,I \quad , \tag{3.97}$$

where the generation of $F_j(a_i)$ is obtained by alternate formulations defined shortly. The domain $\Omega$ and boundary $\Gamma$ are subdivided into K "finite element" zones (Fig.3.19), where $\Omega_k^e$, $\Gamma_k^e$ denote the domain and boundary of a typical $k^{th}$ finite element; thus, the integrals appearing in (3.97) are composed of the sum of the K element contributions

$$\int_\Omega G_j(\hat{\phi})d\Omega = \sum_{k=1}^{K} \int_{\Omega_k^e} G_j(\hat{\phi})d\Omega \qquad \int_\Gamma g_j(\hat{\phi})d\Gamma = \sum_{k=1}^{K} \int_{\Gamma_k^e} g_j(\hat{\phi})d\Gamma \quad ,$$

where the $\sum$ operators refer to the "standard assembly rules" of structural forms [3.71,74].

III) The usual condition, although not absolutely necessary, is imposed that the trial shape functions $N_i(\bar{x})$ be narrowly based (i.e., for an arbitrary i, $N_i(\bar{x})$ are given zero values everywhere, except in elements containing the unknown $a_i$). The narrow based trial functions lead to a set of banded (often symmetric) set of simultaneous equations for the unknown $\{a\}$.

Thus, substituting shape function expansions given by (3.96) into (3.97) (and performing the explicit spatial integrations, numerically if necessary) yields a set of simultaneous equations for the unknowns $\{a\}$. Restriction (III) only affects the

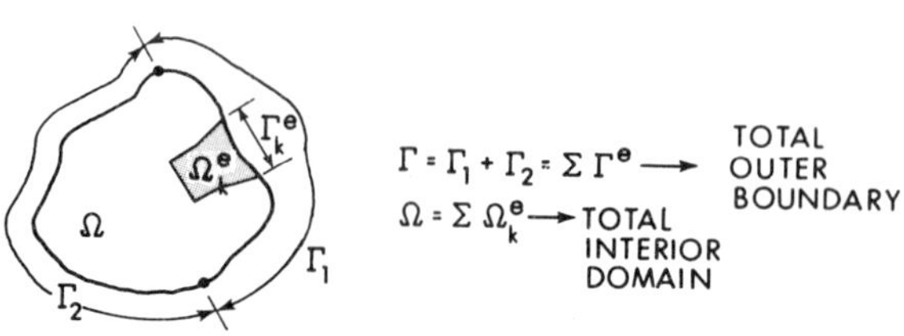


Fig. 3.19. Definition of interior domain and boundary

bandedness of the resulting set of simultaneous equations, and is not a strict requirement of the FEM technique [3.94]. The differences between various finite element approaches arise from the choice of the shape functions $N_i(\bar{x})$ selected at stage (I) and also by the manner in which the approximating equations (3.97) are derived. Specific integral formulations of (3.97) are constructed from weak formulations of the Galerkin, least square or collocation types; or, through the introduction of perhaps the most popular (although not always applicable) "variational principles". In the least square and variational principle formulations, the development of the $F_j(a_i)$ in (3.97) involves a minimization of the functional $I(\hat{\phi})$

$$F_j(a_i) = \frac{\partial I(\hat{\phi})}{\partial a_j} = 0 \quad , \quad j = 1,2,\dots,I \quad , \tag{3.98}$$

over the problem variables {a}, where $I(\hat{\phi})$ is of the form

$$I(\hat{\phi}) = \int_{\Omega} H(\hat{\phi})d\Omega + \int_{\Gamma} h(\hat{\phi})d\Gamma \quad . \tag{3.99}$$

The evaluation of the $F_j(a_i)$ expression (3.97) usually leads to a linear set of algebraic equations of the form[6]

$$[K]\{a\} = \{f\} \quad . \tag{3.100}$$

For either the variational principle, the least square formulation, it follows from (3.98) that the $G_j(\hat{\phi})$, $g_j(\hat{\phi})$ are related to $H(\hat{\phi})$, $h(\hat{\phi})$ by

$$G_j(\hat{\phi}) = \frac{\partial H(\hat{\phi})}{\partial a_j} \quad ; \quad g_j(\hat{\phi}) = \frac{\partial h(\hat{\phi})}{\partial a_j} \quad .$$

The matrix [K] (known as the generalized stiffness matrix of the system for structural applications) is obtained from the volume integrals of (3.97), and the problem loading {f} is determined from the surface integrals of (3.97). From a computational viewpoint, it is desirable to have the resulting [K] banded and symmetric, although not all FEM formulations yield this feature.

The volume integration in (3.97) generates the generalized stiffness matrix [K], thus attention is focused on the various forms for the kernel $G(\hat{\phi})$ that are employed for a specific problem, namely that of solving the steady-state acoustic response in a fluid domain governed by the Helmholtz equation

$$\nabla^2\phi(\bar{x}) + k^2\phi(\bar{x}) = 0 \quad , \tag{3.101}$$

---

[6]In certain cases such as transient solutions [3.72], or with special "continuous coordinate" finite-element formulations [3.95], (3.100) is replaced by an ordinary differential equation in {a}.

where the fluid pressure p is related to the velocity potential $\phi$ by the relation $\phi = -i\omega\rho p$ with $\rho$ being the fluid density, $k = \omega/c$, $\omega$ denoting the circular frequency, and c the piecewise constant (within the element $\Omega_k^e$) sound speed. Paper [3.96] employs a variational principle where

$$G_j(\hat{\phi}) = \frac{\partial H(\hat{\phi})}{\partial a_j} = \frac{\partial}{\partial a_j} (\nabla\hat{\phi}\cdot\nabla\hat{\phi}-k^2\hat{\phi}^2) \tag{3.102a}$$

leads to the development of (3.100); in another approach, [3.89] employs a Galerkin variational form where

$$G_j(\hat{\phi}) = - \nabla\hat{\phi} \cdot \nabla\psi_j + k^2\hat{\phi}\psi_j \tag{3.102b}$$

[with $\psi_j(\bar{x})$ as a weight function having the form of the shape functions $N_j(\bar{x})$] and leads to another alternate development of (3.100). Finally, [3.97] employs a least square approach where

$$G_j(\hat{\phi}) = \frac{\partial H(\hat{\phi})}{\partial a_j} = \frac{\partial}{\partial a_j} (\nabla^2\hat{\phi}+k^2\hat{\phi})^2 \quad , \tag{3.102c}$$

which leads to still another alternate form of (3.100).

b) *Merits/Shortcomings of the FEM*

The merits and shortcomings of numerical solutions employing the FEM are summarized in [3.65,93] and are given below with special emphasis on the ocean-bottom interaction problem.

*Merits of the FEM*

1) Representation of both the ocean (fluid) and bottom (solid) with finite elements is easily done. The total fluid and bottom domain are modeled with individual finite elements (each having their own physical constants), hence, the representation of solid media with multi-material properties (e.g., layers); or representation of the fluid with variable sound speed profile in one, two, or three directions is achievable. The boundaries of the media can be irregular, thus sea mounts etc. can be modeled. A demonstration problem employing some of these features is presented later.

2) Nonlinearities (e.g., a nonlinear representation of the bottom media stress-strain law) can potentially be included in the formulation. This, however, is accomplished at the expense of having to treat the time variation as a transient rather than steady-state problem. Thus, time-dependent ordinary differential equations in $\{a\}$ must be solved.

3) Due to the locally based shape functions, the final equations for the discretized unknowns are banded (and usually symmetric) thus offering certain computational advantages with relation to speed of solution and computer storage capacity.

4) The unknown parameters {a} are physically identifiable (e.g., displacement and pressure variables for the ocean-bottom interaction problems).

*Shortcomings of the FEM*

1) The number I of unknown parameters in {a} is large in that both the volume domain $\Omega$ and surface domain $\Gamma$ are discretized; this is a particularly strong disadvantage for the ocean problems, especially if a total three-dimensional representation is considered. Practically speaking, 2-d planar or rotationally symmetric domains are considered to keep the number of degrees of freedom manageable, although theoretically the method is totally applicable to the general three-dimensional case.

2) Representation of the infinite domain presents a problem wherein only approximate truncation boundary conditions are normally employed. The closely related *B*oundary *S*olution *M*ethodology (BSM) and *B*oundary *I*ntegral *M*ethodology (BIM) offer relief from element modeling, however, often at the expense of requiring the media to be homogeneous [3.65,91,93].

3) For steady-state problems, as many as eight elements per wavelength could be needed to adequately model the domain [3.82,98]; however, [3.90] appears to require less fine modeling (e.g., two elements per wavelength) with their particular approach.

4) Singularities that arise under concentrated loads are troublesome to model, e.g., the representation of a point (or line) source radiating from some location in the ocean.

*c) Transparent Boundary Simulation Techniques*

The chief FEM disadvantage (aside from the obvious large number of degrees of freedom required to model a problem) is the proper treatment of the infinite domain truncation. A typical planar or axis of revolution finite element model for an ocean-bottom problem typically involves four separate boundaries of an overall elongated region that is bounded by four surfaces (e.g., Fig.3.20). The region is a planar section for two-dimensional Cartesian coordinate problems or is a cross-section of a torus for rotationally symmetric (r-z cylindrical problems). The ocean surface boundary is perhaps the only clear-cut boundary, wherein a zero pressure condition is imposed along this side. The bottom boundary condition has several choices open, depending upon the degree of bottom detail imposed by the modeler. It can be taken as rigid; modeled with a prescribed impedance; modeled with finite elements terminated by either of the two previously mentioned options; modeled with finite elements and terminated with a prescribed impedance; modeled with finite elements and terminated with a transparent boundary condition. The remaining two vertical surface boundaries cut through and thus truncate both the fluid and bottom domain. Each of these two boundaries have the common problem of requiring that a proper radiation (i.e., transparent) boundary condition be imposed. In situations where the problem loading is such that

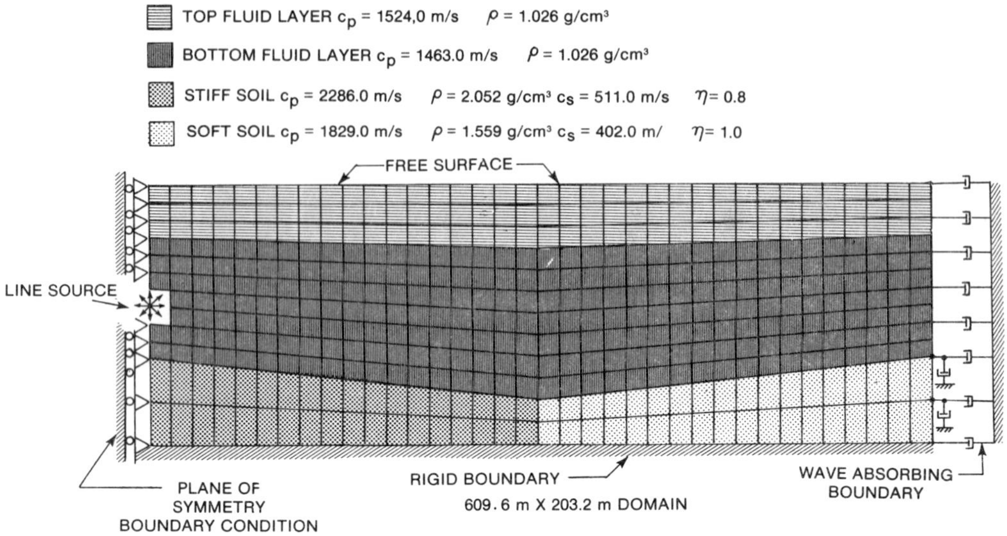


Fig. 3.20. Ocean-bottom interaction finite element method

a plane of symmetry exist (e.g., the sample problem considered later, Fig.3.20) a boundary condition demanding that the particle displacements are zero normal to the plane of symmetry takes care of one of the vertical boundaries. In various forms (in both the field of acoustics and the field of seismology) the remaining down range vertical boundary has received a good deal of attention in the literature. Other than [3.65,99] there appears no single reference that considers the most general case of having the vertical boundary made up of part fluid and part solid; consequently, the transparent boundary element for each of these two types of media is briefly considered separately. The reader is referred to [3.65,99] for a more detailed survey.

*d) Solid Domain Boundaries*

Perhaps the simplest idea for handling the transparent boundary is given by [3.82], wherein a plane wave type boundary condition of the form

$$\sigma_n = \rho c_p v_n \quad , \quad \sigma_t = \rho c_s v_t \quad , \tag{3.103}$$

is applied at the truncation of the solid finite elements, where $\sigma_n$ and $\sigma_t$ are the normal and tangental interface stresses, $\rho$ is the solid media mass density, $c_s$ and $c_p$ are the shear and dilatational wave speeds, and $v_n$ and $v_t$ are the corresponding normal and tangental velocities at a typical point on the solid boundary surface, respectively. The (3.103) conditions are exact if the incident radiating energy impinging on the surface are made up of plane waves of normal incidence to the surface. As pointed out in [3.82,98], they are still reasonably accurate, even when either of the incident waves is substantially off normal incidence. Employing a displacement finite element approach (i.e., the unknown parameters {a} are displacement quantities at the finite element grid work) the transparent boundary appears as though there are viscous dampers applied normal and tangental to the solid domain boundary point (where the $\rho c_n$ (or $\rho c_s$) velocity coefficient multiplied by an appropriate area factor, represents the value of the viscous damping constant). An example application of this type absorber is illustrated in Fig.3.20. The accuracy of this type boundary condition was considered in [3.65,82,100,101].

More accurate methods of treating the transparent boundary exist, but are employed at the expense of a more complicated formulation. KUHLEMEYER [3.102] considered a Rayleigh wave type viscous absorber, similar to [3.82] except the right side of these equations are multiplied by known depth and frequency-dependent parameters. A semi-analytical consistent boundary concept involving "hyperelements" was investigated in [3.87,88] for rotationally symmetric domains and in [3.83,86] for planar domains. These techniques are awkward to apply because they involve finding eigenvalues and associated eigenfrequencies of the solid media beyond the mesh truncation. GUTIEREZ [3.84] considered an approach analogous to [3.83,86-88] except that "finite element substructuring" is employed which enables one to treat a large domain of finite elements representing irregular material and shape variation zones of the solid domain. RUKOS [3.95] advanced an approach viewed as a generalization of the [3.88] hyperelement. A main difference is that the solid domain is modeled with finite elements in all spatial coordinate directions except one, which is represented continuously rather than discretely as in the usual FEM. The result is that even for steady-state problems, the final governing equations are ordinary differential equations (rather than algebraic equations) in the problem unknowns {a}. Finally, [3.93] treats the infinite domain by coupling the FEM with either the boundary solution method or boundary integral method.

*e) Fluid Domain Boundaries*

In theory, one should be able to appropriately reduce the various schemes described above for the solids case into the fluid application case by appropriately discarding the shear wave response portion of the solid media response. As an example of this, consider the viscous type boundary condition given by the first of (3.103). This type boundary condition was employed for pressure element formulation [3.103] and [3.79, 98] considered it for a displacement fluid element formulation. The accuracy of this type boundary absorber, with relation to fluid applications, is discussed in more detail in [3.65,90,98].

Other transparent boundary treatments that are not a limiting case of the solids approaches already considered are considered next. A "generalized radiation condition" of the form $\partial\phi/\partial r = T(\phi)$ was considered in [3.90] which is incorporated into a Galerkin type variational formulation, where $\phi$ is the velocity potential and $T(\phi)$ is a Hankel function expansion representing outward radiating waves with coefficients determined as part of the solution methodology [e.g., for the (3.103) type of absorber, $T(\phi)$ reduces to $i\omega\phi$]. "Infinite elements" were investigated in [3.104-106], which constitute the last (outermost) elements of the domain. The shape functions $N_j$ are of the form

$$N_j = \tilde{g}_j(s)\, e^{-s/L}\, e^{iks} \quad ,$$

where s is a spatial coordinate in the direction of the parametric coordinate extending to infinity, $\tilde{g}_j(s)$ is a Lagrange polynomial allowing for the usual amounts of shape change in the field of interest, exp(-s/L) represents the wave decay with increasing s (where L is an approximate decay length constant), and exp(iks) represents the usual harmonic waveform variation. A "generalized outflow boundary condition" through the application of Fourier transforms in the region exterior to the computational domain was derived in [3.107]. Finally, techniques employing the boundary integral method are often used to handle the infinite fluid domain, however, these are mainly concerned with application of these techniques to radiation or scattering problems related to various types of solids [3.79,108-110] which are submerged in a constant sound speed fluid.

### *f) Combined Solid-Fluid Domain Boundaries*

The ocean-bottom interaction problem has not received much attention in the finite element literature. Consequently, those wishing to employ the FEM to this field must employ a certain amount of ingenuity in applying (or combining) the separate techniques described earlier for treating the solid domain alone and for treating the fluid domain alone.

As an example of making one such combination, consider the problem of solving for the pressure response in a variable sound speed, two-dimensional fluid region which is bounded by a free surface from above and by a multi-material soil media (terminated by rock) from below (Fig.3.20). The input is a steady-state line source, whose axis is normal to the two-dimensional domain. For convenience in modeling, it is assumed that the source lies on a plane symmetry thus enabling one to use an appropriate plane-of-symmetry boundary condition along one of the vertical faces. The remaining vertical face, at the opposite end of the model, is the one requiring the transmitting boundary condition. This boundary, shown at the right and in the finite element model (Fig.3.20), contains part fluid and part solid media. The elementary viscous boundary conditions (3.103) are used for both the solid and fluid (the fluid case omits the unneeded viscous shear dampers). Wave speed parameters are assigned values according to the material they are directly contacting. The 430 element model is only used as a demonstration of the approach, consequently, the size of the mesh modeling, particularly in the solid domain, is rather coarse. The model is selected to emphasize the positive features of the finite element method, namely, the irregular shaped bottom; the variable material bottom; inclusion of dissipative loss factor in the bottom; variable sound-speed fluid domain (note fluid layers need not be flat nor parallel with the global coordinate system). The generalization of the above sample problem to additional solid-material variations and more complicated sound-speed profiles is straightforward. The finite element model is constructed from a displacement formulation approach for both the solid and fluid domain (an alternate approach more often used is to represent the fluid with "pressure type" finite elements, and only the solid domain with "displacement type elements"; the distinction between these basic types of elements is covered in more detail later). The model satisfies the fluid-solid interface boundary condition of equal displacements normal to each point of interface contact. The sample problem physical constants such as mass density $\rho$, dilatational sound speed $c_p$, shear wave sound speed $c_s$, and loss factor $\eta$ are specified directly on Fig.3.20. A 10 Hz line source pressure loading is represented by a set of nodal forces (applied at the open cut in the Fig.3.20 mesh) which corresponds to a unit pressure at the initial wave front (all other response pressure plots are referenced to this value). In order to emphasize the importance of including the effect of the bottom compliance on the down range pressure response, the problem was solved twice; one with a compliant soil sloping bottom, as shown in Fig.3.20, and again with a rigid sloping bottom. The solution time on a Univac 1108 computer is approximately four minutes of CPU time per incident frequency considered.

One method of displaying the pressure response is with the aid of contour plots. Since the solution is complex, plots reflecting both the real and imaginary parts of the solution (or alternatively the amplitude and phase angle) are usually made. The amplitude, phase angle contours are perhaps the most informative for the ocean-bottom interaction class of problems. Contour lines of constant pressure amplitude reveal where possible shadow zones are located, whereas lines of constant phase indicate wavefronts (lines normal to the constant phase contours are analogous to rays indicating the direction of wave propagation). The plot package used to generate the Fig.3.21 results is not sophisticated in that contours approaching the boundary nodes are not as reliable (due to the interpolation scheme employed at the boundary) as contours at or near interior nodes. Consequently, the contours are terminated at the nodes that are just inside the boundary nodes. Further, only pressure amplitude in the fluid (i.e., omitting the solid domain) is plotted. For the phase angle plots, the $0^\circ$ and $-360^\circ$ contours mathematically represent the same plotted contour line, consequently, this non-uniqueness creates some confusion to the contour plotter, particularly for closely packed contours. The computer-plotted phase angle contours in Fig.3.21b were considered unreliable in the shadow zone area, consequently, the contours were sketched in by hand (dashed portion of contours only) based on conjec-

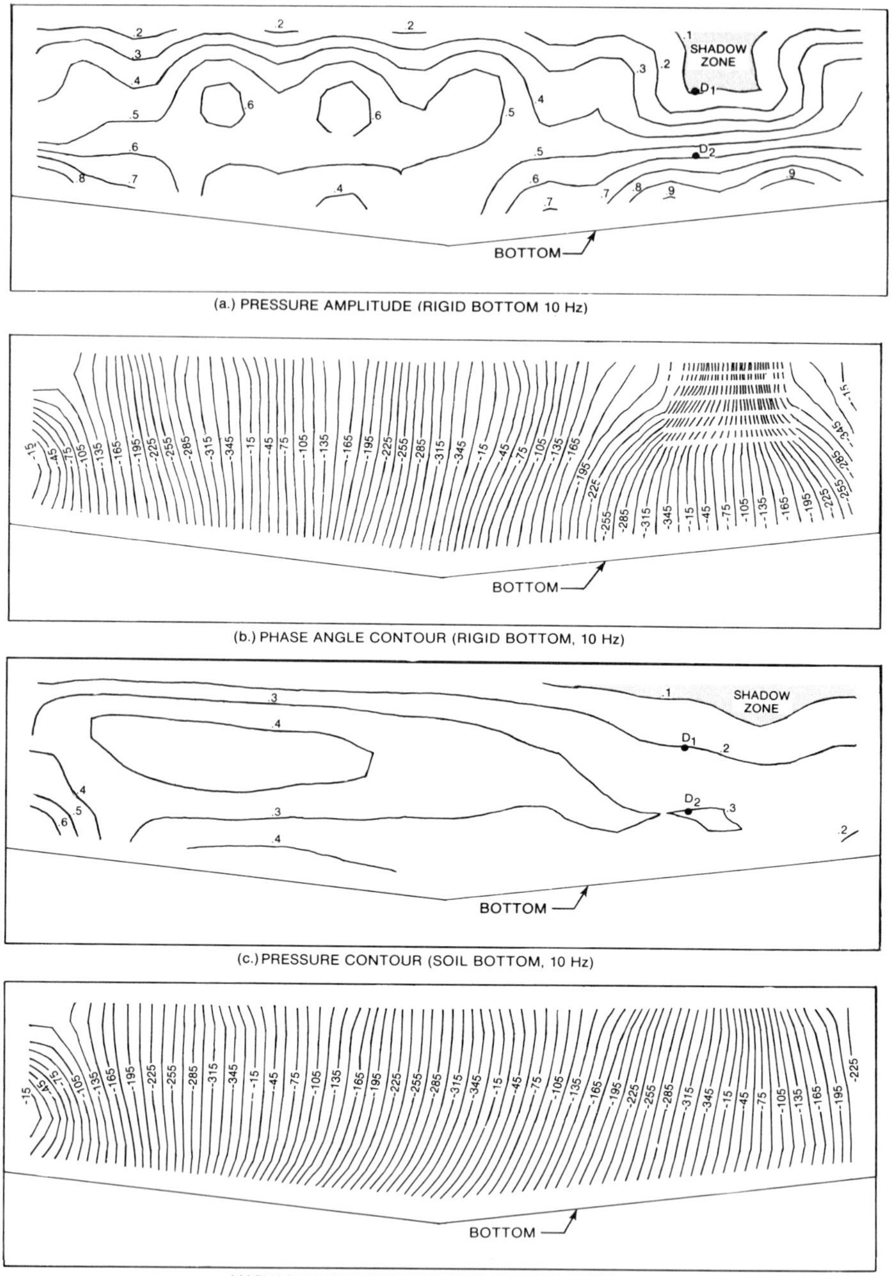


Fig. 3.21a-d. Finite element amplitude-phase contours

ture and by analogy with Fig.3.21d. The importance of modeling the compliance of the bottom in the problem formulation is clear upon making a comparison of the corresponding rigid bottom and soil bottom plots. A closer look at the data reveals how some erroneous conclusions could be drawn from an improperly modeled bottom (e.g., if the bottom were approximated as rigid as a modeling simplification). Consider, for example, the point labeled $D_1$ in Figs.3.21a and c. The rigid bottom pressure amplitude indicates point $D_1$ to be in a shadow zone wherein the soil bottom pressure amplitude for the same point, all other things held constant, is twice as large. Conversely comparing points $D_2$ (same range but roughly 44 meters deeper) in the same plots reveals just the reverse situation, namely, that the rigid bottom pressure amplitude is twice as large as the corresponding soil bottom pressure amplitude.

As a final comment regarding the suitability of the standard viscous transparent boundary in the fluid domain, it is noted that in Fig.3.21d normals to the wavefronts at the right end are very nearly perpendicular ($90^\circ$ is normal incidence) to the right side vertical face. This is the ideal situation for this viscous absorber to work. On the other hand, the plot in Fig.3.21b shows the wavefront normals are as far as $55^\circ$ off normal. Paper [3.98] contains percent error versus angle of incidence plots indicating that at $55^\circ$, approximately 97 percent of the energy is still absorbed by a plane wave obliquely indicent upon a set of fluid viscous absorbers, as employed in the demonstration problem of Fig.3.20. In fact, according to [3.98], 90 percent of the incident energy is absorbed even at a shallow $30^\circ$ angle of incidence. The phase angle contour plots approaching the mesh termination boundary thus provides a secondary function in aiding the modeler to check the validity conditions of the viscous absorbers. In fact, going one step further, it appears possible, although as yet untried, to alter the size of the damper [velocity coefficient in (3.16)] according to the observed angle of incidence as determined from a contour phase angle plot. Portions of the boundary not meeting a prescribed tolerance regarding the deviation from the ideal normal incidence, can have those particular absorbers adjusted by a prescribed amount (the amount depends on the angle of incidence) so that a second iteration computation can be made to improve the fluid portion transparent boundary performance.

### *g) Fluid-Finite Elements*

The solid finite elements are well established and are treated in sufficient detail elsewhere [3.73,75,111]. The fluid elements, on the other hand, deserve additional discussion in that the interface of solid and fluid elements require a certain degree of care. For the class of FEM problems which explicitly model the fluid, there are three approaches that are used in the field: 1) the pressure formulation for the fluid domain wherein there is one scalar (pressure) per discrete point in the FEM mesh [3.103,111,112]; 2) the displacement formulation wherein there are M unknowns per grid point where M corresponds to the number of independent spatial coordinates (1, 2, or 3 dimensions) needed to describe the response [3.79,113-115]; the mixed formulation where both pressure and the displacement vectors (or equivalently by velocity potential $\phi$ and its divergence $\nabla\cdot\phi$) are taken as the problem unknowns

[3.116,117]. The displacement and pressure formulation have certain advantages over each other which were discussed in detail in [3.65,118].

## References


3.1 F.R. DiNapoli, R.L. Deavenport: "Computer Models of Underwater Acoustic Propagation"; Tech. Rpt. 5867 (Naval Underwater Systems Center, New London, Conn. 1978)

3.2 H.F. Baker: Proc. London Math. Soc. (Ser.1) *35*, 333 (1902)

3.3 R.A. Frazer, W.J. Duncan, A.R. Collar: *Elementary Matrices* (Cambridge University Press, London 1960)

3.4 F.R. Gantmacher: *The Theory of Matrices*, Vol.2 (Chelsea Publishing, New York 1959)

3.5 F. Gilbert, G.E. Backus: Geophys. *31*, 326 (1966)

3.6 L.B. Felsen, N. Marcuvitz: *Radiation and Scattering of Waves* (Prentice-Hall, Englewood Cliffs, NJ 1973)

3.7 E.A. Coddington, N. Levinson: *Theory of Ordinary Differential Equations* (McGraw-Hill, New York 1955)

3.8 H.W. Marsh, S.R. Elam: Internal Document, Raytheon Company, Marine Research Laboratory, New London, Conn. (1967)

3.9 F.R. DiNapoli: "A Fast Field Program for Multilayered Media"; Tech. Rpt. 4103 (Naval Underwater Systems Center, New London, Conn. 1971)

3.10 F.R. DiNapoli, M.R. Powers:"Recursive Calculation of Products of Cylindrical Functions", Tech. Memo. PA-83-70 (Naval Underwater Systems Center, New London, Conn. 1970)

3.11 F.R. DiNapoli: "The Collapsed Fast Field Program (FFP)"; Tech. Memo. TA11-317-72 (Naval Underwater Systems Center, New London, Conn. 1972)

3.12 D.C. Stickler: J. Acoust. Soc. Am. *57*, 856 (1975)

3.13 W.M. Ewing, W.S. Jardetzky, F. Press: *Elastic Waves in Layered Media* (McGraw-Hill, New York 1957)

3.14 C.L. Bartberger: ;AP2 Normal Mode Program", Report, Naval Air Development Center, Warminster, Pa. (1978)

3.15 C.L. Pekeris: Geol. Soc. Am. Mem. *27* (1948)

3.16 I. Tolstoy: J. Acoust. Soc. Am. *28*, 1182 (1956)

3.17 S.R. Santaniello, F.R. DiNapoli, R.K. Dullea, P. Herstein: "A Synopsis of Studies on the Interaction of Low Frequency Acoustic Signals with the Ocean Bottom"; Tech. Doc. 5337, Naval Underwater Systems Center, New London, Conn. (1976)

3.18 P. Debye: Phys. Z. *9*, 775 (1908)

3.19 B. Van der Pol, H. Bremmer: Phil. Mag. *24*, 141, 825 (1937)

3.20 D.V. Batorsky, L.B. Felsen: Radio Sci. *6*, 911 (1971)

3.21 H. Weinberg: J. Acoust. Soc. Am. *58*, 97 (1975)

3.22 C.W. Spofford: "The FACT Model"; Tech. Rpt. 109, Maury Center for Ocean Science, Washington D.C. (1974)

3.23 G.A. Leibiger: "The Acoustic Propagation Model RAYMODE: Theory and Numerical Treatment"; NUSC Tech. Rpt., Naval Underwater Systems Center, New London, Conn. (1978)

3.24 M.A. Pedersen, D.F. Gordon: J. Acoust. Soc. Am. *51*, 323 (1972)

3.25 C. Bartberger: "Normal Mode Solutions and Computer Programs for Underwater Sound Propagation"; Tech. Rpt. NADC-72001-AE, Naval Air Development Center, Warminster, Pa. (1973)

3.26 E.C. Titchmarsh: *Introduction to the Theory of Fourier Integrals*, 2nd ed. (Oxford University Press, Oxford 1948)

3.27 H. Bremmer: *Terrestrial Radio Waves* (Elsevier Publishing, Amsterdam 1949)

3.28 H.M. Nussenzveig: Ann. Phys. *34*, 23 (1965)

3.29 F. Gilbert: Geophys. J. R. Astr. Soc. *44*, 275 (1976)

3.30 A. Ben-Menahem: Bull. Seis. Soc. Am. *54*, 1315 (1964)

3.31 C.L. Pekeris: *Proc. Symp. Appl. Math.*, Vol.2 (American Mathematical Society, New York 1950) pp.71-75
3.32 D.S. Ahluwalia, J.B. Keller: "Exact and Asymptotic Representations of the Sound Field in a Stratified Ocean", in *Wave Propagation and Underwater Acoustics*, ed. by J.B. Keller, J.S. Papadakis, Lecture Notes in Physics, Vol.70 (Springer, Berlin, Heidelberg, New York 1977)
3.33 N.A. Haskell: J. Appl. Phys. *22*, 157 (1951)
3.34 L.M. Brekhovskikh: Sov. Phys. Acoust. *2*, 124 (1956)
3.35 R.B. Lauer, B. Sussman: "A Methodology for the Comparison of Models for Sonar System Applications"; Vol.I, Naval Sea Systems Command Tech. Rpt. SEA 06H1/036-EVA/MOST-10, Naval Underwater Systems Center, New London, Conn. (1976)
3.36 R.B. Lauer, B. Sussman: "A Methodology for the Comparison of Models for Sonar System Applications - Results for Low Frequency Propagation Loss in the Mediterranean Sea"; Vol.II, Naval Sea Systems Command Tech. Rpt. SEA 06H1/036-EVA/MOST-11, Naval Underwater Systems Center, New London, Conn. (1978)
3.37 D.F. Yarger: "The User's Guide for the Raymode Propagation Loss Program"; Tech. Memo. 222-10-76, Naval Underwater Systems Center, New London, Conn. (1976)
3.38 C.L. Bartberger: "PLRAY, A Ray Propagation Loss Program"; Tech. Rpt. Naval Air Development Center, Warminster, Pa. (1978)
3.39 D.F. Gordon: "Normal Mode Computation of Propagation Loss for an Arbitrary Number of Layers"; Tech. Pub. 236, Naval Undersea Center, San Diego, Calif. (1971)
3.40 D.W. Hoffman: "LORA, A Model for Predicting the Performance of Long-Range Active Sonar Systems"; Tech. Pub. 541, Naval Undersea Center, San Diego, Calif. (1976)
3.41 W.H. Watson, R. McGirr: "Raywave-II, A Propagation Loss Model for the Analysis of Complex Ocean Environments"; Tech. Note 1516, Naval Undersea Center, San Diego, Calif. (1975)
3.42 E.B. Wright: "Acoustic Transmission Loss by Single-Profile Ray Tracing, Program RTRACE"; Tech. Rpt. 7815, Naval Research Laboratory, Washington D.C. (1974)
3.43 I.M. Blatstein: "Comparisons of Normal Mode Theory, Ray Theory, and Modified Ray Theory for Arbitrary Sound Velocity Profiles Resulting in Convergence Zones"; Tech. Rpt. 74-95, Naval Ordnance Laboratory, White Oak, Silver Spring, Md. (1974)
3.44 H. Weinberg: J. Acoust. Soc. Am. *50*, 975 (1971)
3.45 R.P. Porter, H.D. Leslie: J. Acoust. Soc. Am. *58*, 812 (1975)
3.46 F.R. DiNapoli: "The Inverse Fast Field Program (IFFP): An Application to the Determination of the Acoustic Parameters of the Ocean Bottom"; Tech. Memo. 771160, Naval Underwater Systems Center, New London, Conn. (1977)
3.47 H.W. Kutschale, F.D. Tappert: J. Acoust. Soc. Am. *62*, S518 (1977)
3.48 H.W. Kutschale, F.R. DiNapoli: J. Acoust. Soc. Am. *62*, S518 (1977)
3.49 F.D. Tappert: "The Parabolic Approximation Method", in *Wave Propagation and Underwater Acoustics*, ed. by J.B. Keller, J.S. Papadakis, Lecture Notes in Physics, Vol.70 (Springer, Berlin, Heidelberg, New York 1977)
3.50 R.H. Hardin, F.D. Tappert: SIAM Rev. *15*, 423 (1973)
3.51 F.D. Tappert, R.H. Hardin: Proc. 8th Intern. Cong. on Acoustics, Goldcrest, London, Vol.2, 452 (1974)
3.52 R.M. Wilcox: J. Math. Phys. *8*, 962 (1967)
3.53 F. Jensen, H. Krol: "The Use of the Parabolic Equation Method in Sound Propagation Modeling"; Tech. Rpt. Sm 72, NATO SACLANTCEN, La Spezia, Italy (1975)
3.54 J.S. Papadakis, D. Wood: "A Parabolic Decomposition of Helmholtz Equation"; Tech. Rpt., Naval Underwater Systems Center, New London, Conn. (1978)
3.55 M.C. Smith: J. Acoust. Soc. Am. *46*, 233 (1969)
3.56 S.T. McDaniel: J. Acoust. Soc. Am. *58*, 1178 (1975)
3.57 D. Lee, J.S. Papadakis: "Numerical Solutions of Underwater Acoustic Wave Propagation Problems"; NUSC Tech. Rpt. 5929, Naval Underwater Systems Center, New London, Conn. (1978)
3.58 D. Lee: "Nonlinear Multistep Methods for Solving Initial Value Problems in Ordinary Differential Equations"; Ph.D. dissertation, Polytechnic Institute of New York, N.Y. (1977)
3.59 P. Henrici: *Discrete Variable Methods in Ordinary Differential Equations* (Wiley and Sons, New York 1962)

3.60 W.G. Kanabis: "A Shallow Water Acoustic Model for an Ocean Stratified in Range and Depth"; NUSC Tech. Rpt. 4887-1, Naval Underwater Systems Center, New London, Conn. (1975)
3.61 W.G. Kanabis: "Computer Programs to Calculate Normal Mode Propagation and Applications to Analysis of Explosive Sound Data in the BIFI Range"; NUSC Tech. Rpt. 4319, Naval Underwater Systems Center, New London, Conn. (1972)
3.62 C.W. Spofford, H.M. Garon: "Deterministic Methods of Sound-Field Computation"; in *Proc. NATO Conf. on Oceanic Acoustic Modeling*, ed. by W. Bachmann, R.B. Williams (SACLANTCEN, La Spezia, Italy, 1975) pp.40-1 to 43
3.63 J.J. Cornyn: "Grass, A Digital-Computer Ray-Tracing and Transmission-Loss-Prediction System"; NRL Rpt. 7621, Vol.I, Naval Research Laboratory, Washington D.C. (1973)
3.64 H.P. Bucker: "The RAVE (Ray Wave) Method"; in *Proc. NATO Conf. on Geometrical Acoustics*, ed. by B.W. Conolly, R.H. Clark (SACLANTCEN, La Spezia, Italy 1971) pp.32-36
3.65 A.J. Kalinowski: The Shock and Vibration Digest, *11*, 12 (March 1979)
3.66 M.J. Turner, R.W. Clough, H.C. Martin, L.J. Topp: J. Aeronaut. Sci. *23*, 805 (1956)
3.67 J.T. Oden: *Finite Elements of Nonlinear Continuum* (McGraw-Hill, New York 1972)
3.68 O.C. Zienkiewicz, Y.K. Cheung: "Finite Elements in the Solution of Field Problems", The Engineer *220*, 507 (1965)
3.69 R.H. Gallagher, J.T. Oden, C. Taylor, O.C. Zienkiewicz: *Finite Elements in Fluids* - Vol.1, *Viscous Flow and Hydrodynamics* (Wiley and Sons, New York 1975)
3.70 R.H. Gallagher, J.T. Oden, C. Taylor, O.C. Zienkiewicz: *Finite Elements in Fluids*, Vol.2, *Mathematical Foundations, Aerodynamics and Lubrication* (Wiley and Sons, New York 1975)
3.71 J.T. Oden, O.C. Zienkiewicz, R.H. Gallagher, C. Taylor: *Finite Element Methods in Flow Problems* (University of Alabama Huntsville Press, Huntsville, Ala. 1974)
3.72 O.C. Zienkiewicz, Y.K. Cheung: *The Finite Element Method in Structural and Continuum Mechanics* (McGraw-Hill, New York 1967)
3.73 L.J. Segerland: *Applied Finite Element Analysis* (Wiley and Sons, New York 1976)
3.74 K.H. Huebner: *The Finite Element Method for Engineers* (Wiley and Sons, New York 1975)
3.75 R.D. Cook: *Concepts and Applications of Finite Element Analysis* (Wiley and Sons, New York 1974)
3.76 J.T. Oden, J.N. Reddy: *An Introduction to the Mathematical Theory of Finite Elements* (Wiley and Sons, New York 1976)
3.77 G. Strang, G. Fix: *An Analysis of the Finite Element Method* (Prentice-Hall, Englewood Cliffs, N.J. 1973)
3.78 D. Norrie, Gerard de Vries: *Finite Element Bibliography* (IFI/Plenum Data Company, New York 1976)
3.79 A.J. Kalinowski: "Fluid-Structure Interaction", in *Shock and Vibration Computer Programs, Review and Summaries*, ed. by Pilkey and Pilkey, SVM-10, The Shock and Vibration Information Center, Washington D.C. (1975)
3.80 L.H. Chen, V.M. Pierucci: "Underwater Fluid-Structure Interaction", Parts I and II, The Shock and Vibration Digest (April issue, pp.23-24; May, pp.17-21, 1977)
3.81 R.L. Kuhlemeyer, J. Lysmer: " J. Soil Mech. Found. Div., ASCE, *99*, 421 (1973)
3.82 J. Lysmer, R.L. Kuhlemeyer: J. Engr. Mech. Div., ASCE , *95*, 859 (1969)
3.83 J. Lysmer, L. Drake: "A Finite Element Method for Seismology", *Methods in Computational Physics*, ed. by B. Bolt (Academic Press, New York 1972)
3.84 J.A. Gutierrez: "A Substructure Method for Earthquake Analysis of Structure-Soil Interaction", Earthquake Engineering Research Center, Rpt. No. 76-9 (April 1976)
3.85 J. Lysmer: Bull. Seism. Soc. Am. *60*, 89 (1970)
3.86 J. Lysmer, G. Waas: J. Eng. Mech. Div., ASCE, *98*, 85 (1972)
3.87 E. Kausel, J. Roesset, G. Waas: J. Eng. Mech. Div., ASCE, *101*, 679 (1975)
3.88 E. Kausel, J. Roesset: J. Eng. Mech. Div., ASCE, *103*, 569 (1977)
3.89 P. Chakrabarti, A.K. Chopru: Earthquake Engineering and Structural Dynamics *2*, 107 (1973)
3.90 G.J. Fix, S.P. Marin: "Variational Methods for Underwater Acoustics Problems", ICASE Rpt. 77-16 (August 1977)

3.91 A. Carlson, A.J. Kalinowski, J. Patel: "Solutions of a General Class of Fluid Structure Problems by the Finite Element Method", Naval Underwater Systems Center TR-5361 (Conf.) (June 1976)
3.92 "The NASTRAN User's Manual", NASA SP-222(03) (March 1976)
3.93 O.C. Zienkiewicz, D.W. Kelly, P. Bettess: International Journal of Numerical Methods in Engineering, *11*, 355 (1977)
3.94 C.D. Mote, Jr.: Int. J. Numerical Methods in Engineering, *3*, 565 (1971)
3.95 E.A. Rukos: International Journal for Numerical Methods in Engineering,*12*, 11 (1978)
3.96 M. Petyt, J. Lea, Koopman: Journal of Sound and Vibration, *45*, 495 (1976)
3.97 G.J. Fix, M.H. Gunzburger: "On the Use of Modern Numerical Methods in Acoustics", ICASE Rpt., no date
3.98 A.J. Kalinowski: Shock and Vibration Bulletin, *48*, 62 (1977)
3.99 A.J. Kalinowski: "Application of the Finite Element Method to Acoustic Propagation in the Ocean"; NUSC Tech. Rpt. S891, Naval Underwater Systems Center, New London, Conn. (1979)
3.100 R. Dungar, P.J.L. Eledred: Earthquake Eng. Struct. Dynamics *6*, 123-138 (1978)
3.101 J.M. Rosset, M.M. Ettouney: International Journal for Numerical and Analytical Methods in Geomechanics, *1*, 151 (1977)
3.102 R. Kuhlemeyer: "Vertical Vibrations of Footings Embedded in Layered Media", Ph.D. Thesis, University of Calif., Berkeley (1969)
3.103 O.C. Zienkiewicz, R.E. Newton: "Coupled Vibrations of a Structure Submerged in a Compressible Fluid", Proc. Int. Symp. on Finite Element Techniques (Stuttgart, F.R. Germany 1969)
3.104 P. Bettess, O.C. Zienkiewicz: International Journal for Numerical Methods in Engineering, *11*, 1271 (1977)
3.105 P. Bettess: International Journal for Numerical Methods in Engineering, *11*, 53 (1977)
3.106 O.C. Zienkiewicz, P. Bettess: 2nd Intern. Symp. Computing Meth. Appl. Science and Engng. (Versailles, France 1975)
3.107 B. Engquist, A. Majda: Mathematical of Computation, *31*, 629 (1977)
3.108 J.T. Hunt, M. Knittel, C.S. Nichols, D. Barach: J. Acoust. Soc. Am. *57*, 287 (1975)
3.109 J.T. Hunt, M.P. Knittel, D. Barach: J. Acoust. Soc. Am. *55*, 269 (1974)
3.110 H.A. Schenck: J. Acoust. Soc. of Am. *44*, 41 (1967)
3.111 O.C. Zienkiewicz: *The Finite Element Method in Engineering Science* (McGraw-Hill, London 1971)
3.112 G.C. Everstine, E.M. Schroeder, M.S. Marcus: "The Dynamic Analysis of Submerged Structures", 4th NASTRAN User's Colloquium, Langley Research Center, Hampton, Virginia (1975)
3.113 D. Shantaram, D.R.J. Owen, O.C. Zienkiewicz: Earthquake Engineering and Structural Dynamics, *4*, 561 (1976)
3.114 E.P. Sorensen, D.V. Marcal: "A Solid Mechanics Approach to the Solution of Fluid-Solid Vibration Problems by Finite Elements", Brown University Tech. Rpt. N00014-0007/13 (May 1976)
3.115 L. Kiefling, G.C. Feng: AIAA J. *14*, 199 (1976)
3.116 G.J. Fix, M.D. Gunzburger, R.A. Nicolaides: "On Mixed Finite Element Methods I, The Kelvin Principle", ICASE Rpt. (Dec 1977)
3.117 G.J. Fix, M.D. Gunzburger, R.A. Nicolaides: "On Mixed Finite Element Methods II, The Least Squares Method", ICASE Rpt. 77-18 (Dec 1977)
3.118 A.J. Kalinowski: "Transmission of Shock Waves into Submerged Fluid Filled Vessels", in *Fluid Structure Interaction Phenomena in Pressure Vessel and Piping Systems*, ed. by M. Wang, S.J. Brown (ASME, New York 1977) pp.83-105

# 4. Physical Modeling of Underwater Acoustics

J. G. Zornig

With 17 Figures

In the attempt to explain underwater acoustic phenomena and to exploit these phenomena in useful ways a number of techniques for analysis have been developed. These include, but are not limited to, mathematical analysis, numerical modeling, in situ empirical description, and physical modeling. All have made substantial contributions to our understanding of the way in which sound behaves in water. The purpose of this chapter is to review the methods and results of physical modeling and to discuss its role in underwater acoustic research. To do this an attempt will be made to define the technique and to set in perspective its relationship to the other methods of analysis. Some attention will be paid to the use of small natural bodies of water, lakes and bays, for example, as model facilities. However, the main thrust of the discussion will be concerned with artificially constructed model tanks and particularly with the study of surface scattering in such facilities. Physical model research has been conducted for some time in a number of contexts. Unfortunately, until the present there has been little communication among various investigators on matters of technique. By discussing the development of physical model experimental methods and illustrating the degree of precision and flexibility possible in model tanks this chapter seeks to encourage and facilitate a more widespread and systematic employment of physical modeling in future underwater acoustic research.

## 4.1 Background Information

### 4.1.1 Definition and Purpose

Physical modeling in this context refers to the empirical investigation of underwater acoustic phenomena in physical situations other than the ocean. This somewhat imprecise definition is intended to include acoustic experimentation in water tanks, lakes, bays and so forth; physical environments which, by virtue of their stability, convenience and flexibility, provide the experimenter with opportunities for precision, repeatability, and economy not normally available in ocean experimentation. Physical modeling is, in general, useful for two broad classes of application. The first, and probably more significant of these roles is in the validation of analytical models. It is in this application that the inherent precision of physical modeling has made its biggest impact. A more secondary role, although one which has resulted in important contributions, is in the empirical characterization of phenomena for which no unambiguous ocean measurements are available. The relative economy of the model tank has frequently made it the sole source of data of a particular type.

### 4.1.2 Relationship to Ocean Experimentation

Perhaps the single most important consideration in the planning and evaluation of physical model experiments is that the physical apparatus is not an exact scale model of the ocean. Many misunderstandings have resulted from failures to recognize this fact and to account for its implications. Unlike many scale models used in electromagnetic studies, hydroacoustic models invariably involve physical quantities which are difficult or impossible to scale accurately. For example, the overwhelming majority of physical models use water as a medium. The use of water forces the experimenter to account for the fact that the properties of the fluid, viscosity, surface tension, density and absorption among others, are not scaled. On the other hand, because of the reduced dimensions of the apparatus, acoustic wavelength and target scale are normally reduced. This results in a relationship between ocean phenomena and those observed in the physical model that is far more complex than a simple scaling factor. Normally that relationship is accounted for by analytical and numerical models in which the assumptions are such that they are equally valid in both cases. In such a situation, the physical model is used to validate the mathematical model which then requires only calibration to be useful as a predictor of ocean phenomena. This reduces the requirement for expensive ocean data to a relatively few calibration checkpoints, the much more extensive validation having been performed comprehensively using physical model data at low cost and high precision.

### 4.1.3 Historical Overview

The applications of physical modeling in underwater acoustics extend back some thirty years to early investigations of diffraction and scattering from solid targets. The first major impact, however, was made by LACASCE and TAMARKIN [4.1] in 1956 in a study of scattering from sinusoidal surfaces. This study came as a result of a considerable amount of theoretical activity in the early 1950's in attempts to explain acoustic and electromagnetic scattering from rough surfaces. It was followed closely by an investigation of scattering from periodic sawtooth surfaces [4.2] and spatially random but fixed surfaces [4.3] at the same Brown University facility. During this period, LEPORSKII [4.4,5] in the USSR conducted model studies of the spatial distribution of scattering from periodic surfaces of a similar nature. In both cases the surfaces varied in only one dimension. Similar work was initiated at this point by HORTON and others at the University of Texas [4.6-8]. This effort continued for a number of years and extended the surfaces studied to a set of two-dimensional polystyrene surfaces based on aeromagnetic contour maps of the Canadian Shield. As before the signal sources used were pulsed sinusoids and the results of the experimental series were analysed primarily in terms of various approximate solutions to the Helmholtz integral. Other researchers have continued work on solid surfaces including a variety of improvements in technique [4.9-13].

While the experiments with solid surfaces were of tremendous value in the validation of theory for some relatively simple statistics, they obviously were not time varying. The inclusion of time variation into the physical model was necessary in order to achieve the statistical sufficiency necessary to develop high precision results and to explore such scattering phenomena as frequency spread (Doppler) and time spread. This time variation was included in a physical model in 1965 in a series of experiments on short fetch wind driven surfaces at the University of Rhode Island by D'ANTONIO and HILL [4.14]. This study produced some rather rough results for the envelope statistics of high frequency (1.45 MHz) cw signals at a variety of source-receiver ranges. At roughly the same time other experiments with wind driven surfaces were begun by MEDWIN et al. [4.15-18], RODERICK and CRON [4.19]. and KINGSBURY [4.20]. These experiments contributed a variety of data on wind driven surfaces using pulsed cw sources. SPINDEL and SCHULTHEISS [4.21] later used impulsive sources to obtain estimates of the impulse response of wind driven surfaces in a variety of specular orientations. This work led to a continuing series of experiments by the author [4.22-26] using a variety of computer synthesised broadband probing signals to make measurements of a number of second-order temporal and spatial statistics of model surface scattering process.

A number of researchers [4.27 30] have conducted experiments in lakes and bays as a compromise between the desire to have the most realistic possible surface and the desire to experiment at the high level of precision made possible by long term fixed installations. These experiments, while they have the advantage of a physical surface of larger scale and longer fetch, have in general suffered from the difficulties with stationarity, repeatability, and surface characterization that afflict ocean experiments. In addition, a large amount of work in model facilities has concerned itself with solid targets of various sorts [4.31-40], bubbles [4.41], particles [4.42], and temperature microstructure. This work is in general of very high precision and uses narrow band test signals.

## 4.2 Water Facilities

### 4.2.1 Model Tanks

Physical model tanks used in hydroacoustic studies have ranged in size from approximately 1.3m × 1.3m × 3.7m to 4.6m × 3.7m × 18.3m. In general, the requirement for this scale of apparatus comes from two basic considerations. The first of these is that the facility makes it possible to place projector and hydrophone(s) a large number of wavelengths from the scattering target (100 wavelengths at 100 KHz is approximately 1.5 m). The second is a limitation on acoustic wavelength imposed by data acquisition electronics and surface scale. Above 1.0 MHz, the cost of electronic recording and digitization becomes increasingly significant. In addition, the production of wind driven surfaces with RMS waveheights of much less than 1 mm and practically useful statistics is quite difficult. Whereas solid, perfectly reflecting targets can be made at quite small scales, wind driven water surfaces that are spatially stationary and exhibit some statistical resemblance to ocean waves are difficult to produce with fetches much smaller than two meters or wind speeds less than 3 $m\,s^{-1}$. For these reasons, signals used in physical models rarely exceed 2 MHz and are more commonly in the 50 to 500 KHz range.

Model tanks have been constructed of materials varying from permanent concrete installations to those based on ordinary backyard swimming pools. Anechoic linings are in some cases used, but in many cases it has been found that the use of pulsed signals and narrow projector beam patterns makes this unnecessary. Projector and hydrophone(s) are normally positioned using a manual or remotely controlled apparatus. Both translators [4.14] and goniometers [4.25] have been constructed for this purpose, the primary requirement being that the transducers be repeatably positionable to within a small fraction of an acoustic wavelength (for coherent measurements).

Wind sources are normally fan-driven wind tunnels. In order to insure symmetry of the wind-driven surface a collimator and/or plenum is used between fan(s) and tunnel. The tunnel may extend over the area of the surface used for scattering [4.18,20-22] or it may stop short of that area [4.15]. The consequence of the latter is a drastically reduced local wind interaction in the scattering region. Surfaces of this sort tend to have a higher proportion of sinusoidal, swell-like components than do locally wind driven surfaces. The skirts of tunnels are in most cases smooth and penetrate the water. This has the effect of producing quite a long water surface correlation function in the crosswind direction. In order to more closely model typical ocean surfaces, the author has modified the skirts in his facility such that they ripple horizontally. This has had the desired effect of decreasing the crosswind correlation distance substantially [4.23]. It has been found to be desirable, when possible, to isolate the wind source mechanically from the tank and to use a "beach" downwind of the scattering area so as to control the formation of standing water waves in the tank.

The principal requirement of physical models, as was stated above, is to exercise the assumptions of theory sufficiently to test validity. For this reason the addition of controlled swell to wind driven surfaces has been found desirable in some circumstances. To some extent, the removal of the "beach" has the effect of generating long standing waves which resemble swell. A preferable method is to use a mechanical wave generator, a cylinder moved mechanically on the surface. Such generators have been used by RODERICK and CRON [4.19], CLAY et al. [4.18], and SANDNESS [4.44].

The medium in essentially all hydroacoustic model tanks is fresh water. It has been suggested to the author that more exotic fluids both above and below the scattering interface might make it possible to more closely emulate ocean surface behavior. The difficulty of pursuing this approach, however, is substantial and the likelihood of improvement in modeling accuracy sufficient to justify the expense seems remote. Maintenance of clean fresh water over extended periods, while not without its problems, is a relatively simple task; the physics of sound propagation in fresh water is relatively well understood; and the surface statistics produced by model tanks are adequate. Therefore it remains the medium of choice in most studies.

### 4.2.2 Lakes and Bays

As a compromise between the restrictions on scale imposed by model tanks and the imprecision and nonstationarity of ocean situations several researchers have turned to small natural bodies of water. Lakes and bays have the advantage of substantially longer fetch and greater depth than are possible in even rather large model tank facilities. Unfortunately, the price for these benefits is that to a significant extent the physical conditions in such natural bodies are unpredictable and nonstationary. PLEMONS et al. [4.27,28] reported a representative series of experiments performed at the Lake Travis Test Station of the University of Texas in 1970-1971. This example of a careful and thorough lake study gives an idea of the nature of the compromises inherent in that approach. Waveheights reported were almost an order of magnitude greater than those produced in tanks. This permitted the use of rather lower acoustic frequencies and longer duration signals than are possible otherwise. Stationarity and homogeneity, on the other hand, proved to be a sufficiently difficult problem, in spite of the fact that data collection was performed during relatively short periods, that a major portion of the analysis was devoted to statistical testing of stationarity and homogeneity. Because of the unpredictability of weather in the outdoor facility, environmental conditions including surface conditions and temperature gradient could not be maintained constant from experiment to experiment. Nevertheless, a stable instrumentation platform and a generally less complicated environmental situation than is found in the ocean made a detailed analysis possible which was useful in the evaluation of the theoretical models.

A good example of what can be accomplished in a sheltered bay can be found in [4.45]. In presenting the results of a series of backscatter experiments made in

Dabob Bay, Washington, in 1956-1957 the authors pointed out that the absence of ship traffic and the relative predictability of surface conditions together with fetches of up to 40 km made it possible to obtain high quality data over an extended period of time. In addition, because of easy access to the bay and the existence of an establishment of facilities, extensive environmental data was collected which made calibration of the data collected possible. Nevertheless, the precision obtained did not begin to approach that of a model facility and considerable effort had to be expended in order to account for uncontrollable variations in the experimental conditions, changing wind speed and direction and rain for example. The analysis of data was principally phenomenological.

As can be seen from these examples, lake and bay experiments are useful for some purposes, but have many of the same disadvantages as ocean measurements. The principal contributions from this class of facilities have been in the form of extensive characterizations of phenomena for which precise physical theory is not available. For this sort of measurement, the physical scale of the environment makes possible a high level of confidence that the empirical results are applicable to the ocean. The lack of precision and statistical sufficiency, on the other hand, is generally not a great handicap when the purpose of the experiment is empirical characterization rather than validation of a predictive model.

## 4.3 Targets

### 4.3.1 Wind Driven Surfaces

One of the most challenging problems in underwater acoustics is the explanation of scattering of hydroacoustic waves from rough surfaces. A great deal of analytical effort has gone into this problem and the result has been a large body of approximate theory attempting to model the scattering process. The effort to test the validity of these theories and to characterize scattering phenomena for which no models exist has been one of the principal applications of physical modeling in underwater acoustics. A great variety of surfaces have been used as targets in physical models. These tend to fall into three general categories: wind driven, mechanically excited, and solid. Each category of surface has found application in the testing of theory, however all have limitations. In this section the characteristics of physical model surfaces will be reviewed briefly.

By far the most complex and phenomenologically rich of the physical model surfaces used as scattering targets is the short fetch wind-driven air-water interface. Wind-driven surfaces have been used in a number of investigations over the past 20 years as a target which best tests theories of ocean-surface scattering. It is important to emphasize, however, that short fetch model-tank surfaces are not scale models of the ocean surface. The model surface is a time varying randomly rough surface, as is the ocean, which can be represented by a two-dimensional random process, as can the ocean surface. In both cases there are three physical forces acting to produce the surface roughness: gravity, surface tension, and the complex wind-water interaction (neglecting the coriolis force, which is of small importance in acoustic scattering). The major difference between model surfaces and ocean surfaces is in the relative importance of these interactions and in boundary constraints. KINSMAN [4.46] is an excellent source of information on the properties of fetch and time-limited wind-driven water surfaces. The model-tank surface is extremely fetch limited and not at all time limited and no short fetch surface has all of the properties of a more fully developed wind-driven water surface. On the other hand, few, if any, of the analytical models applicable to underwater acoustic scattering depend on the scattering interface to be a large scale sea surface. What is important is that the physical model surface be stationary and well enough developed to exhibit statistical properties that are sufficiently similar to those of the ocean to yield useful information.

The easiest requirement to fulfil is temporal stationarity. With an effective beach, a model tank of the scale of those mentioned in this chapter will stabilize in a few minutes. Some investigators have allowed as much as an hour of wind to insure that equilibrium has been reached, but the author has found that measured surface statistics are stable in 5 to 10 minutes at most. The limiting process seems to be the rate at which the small standing waves generated in the tank, both inside and outside of the wind tunnel, come to equilibrium. These standing waves can be suppressed, but to some extent they are always present and take some time to stabilize.

Spatial stationarity is somewhat more difficult to achieve. A fetch limited surface becomes rougher as a function of downwind distance. This change occurs at a rate which is rather slow, however, with respect to the temporal and spatial roughness of the surface itself. Therefore, depending on the degree of homogeneity desired, a surface developed by a model-tank wind tunnel can be considered locally homogeneous over a small area. The experimenter is inevitably forced to design his experiments such that the area insonified on the surface is not larger than this region of local homogeneity. Moreover, if the force of the wind is slackened somewhat in the vicinity of the scattering area, this region of local homogeneity can be enlarged somewhat in the downwind direction. This can be accomplished by truncating the wind tunnel upwind of the scattering area. However, as was mentioned above, the surface so produced will necessarily have a smaller content of small capillary waves. The author of this chapter has had some success with a tunnel which diffuses the air flow vertically starting at the scattering site. This technique seems to have the effect of maintaining the capillary waves, but diminishing the extent to which they are increased over the extent of the scattering region.

In the crosswind direction, homogeneity is somewhat more easily maintained. Except in the near vicinity of the skirts of the wind tunnel, the wind and water conditions are uniform across the width of the tunnel and a high degree of homogeneity is observed. Irregularities in the shape of the skirts also tend to improve the homogeneity of the surface. In fact, however, no rough-water surface produced by a wind tunnel is completely homogeneous.

Of paramount importance in the generation of wind-driven surfaces is the degree to which the surfaces produced represent the models proposed for them in analytical theories. Fortunately, analytical models tend only to require a statistical specification of the surface sufficient for the specific calculations being performed. For example, a common requirement is that the probability density function of surface waveheights be Gaussian in one dimension. In order to test a theory making this assumption, the physical model ought to be able to generate two sorts of surfaces. The first is an ensemble of very nearly Gaussian surfaces useful for testing the accuracy of the mathematics. The second is a set of somewhat non-Gaussian surfaces to test the sensitivity of the theory to its assumptions. Preferable still is a variety of documented surfaces which depart in a variety of ways from the assumptions. Sensitivity testing, however, is an area which has received little attention in physical model research. Figures 4.1 and 4.2 illustrate the degree to which short fetch wind

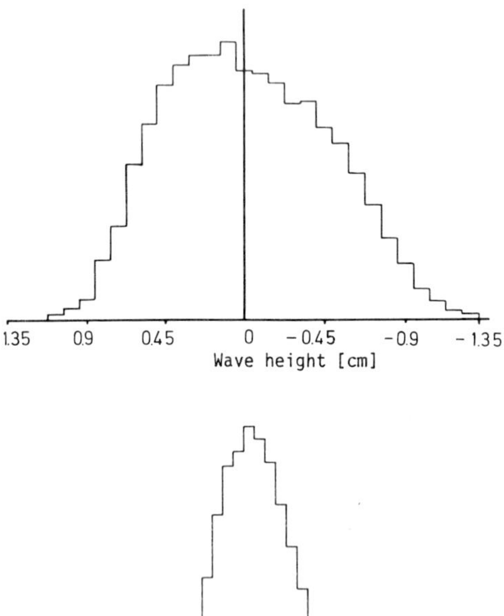


Fig. 4.1. Measured histogram of waveheights of the wind-driven surface used by CLAY et al. [4.18]. The surface was generated by a set of five fans (positive values of waveheight denote depression with respect to the mean)

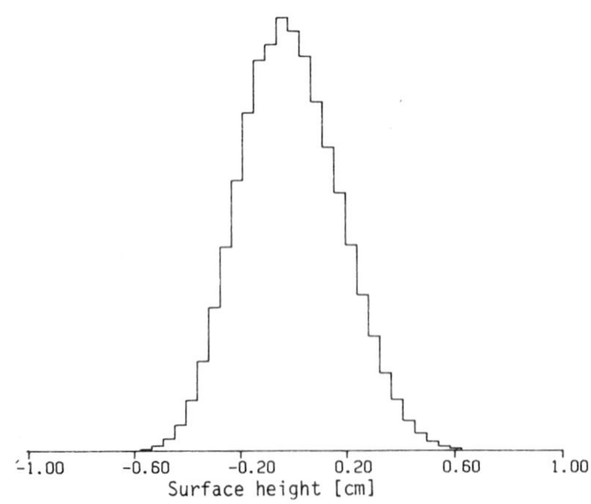


Fig. 4.2. Measured histogram of waveheights of the wind-driven surface of ZORNIG and McDONALD [4.22]. The surface was generated by a single fan and collimator driving a wind tunnel with irregular skirts (positive values of surface height denote elevation with respect to the mean)

driven surfaces have satisfied the assumption of a Gaussian probability density function (PDF). SPINDEL and SCHULTHEISS [4.47] have made some measurements of the joint distribution of waveheights in two dimensions which indicate that the Gaussian assumption may in some circumstances be very difficult to justify.

A number of methods for measurement of surface waveheights have been employed with satisfactory results. These include optical and acoustic methods as well as the measurement of the capacitance or resistance of wires penetrating the surface. For small scale models and for acquisition of records of waveheight time series, however, the capacitance wire has enjoyed the widest use. Capacitance detectors can, if improperly adjusted or designed, exhibit a nonlinearity in transfer characteristic similar to an improperly biased vacuum tube or transistor circuit. The effect on the resulting histogram is to compress the tails relatively. While this distortion might very well escape detection in the histogram, the effect on a characteristic function computed from that histogram can be quite noticeable at high spatial frequencies. Since a great deal of the interest in coherent scattering is centered around high acoustic wavenumbers, these errors, if present, could be important.

As mentioned above the statistic of the surface normally used to predict coherent scattering is the characteristic function of the surface height distribution, the Fourier transform of the PDF. Unfortunately, this complicates the problem of estimation of the PDF. Normally surface-heights are measured using either resistance or capacitance sensing wires of small diameter which penetrate the surface of the water. The sensing electronics produce a voltage which varies in proportion to the waveheight. This voltage is then sampled and converted to digital form. A histogram is accumulated from these data samples either in a multichannel analyser or a general purpose computer until the desired degree of statistical sufficiency is achieved. In principle, this histogram is a good estimate of a sampled version of the surface height PDF. There are, however, two sources of error which can tend to distort the characteristic function a great deal and are not obvious from an inspection of the unprocessed histogram. These are amplitude nonlinearity in the detection electronics and nonuniform quantization in the analog-to-digital conversion process.

Measurements of surface slope statistics have been made primarily using glitter photographic techniques [4.15,48,49]. This method has the advantage of not disturbing the surface, but the disadvantage of being difficult to automate so as to collect large statistical ensembles. Another approach is to use pairs of capacitance or resistance probes to obtain two-time series of waveheights at adjacent points. The basic measurement is performed at a variety of horizontal orientations of the pairs in order to obtain directional information. Probe-pair measurement, unfortunately, is not easy. It requires very fine probe wire, which must be held stiff and vertical in spite of water and air turbulence. Minimum probe spacing, hence the smallest area observable, is limited primarily by the requirement to have negligible meniscus between the wires. Measurements indicate that several diameters of the probe wires is a minimum for accurate estimation. In addition, electromagnetic coupling between probes must be eliminated. In some probe systems this has been accomplished by using separate detector circuits operating at different frequencies.

### 4.3.2 Fixed Surfaces

Much of the study of scattering from rough surfaces has been conducted using solid-carved surfaces of various materials and shapes. The most obvious advantage of such an approach is that it involves a non-time-varying target of precisely defined shape. Since much of the early theoretical work in this area concerned itself with idealized deterministic surfaces, there was an obvious motivation to conduct confirming measurements on physical surfaces of the same sort. One of the first studies was made by LACASCE and TAMARKIN using corrugated cork surfaces floated on the water surface [4.1]. Cork was used to provide an approximately pressure release surface. The surfaces were of sinusoidal cross section in one dimension and had RMS roughnesses of between 0.15 and 0.32 cm. Evidently, the fabrication of the surfaces was somewhat rough as the authors pointed out that some of the surfaces "departed considerably from sinusoidal form". A plane cork surface was used as a reference for calibration. A subsequent study [4.2] used an echelette grating of the same material to test a somewhat different theoretical approach. In this case the upper frequency range was extended to 300 KHz and it was discovered that cork was an imperfect reflector in short acoustic wavelengths. Nevertheless, cork was again used in a study of spatially random surfaces which followed [4.3]. These were constructed of flat cork strips laid down in such a way as to give a roughly Gaussian distribution of heights and a regular correlation function. These surfaces were insonified at longer acoustic wavelengths and seemed to give quite satisfactory results for that case. Later, in response to a new theoretical approach, BARNARD et al. [4.7] fabricated a styrofoam surface by milling with a cutter of sinusoidal cross section. This surface was probably somewhat more ideal than the cork surfaces. Apparently the experiment produced very precise data.

Another motivation for the study of fixed surfaces is that the bottom of the ocean is a relatively time-invariant rough surface which reflects and absorbs sound. A study of this phenomenon was made by BARNARD et al. [4.6]. For these experiments three surfaces were constructed. These consisted of layered combinations of styrofoam, sand, a sand and clay mixture, and limestone having a smooth surface. These surfaces were used to measure reflection coefficients as a function of grazing angle at 100 KHz. Subsequently, that project was extended to the measurement of two-dimensionally rough surfaces. HORTON et al. [4.8] introduced a set of surfaces made of expanded polystyrene with relief generated by a two-dimensional random process, some aeromagnetic contours of the Canadian Shield. Four such surfaces were used in a variety of studies [4.9-13] of scattering phenomena.The surfaces had RMS roughnesses

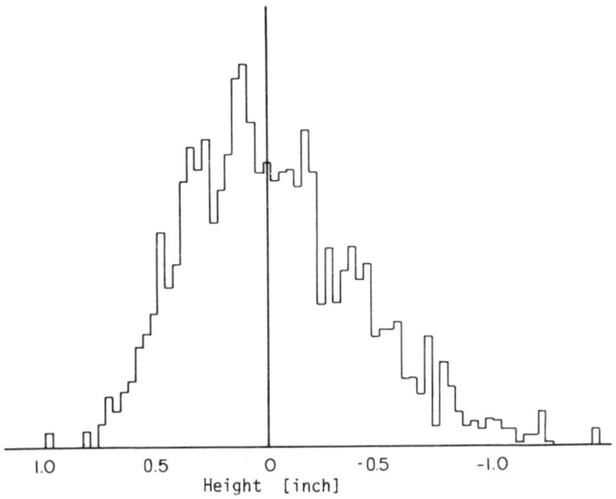


Fig. 4.3. Measured histogram of surface heights of the Canadian Shield surface III [4.11]. The histogram contains 1089 samples and has 75 bins. Attempts were made to fit this distribution to Gaussian, exponential, and a modified exponential distribution but the results were ambiguous

of between 2.3 and 9.2 mm and were fabricated to a precision of 0.19 mm. This is roughly the same step size that was used in the fabrication of the surfaces of [4.1].

An unfortunate property of fixed random surfaces, such as the Canadian Shield models, is that the statistical ensemble is incomplete. These surfaces were 81 cm. square and had correlation distances of up to 7.1 cm. With such a limited extent it would be unreasonable to expect statistical sufficiency great enough to permit stable characterization of either the surface statistics or those of the scattered field. Fig. 4.3 gives an indication of the degree of this insufficiency. Nevertheless, these surfaces played a major role in the measurement of rough surface scattering phenomena by several investigators.

### 4.3.3 Volume Targets

Volume targets used in model tank measurements are so many and varied that it is essentially impossible to enumerate them. A great number of studies of idealized structures such as cylinders, plates, disks, spheres and so forth have been performed. These studies have been directed largely at either target strength measurement or the characterization of local phenomena such as circumferential waves on cylinders. Both hydrophones and Schlieren photography [4.35] have been used with considerable success. In general, continuous or pulsed cw signals are used having wavelengths on the order of a centimeter. These experiments are characterized by relatively simple instrumentation and very high precision.

Another class of targets of considerable interest has been air bubbles. A representative study containing references is MOLE et al. [4.41]. These studies have employed methods roughly similar to those used with solid targets. The methods of production of bubbles, however, have varied widely and include among others the suspension of particles in gelatin [4.42]. Acoustic wavelengths have varied from several centimeters to less than a millimeter and in most cases only scattering strength has been measured. Acoustic propagation in water with a random temperature microstructure has been studied in physical model by CAMPANELLA and FAVRET [4.43]. They used a resistance heater to generate temperature variations in a model tank less than 1 m deep. Relatively random temperature fluctuations on the order of a few tenths of a degree Celsius were measured using a pair of thermistors. The authors presented evidence that it is possible to generate statistically stable microstructure, both spatially and temporally, in such a model. This "target" was used in a measurement of time autocorrelation of acoustic pulses at 480 KHz and 1.42 MHz which showed that the time autocorrelation of pulses transmitted through the medium was equal to that of the medium itself. In spite of the success of this experiment, however, not a great deal of work has apparently been done with temperature microstructure targets in physical models.

## 4.4 Instrumentation and Signal Processing

### 4.4.1 Sources

The vast preponderance of sound sources used in model experiments have been piezoelectric transducers of cylindrical shape operating in thickness mode. These projectors have the advantage of being relatively easy to make in the frequency range used in such studies, and of projecting a more or less regular beam pattern, that of a baffled piston (Fig.4.4). Some use has been made of larger electrostatic transducers [4.15], but the difficulty of driving these projectors has not led to their widespread use. SPINDEL [4.50] reported the consideration of several alternative broadband sources including pulsed laser, spark gap, explosive, and mechanical impact. Whereas some of these sources are in widespread use in ocean work, problems of cost, scale, repeatability and calibration among others led to the adoption of a piezoelectric source in that study.

In most cases, the signal source is a single frequency sinusoid. In order to resolve paths to eliminate multiple reflections the sinusoid is normally gated. This source has the advantage of ease of production, but yields relatively little information about the target in any single experiment. Recognizing this shortcoming, MEDWIN [4.15] used a signal consisting of a sequence of sinusoids of different frequencies in rapid succession, a "multiburst signal". The signal source (Fig.4.5)

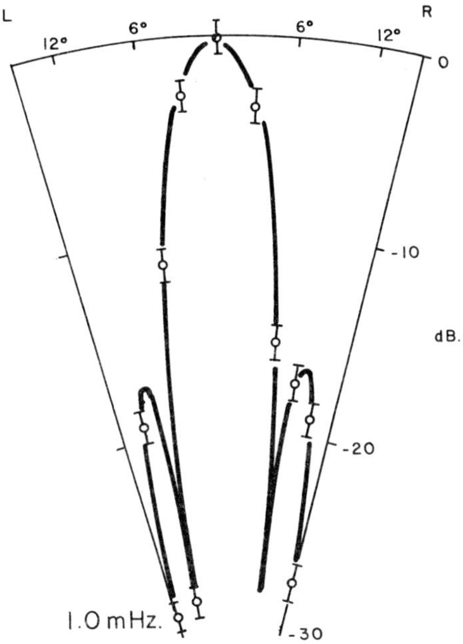


Fig. 4.4. The measured beam pattern of a Branson Z803B pressure transducer measured at 1 MHz in the far field. Even simple practically realizable patterns such as this one seldom match the assumptions of theory

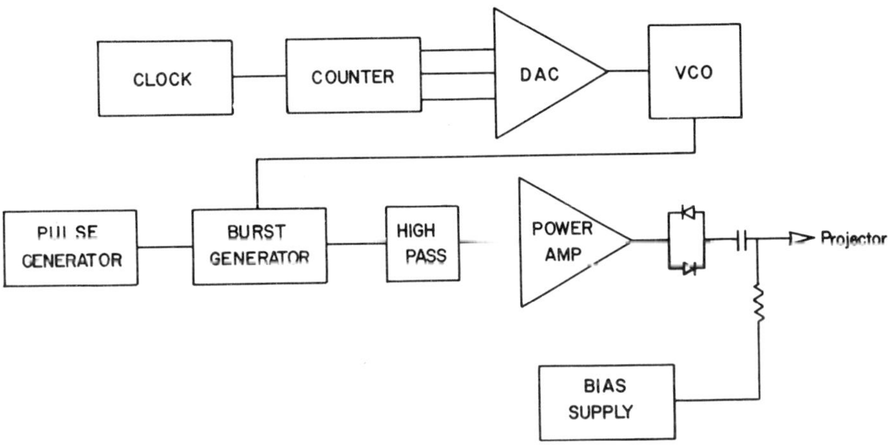


Fig. 4.5. The apparatus used by MEDWIN [4.15] to generate "multiburst" test signals. This approach produced a broadband test signal of short duration but without coherence between frequency components

was used to drive a Mylar electrostatic transducer of adequate efficiency and relatively broad frequency response. This approach made possible the collection of data at a variety of acoustic wavelengths for the same statistical ensemble of target surfaces. Since the sinusoids were separated in time, oscilloscope photography could be used as a data collection method. The resulting data had the relatively unique property that the surface slopes were measured (by glitter photography) at the same time as each acoustic experiment.

SPINDEL and SCHULTHEISS [4.21], in a series of experiments aimed at measurement of coherence, used a source consisting of a short impulsive signal applied to a resonant but relatively low Q piezoelectric transducer. The resulting acoustic signal had useful energy over a broad band but had a very uneven power spectrum. The result was that the scattered signal-to-noise ratio in the data varied greatly. The advantage of this procedure, however, was that the signal was not only broadband but also coherent. That is, all components of the source signal had a fixed phase relationship. This made possible the estimation of statistics of the received signal that would not otherwise have been possible.

As an extension of this approach a method of direct digital synthesis of probing signals in real time by a general purpose computer was developed [4.22,51]. In order to compensate for the transfer functions of the transducers and other components of the data acquisition system, a preemphasized probing pulse was designed. This signal design was based on the measured transfer function of the measurement channel as follows. With the water surface flat, a broadband (about 3 octaves) signal was applied to the projector (Fig.4.6). The received signal was then Fourier transformed digitally. The complex reciprocal of the transform was then inverse transformed to produce a test signal. This signal when applied to the same measurement system would then result in a nearly flat spectrum received signal (Fig.4.7). In practice, small nonlinearities in the projector and quantization errors made it necessary to repeat the process several times in order to get an acceptably smooth received spectrum.

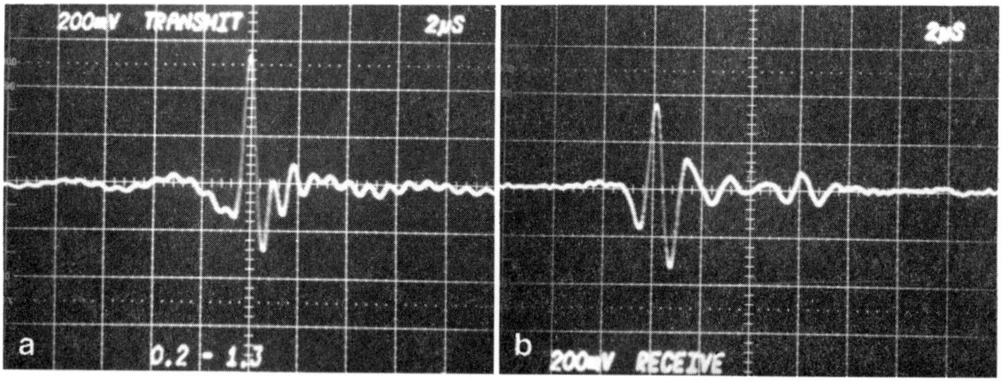


Fig. 4.6. Coherent broadband pulse used in the channel calibration phase of pulse preemphasis (a) and the flat surface response of the measurement system (b)

The resultant amplitude and phase spectra for a transducer response range of nearly 40 dB is shown in Fig.4.8. This signal design technique has expanded the variety of scattering statistics measurable in the model tank substantially in recent years.

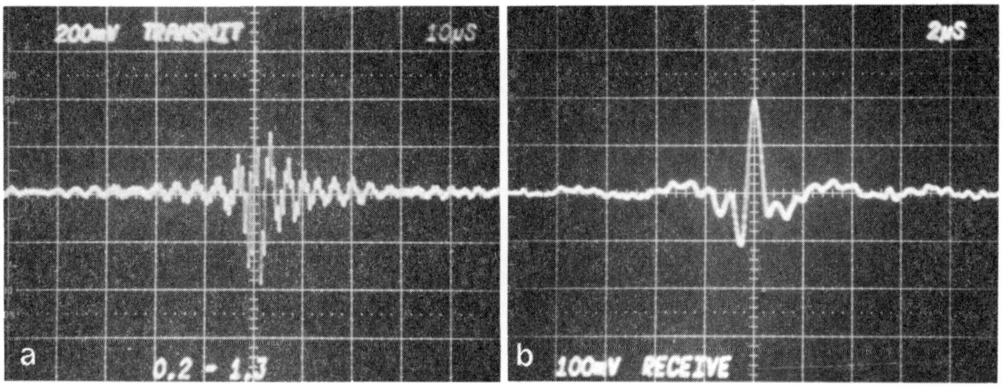


Fig. 4.7. Preemphasized probing pulse computed from Fig.4.6b (a) and the flat surface system response (b)

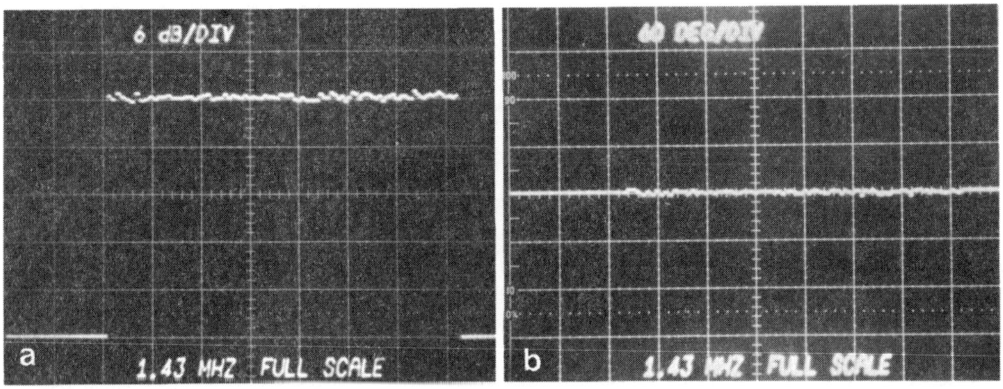


Fig. 4.8. Flat surface system frequency (a) and phase (b) response after using a preemphasized test pulse, the Fourier transform of Fig.4.7b. The precision for which the data acquisition system response can be precompensated substantially simplifies data analysis

### 4.4.2 Data Acquisition

Model tank experiments have as one of their distinct advantages the opportunity for the experimenter to take advantage of a stable situation to acquire as much data as he wants. In most cases this has been done using either analog tape recording or direct measuring instruments. Exceptions include Schlieren and oscilloscope photography in which the cost of acquiring and reducing large amounts of data can be prohibitive. For single frequency experiments, heterodyning has been employed to reduce bandwidth requirements to less than 1 KHz. In other cases, relatively high speed analog tape recorders have been used to record data directly for later conversion at a lower playback speed. Recently, however, the speed and cost of on-line digital data acquisition have improved to the point that it is feasible to digitize and store the receiving hydrophone output(s) directly. Figure 4.9 illustrates a data acquisition system in use which is capable of generating and acquiring signals with data rates of up to 32 megabaud (4 million $s^{-1} \times 8$ bits/sample). Test statistics are immediately available to the experimenter for use in calibration and alignment and the system is capable of recording data for hours without attention. As a result, this system has been used to obtain large volumes of data on a variety of scattering processes and with a variety of test signals that would have been prohibitively time-consuming and expensive using more primitive methods.

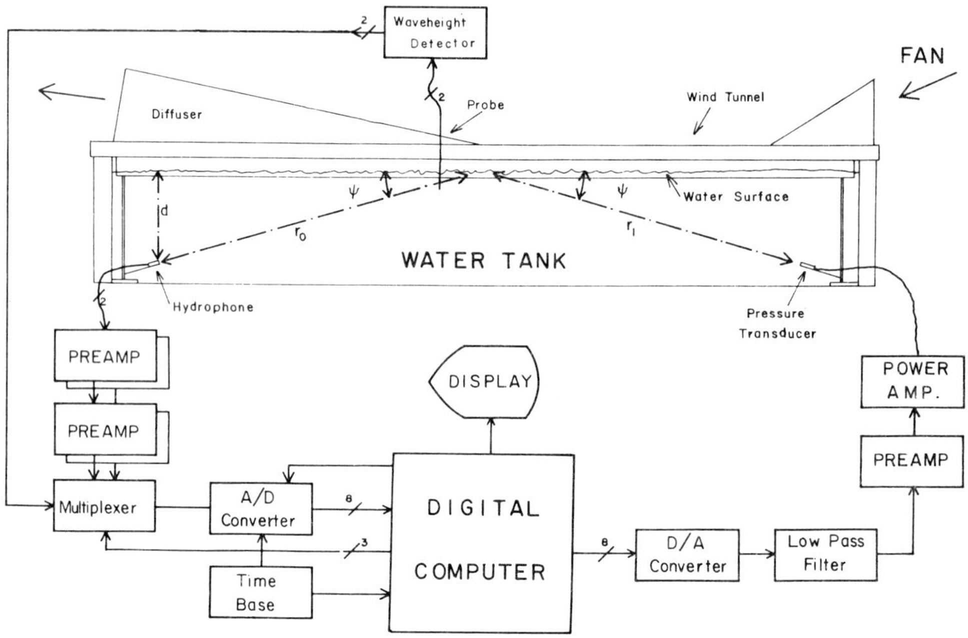


Fig. 4.9. The data acquisition system used by the author in recent investigations. Direct synthesis and acquisition of signals are accomplished by the computer at sample rates up to 4 MHz. On-line analysis of even relatively complicated phenomena makes it possible to correct small errors and drifts precisely

## 4.4.3 Signal Processing

The basic purpose of model tank experimentation makes it generally desirable to produce results which are representative of the process under investigation and not of the particular experimental configuration. In order to more nearly achieve this goal the data obtained is adjusted and averaged in some way. Normally the methods employed fall into two broad classifications: correction, and the computation of a statistic or functional of the raw data. A great variety of methods have been used, most of which are also employed in the processing of ocean data. In the model tank, however, there are some special situations which can be exploited to obtain results of higher precision.

One such advantage is the fact that the acoustic path is very precisely known. The environment in a model tank is exceptionally simple. The water is essentially isothermal; there are no inhomogeneities in it; and projector and receiver positions are stable. The consequence of these conveniences is that it is possible to precisely place the arrival time of received signals with respect to each other and the transmitted pulse. In the computation of coherent statistics this has been an enormous aid, but has imposed a rather tight requirement on instrumentation which would not otherwise be the case. This requirement is that the time reference used to measure arrivals with respect to transmission must be precise to within a small fraction of an acoustic period at the highest frequency employed. For example, if the path length is 6 m and frequencies up to 1.0 MHz are used then a time base stable to better than $1:10^5$ must be used. Normally, this requires an oven controlled crystal or frequency synthesizer.

An advantage of equal importance is the opportunity to periodically recalibrate. Ocean, and even lake and bay experiments are rather desperate affairs. A measurement is being made on a naturally caused system over which the experimenters have little or no control. Once the measurement has started, there is no opportunity to stop for calibration, adjustment or repair. In contrast, the model tank affords nearly unlimited opportunity for periodic recalibration. In the case of a computer controlled apparatus, the recalibration can be frequent and automatic. The author has found that in the course of extended experiments on wind-driven water surfaces, it is advantageous to stop the wind, allow the water to become smooth, and recalibrate at regular 1 to 2 hour intervals. The recalibration can be as simple as trimming the water level in the tank or may involve a complete remeasurement of the composite transfer function of the data acquisition channel or channels. The periodic measurement of systematic errors not only reduces the possibility of loss of data due to undetected equipment failure, but also provides an opportunity to detect and compensate for drifts with time constants shorter than the time required to acquire adequate data.

A further advantage of the small scale model is the fortunate fact that the combination of high frequencies and short distances makes it possible to perform individual acoustic measurements in a time much shorter than that characterizing the motion of the target surface. Ocean measurements of time varying scattering and reverberation processes are often difficult to analyze because the rate of time variation exceeds the reciprocal of the duration of scattered signals. In the model tank the geometries and test signal durations can be arranged so that the entire interaction of a signal with the surface takes place before the surface changes. This situation makes it possible to characterize the scattering process by a time varying transfer function and to measure it conveniently.

## 4.5 Results of Physical Model Experimentation

In the foregoing sections a number of aspects of underwater acoustic physical scale modeling have been discussed. These techniques are still very much developmental and these discussions should be seen as an interim report on the history and present state of the art. In this section several recent results of model experimentation will be discussed. This discussion will be aimed at clarification of the role physical modeling plays in underwater acoustics and at the strengths and weaknesses of various techniques.

### 4.5.1 Scattered Intensity

Perhaps the simplest and in many ways most useful statistic of the sound field scattered from a random surface is the mean intensity. A great amount of work has been done both on fixed and wind-driven surfaces which has yielded a large base of data. The experiments at the University of Texas [4.7,8,11,13] thoroughly studied the scattered intensity from sinusoidal and random rough surfaces in both specular and backscatter directions. These studies used the Canadian Shield surfaces described in Sect.4.3.2 with acoustic signals in the frequency range of approximately 100 to 200 kHz. The results of these measurements in general supported the mathematical models proposed to explain them, but exhibited in nearly every case a large fluctuation probably due to statistical insufficiency. This made it difficult to use the results to sort out fine points in the mathematical models, but was otherwise a minor drawback. A more major problem was the difficulty encountered in characterizing the surface height probability density functions. The surfaces were of such small extent that histograms (Fig.4.2 for example) would fit a variety of distributions nearly equally well. This is a problem that seems to be characteristic of fixed random surfaces and is probably one of the reasons that more work in this area has not been attempted.

D'ANTONIO and HILL [4.14], MEDWIN [4.15], and others have investigated the intensity of sound scattered from wind-driven model tank surfaces. These studies have concentrated on scattered intensity in the forward or specular direction and have produced varied results. More recently ZORNIG [4.25] measured scattering strength at a variety of bistatic azimuths, grazing angles, and surface roughnesses at short wavelengths. This study was made using the computer based data acquisition system described in Sect.4.4. The test signals were cosine modulated sinusoidal pulses at 1.3 MHz. Because the goal of the experiment was empirical characterization and not the confirmation of a specific theoretical model, great care was taken to compute the exact level of insonification of the surface as a function of horizontal location for each bistatic orientation (Fig.4.10). The insonification function was then used to correct the resulting data to the normal reference area, one square

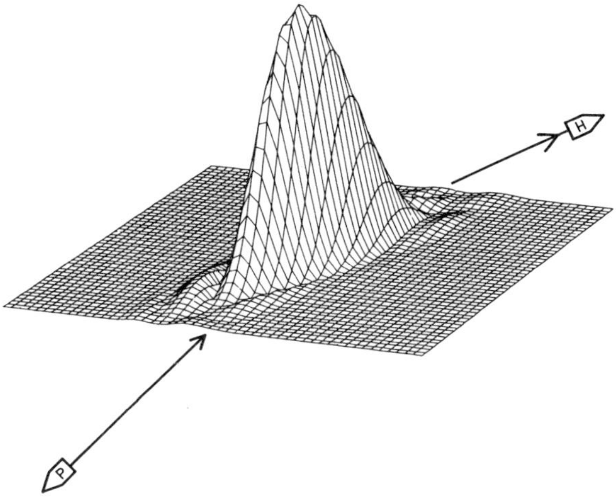


Fig. 4.10. The computed level of effective insonification of the surface over an area 2 $m^2$ due to projector and hydrophone beam patterns and time gating at differential azimuth = 150° and grazing angles = 12°(differential azimuth is the difference between the horizontal components of projector and receiver orientations; grazing angle is the vertical component measured from the mean surface plane). The precision for which such experimental parameters can be accounted significantly enhances the precision of physical model studies

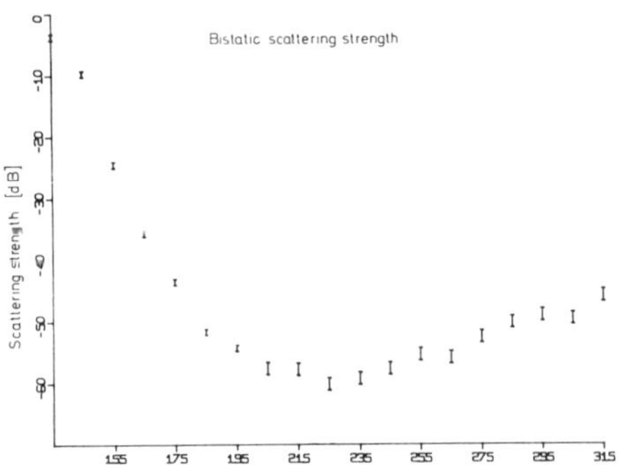


Fig. 4.11. Measured values of surface scattering strength [4.25] for grazing angle 17° and projector azimuth 315° with respect to wind direction. The measurements were taken at 1.3 MHz at an RMS surface roughness of 2.6 mm. These measurements represent data for which no complete analytical model or ocean measurements were available

yard (0.836 $m^2$) uniformly insonified, and receiver range, one yard (0.914 m). An example of a scattering strength measurement made in this way is given as Fig.4.11. Unfortunately, the study did not resolve unequivocally the question of the range dependence of scattering strength and was limited to very short acoustic wavelengths. However, it is a good example of the way in which empirical characterizations of high precision are possible using physical models.

### 4.5.2 Coherence Studies

The study of coherence in scattering from rough surfaces is almost made to order for physical modeling. Coherence in this context means the ensemble average of scattered acoustic waves measured at a distance for a single insonifying wave. If $p_0(\underline{r},t)$ is the pressure incident on a randomly rough surface and $p_1(\underline{r}',t)$ is a random process representing the scattered field, then the ensemble average $\langle p_1(\underline{r}',t)\rangle$ can be referred to as the coherent component of the scattered field. Coherence has also been measured as the average transfer function or average impulse response of the scatter channel defined by a source and receiver pair, the specular component of the scattered field, and the cross correlation between incident and scattered fields. All of these statistics are more or less equivalent and have a common requirement for measurement, a time reference. It is the ability of the physical model to maintain an absolute temporal relationship between transmitted and received signals and to eliminate inhomogeneities in the acoustic medium that has made possible the precise and detailed studies of this effect.

BOYD and DEAVENPORT [4.13] studied the specular coherence at 95.8 kHz of sound scattered from the University of Texas solid surfaces as a function of grazing angle. These measurements were consistent with the notion that scattered field coherence was a function of geometry, acoustic wavelength, and the PDF of surface heights. Unfortunately, as was mentioned above, statistical fluctuations in the data and difficulty in establishing the PDF from measured histograms made detailed analysis difficult. CLAY et al. [4.18] at about the same time measured the coherent component of sound scattered from a wind-driven surface of known PDF (Fig.4.1). An analog correlator was used to measure the incident-reflected covariance and the result was normalized by measuring the same statistic for an undisturbed surface. The results of these measurements are shown in Fig.2.24. In both studies, substantial differences were observed between the theoretical predictions and measured data. However, in the latter case the shape the characteristic function of the surface PDF predicted a dip in coherence which was present to a limited extent in the measured coherence and had been observed in the solid surface data [4.11] (Fig.4.6).

During this period, SPINDEL and SCHULTHEISS [4.21] performed a set of experiments using impulsive signals and a wind-driven surface designed to measure the average impulse response of the surface. These experiments used a novel technique of slew sampling to permit high speed data acquisition. Following this line of development ZORNIG and McDONALD [4.22] made measurements of the same sort using preemphasized pulses and on-line data analysis. A typical measurement from this series is shown in Fig.4.12. Again the results failed to show complete agreement between theory and measurement, but increasing precision and the agreement of a number of independent experiments indicated that a modification of theory for short wavelengths and low grazing angles was in order. Section 2.4 discusses some theoretical approaches which may resolve this disagreement.

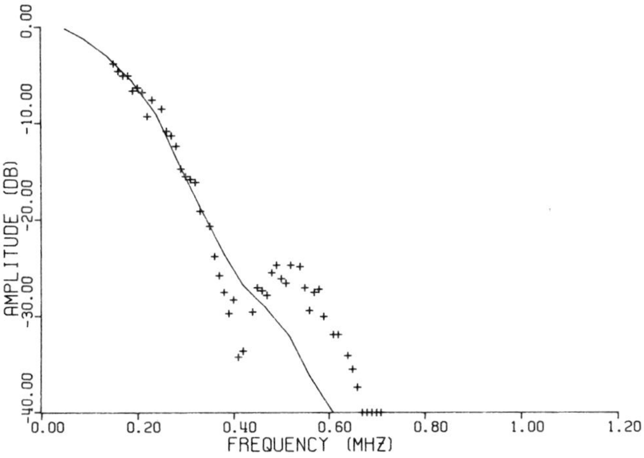


Fig. 4.12. The amplitude component (++++) of a surface scatter average transfer function, the Fourier transform of the average temporal impulse response, measured by broadband pulse probing [4.20]. RMS roughness was 2.5 mm and grazing angle was 16°. A scaled version of the characteristic function of the surface height PDF is plotted for comparison (——)

## 4.5.3 Second-Order Coherence

As a result of experience gained in measurement of the frequency dependence of coherence, and the development of a mathematical prediction of the cross correlation between scattered frequencies [4.52], a series of experiments using the broadband pulse technique were designed to measure the interfrequency correlation of the forward scattered field [4.23]. This function, which is of interest in the design of communication systems is defined as

$$\Phi(\omega,\omega') = \langle H(\omega,t)H^*(\omega',t)\rangle \quad , \tag{4.1}$$

where

$$H(\omega,t) = \int_{-\infty}^{\infty} h(\tau,t)\, \exp(-j\omega\tau)d\tau \quad , \tag{4.2}$$

and $h(\tau,t)$ is the signal scattered from the surface measured at time $t+\tau$ due to a temporal impulse arriving at time t. The average in (4.1) is taken over a statistically sufficient ensemble of incident impulses assumed to be ergodically equivalent to a time average in t. The actual measurements were made using approximate impulses and the results were valid over a frequency range of approximately 3 octaves. The predictive theory in this case was based on the Helmholtz-Kirchoff integral and resulted in an extremely complicated solution. In spite of the complexity of the formulation and limited extent to which the water surface conformed to the assumptions of the theory, extremely good agreement was observed (Figs.4.13 and 4.14). It is

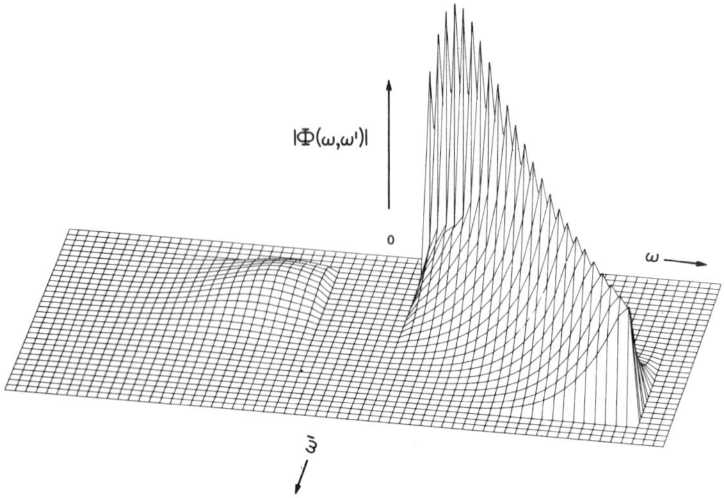


Fig. 4.13

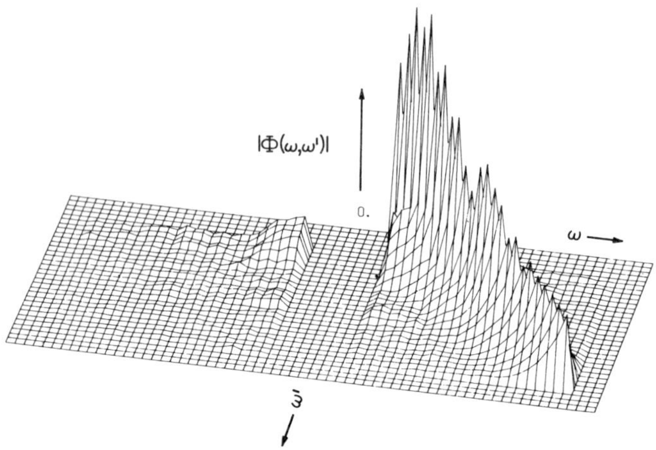


Fig. 4.14

Fig. 4.13. Theoretical prediction of interfrequency correlation for a surface roughness of 1.9 mm, grazing angle 16°, and frequencies up to 1.4 MHz. The mathematical formulation was based on the Helmholtz integral [4.45]

Fig. 4.14. Interfrequency correlation measured by the author [4.23] in a model tank for the same parameters as Fig.4.13

interesting to note that extremely large ensembles of data had to be collected for this experiment. Continuous runs of 6 to 8 hours were the rule during which the apparatus required little or no adjustment except for trimming water height to compensate for evaporation.

In all of the work which has been discussed so far, no information about the statistical form of the scattered wavefront had been obtained. The spatial coherence of scattered sound is, however, a property which is useful in the design of sonar systems and in the prediction of array performance. MEDWIN and CLAY [4.16] measured, among other statistics, the spatial correlation in the vertical direction of sound scattered from a wind-driven surface of very nearly Gaussian PDF and known spatial correlation function. The theoretical formulation was a unique accommodation of a divergent or Fresnel incident wavefront by solving the Helmholtz integral in the Frauenhofer approximation over an ensemble of small areas. Acoustic frequencies of 60, 100, and 160 kHz were used at a grazing angle of $35^{\circ}$. Hydrophone separations of up to 40 cm were used and the results were compared to a scaled version of the surface spatial correlation. Some features of the surface correlation could be seen in the measured data which seemed to be self-consistent. The theory, however, did not predict spatial cross covariance as well as might have been hoped.

Taking an approach similar to that used for interfrequency correlation, McDONALD et al. [4.24] predicted the joint spatial-frequency correlation function for forward scatter. This approach characterized the surface in terms of the same statistics as had MEDWIN and CLAY, the one-dimensional PDF of waveheights and the spatial correlation functions. However, as with the previous model the Fresnel term was included explicitly in the integral. Performing two-hydrophone experiments on an apparatus with a single analog-to-digital converter required that pairs of probing pulses be used as the test signal. The pulses were transmitted in rapid succession and the received signals were digitized in succession. The entire procedure required only a few milliseconds during which period the surface was assumed not to have changed. This assumption was in turn consistent with measurements of the temporal correlation function of the surface. The results were consistent with those of MEDWIN and CLAY in that correlation between receivers decreased with receiver spacing and frequency. The hydrophones were spaced horizontally, however, and no attempt was made to relate acoustic correlation to surface height correlation in any simple way. The results of the experiments compared favorably to a generalized form of the mathematical model of [4.52] (Figs.4.15 and 4.16) and again seem to justify at least some measure of confidence both in the Fresnel corrected Helmholtz-Kirchoff integral and in broadband pulse measurements.

### 4.5.4 Frequency Spreading

In recent years considerable ocean experimentation has centered around measurement of frequency spreading due to scattering from the water surface. D'ANTONIO and HILL [4.14] commented in the discussion of their early measurements of amplitude and phase distributions of sound scattered from a wind-driven surface that these fluctuations gave rise to frequency spreading. They correctly characterized the process as a combination of wide band phase modulation and amplitude modulation. Their data demonstrated that the bandwidth of the power spectrum of the scattered pressure exceeded the surface height temporal spectrum as would be expected for a high modulation index. RODERICK and CRON [4.19] subsequently conducted a set of experiments with a mechanically excited gravity wave surface having a nearly sinusoidal form to

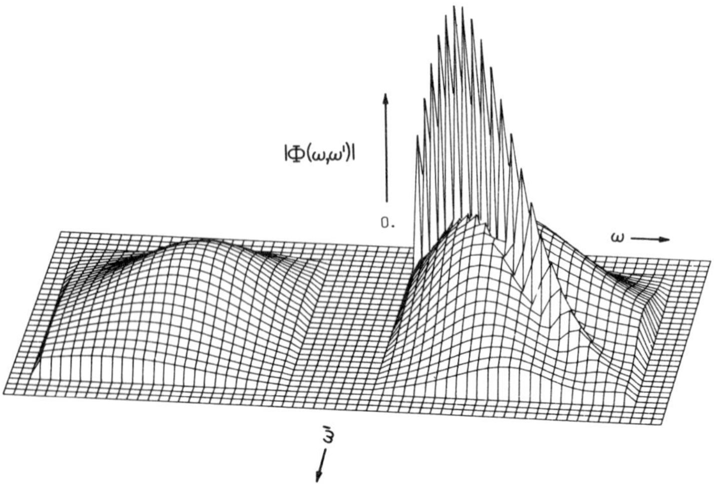


Fig. 4.15

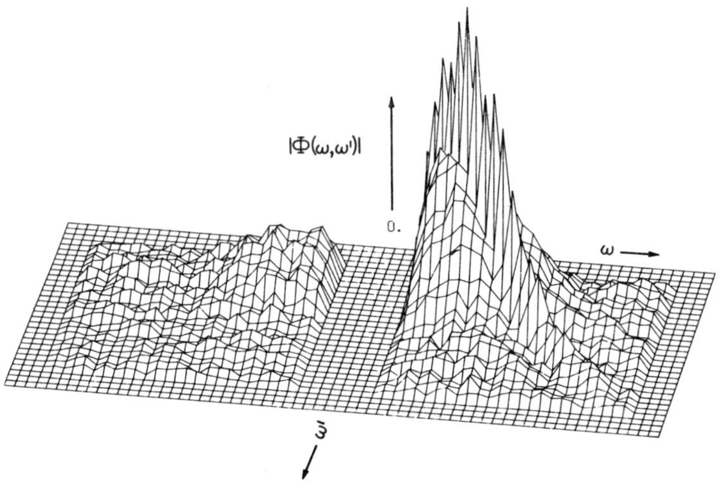


Fig. 4.16

Fig. 4.15. Theoretical prediction of the space-frequency correlation of scattered signals computed by McDONALD et al. [4.22] for a surface roughness of 1.1 mm, grazing angle of 16°, and receiver separation of 6.4 cm

Fig. 4.16. Space-frequency correlation measured by ZORNIG and co-workers [4.24] for the parameters of Fig.4.15

test a mathematical model which predicted higher-order sidebands of the surface temporal frequency. These experiments used cw signals and a spectrum analyser and produced very convincing evidence consistent with their theory.

However, a shortage of precise data and a lack of agreement among several theoretical approaches [4.53-56] left the distinct impression that more needed to be done. The author, therefore, conducted a series of physical model experiments using an adaptation of the pulse probing method to the measurement of frequency spreading

[4.26]. In general, measurements of the frequency spread of a slowly varying process require extremely long observation times with respect to the incident acoustic frequency. In simple cw experiments multiple reflections can cause intolerable interference. In the case of a scattering experiment in which the time and frequency spreading are small with respect to the incident signal duration and frequency, however, the slow variation of the process can be sampled and the frequency spread can be computed from the resulting data. The basic experiment consisted of a rapid series of measurements of the response of the surface to the short preemphasized pulse. The pulses themselves were much shorter than any variation of the surface as was pointed out earlier. The sequence of pulses was transmitted at a rate in excess of the Nyquist sampling rate of the scattering process, that is at least twice as often as the maximum spreading frequency. The actual pulse rate is determined by trial and error. In this way a sequence of responses is obtained that are a complete record of the time variation of the scattered field. The frequency spread as a function of carrier frequency was obtained by a double transform

$$\Gamma(\omega_n,\nu_k) = \left\langle \Delta t \Delta\tau \sum_{m=0}^{M-1} \sum_{i=0}^{I-1} s(\tau_i,t_m) \cdot \exp(-j\omega_n i\Delta\tau - j\nu_k m\Delta t) \right\rangle \quad , \tag{4.3}$$

where $s(\tau_i,t_m)$ is the sequence of pulse responses, $\Delta\tau$ is the sampling interval and $\tau_i$ are the sampling instants for each pulse response, $\Delta t$ is the interval between incident pulses and $t_i$ are the incident pulse arrival times. $\omega_n$ and $\nu_k$ are then effectively the incident carrier and Doppler-shift frequencies, respectively. In practice $\Gamma(\omega_n,\nu_k)$ was windowed to minimize the truncation effect of the finite record of pulse responses. Figure 4.17 gives examples of spreading measured using this technique.

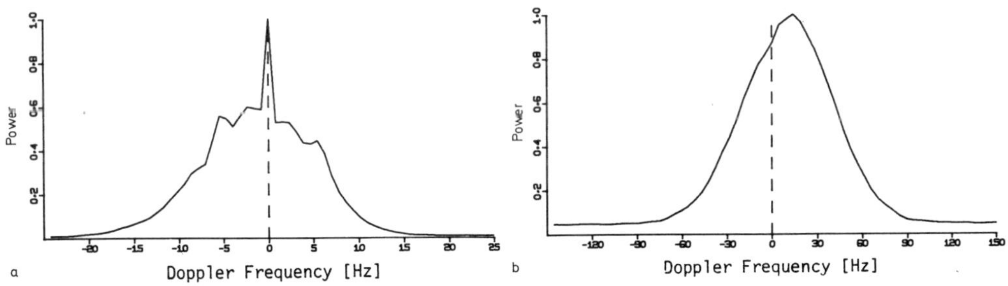


Fig. 4.17. Frequency spreading functions measured using sequences of broadband pulses at moderate (a) and high (b) values of effective surface roughness [4.26]

## 4.6 Remarks

As can be seen from this and the other chapters, there is much to be done. Hydroacoustic propagation is a wonderfully complicated phenomenon which is far from perfectly understood. A great many analytical models which attempt to explain a variety of more or less idealized cases are available. Large and complicated numerical models which seek to provide more global predictions are being developed at several laboratories. Vast amounts of evidence is collected daily in the oceans which helps to shed light on the mystery. Still, there remain much uncertainty and contradiction among researchers and designers of underwater acoustic systems. In order to sort out the growing mass of model and observation, we require among other things conclusive empirical evidence of the correctness and accuracy of our explanations. During the past two decades these explanations have advanced to a level of detail and sophistication that is difficult to evaluate using ocean data alone. The reasons for this are more or less self-evident having to do with separation of effects and precision of measurement. They have been discussed at great length. Fortunately, as has been seen in the overview presented in this chapter, the careful selection and execution of physical model experiments has in many cases been able to provide empirical evidence of such high precision and repeatability that evaluations of theory can be based upon it. The experimental methods and techniques have advanced considerably in recent years and a larger number and variety of results are available. However, there remains much to be done.

Perhaps the most pressing need in future work is a more comprehensive description of the statistical description of random surfaces used in experiments. Modern theoretical approaches are becoming more demanding. They require not only simple first-order statistical characterizations, but also time-space correlations, two-dimensional PDF's, slope distributions, and even slope-waveheight correlations. The measurement of these statistics in the physical model is feasible but is difficult and time consuming. Unfortunately, little ocean data of this kind is available which complicates the problem of either solid or time-varying surface design. On the other hand, interest is increasing and there seems to be reason to expect substantial progress in the coming years.

It seems likely also that physical scale modeling will be extended to underwater acoustic phenomena other than scattering and reverberation. Recently, a technique [4.57] has been developed which attempts to model long-range propagation in refractive media by a laser optical method. The technique itself is still in an early stage of development, but it is illustrative of the sorts of approaches to physical modeling which are becoming possible using contemporary technology. With the increase in speed, precision, and information handling capacity of modern instrumentation it is likely that it will become possible to study in miniature an increasing variety of phenomena which now perplex us. Doing so will not only make it possible to refine other techniques of analysis, but may ultimately give us a better chance to figure out *what is going on in the ocean.*

References


4.1 E.O. LaCasce, P. Tamarkin: J. Appl. Phys. *27*, 138-148 (1956)
4.2 J.M. Proud Jr., P. Tamarkin, W.C. Meecham: J. App. Phys. *28*, 1298-1301 (1957)
4.3 J.M. Proud Jr., R.T. Beyer, P. Tamarkin: J. App. Phys. *31*, 543-552 (1960)
4.4 A.N. Leporskii: Sov. Phys.-Acoust. *1*, 50-59 (1955)
4.5 A.N. Leporskii: Sov. Phys.-Acoust. *2*, 185-189 (1956)
4.6 G.R. Barnard, J.L. Bardin, W.B. Hempkins: J. Acoust. Soc. Am. *36*, 2119-2123 (1964)
4.7 G.R. Barnard, C.W. Horton, M.K. Miller, F.R. Spitznogle: J. Acoust. Soc. Am. *39*, 1162-1169 (1966)
4.8 C.W. Horton, S.K. Mitchell, G.R. Barnard: J. Acoust. Soc. Am. *41*, 635-643 (1967)
4.9 D.R. Melton, C.W. Horton: J. Acoust. Soc. Am. *47*, 290-298 (1970)
4.10 C.W. Horton, D.R. Melton: J. Acoust. Soc. Am. *47*, 299-303 (1970)
4.11 P.J. Welton, H.G. Frey, P. Moore: J. Acoust. Soc. Am. *52*, 1553-1563 (1972)
4.12 P.J. Welton, K.E. Hawker, H.G. Frey: J. Acoust. Soc. Am. *54*, 446-450 (1973)
4.13 M.L. Boyd, R.L. Deavenport: J. Acoust. Soc. Am. *53*, 791-801 (1973)
4.14 R.A. D'Antonio, R.F. Hill: J. Acoust. Soc. Am. *38*, 701-706 (1965)
4.15 H. Medwin: J. Acoust. Soc. Am. *41*, 1485-1495 (1967)
4.16 H. Medwin, C.S. Clay: J. Acoust. Soc. Am. *47*, 1419-1429 (1970)
4.17 H. Medwin, R.A. Helbig, J.D. Hagy Jr.: J. Acoust. Soc. Am. *54*, 99-109 (1973)
4.18 C.S. Clay, H. Medwin, W.M. Wright: J. Acoust. Soc. Am. *53*, 1677-1682 (1973)
4.19 W.I. Roderick, B.F. Cron: J. Acoust. Soc. Am. *48*, 759-766 (1970)
4.20 F.J. Kingsbury: "An Experimental Model Study of Underwater Acoustic Scattering from a Wind-Driven Surface", PhD. Dissertation, Yale University, New Haven, CT (1970)
4.21 R.C. Spindel, P.M. Schultheiss: J. Acoust. Soc. Am. *51*, 1812-1824 (1972)
4.22 J.G. Zornig, J.F. McDonald: J. Acoust. Soc. Am. *55*, 1205-1211 (1974)
4.23 J.G. Zornig, J.F. McDonald: IEEE Trans., COM-*23*, 341-347 (1975)
4.24 J.F. McDonald, F.B. Tuteur, J.G. Zornig: J. Acoust. Soc. Am. *59*, 1284-1293 (1976)
4.25 J.G. Zornig: J. Acoust. Soc. Am. *63*, 758-767 (1978)
4.26 J.G. Zornig: J. Acoust. Soc. Am. (to be published 1978)
4.27 T.D. Plemons, J.A. Shooter, D. Middleton: J. Acoust. Soc. Am. *52*, 1487-1502 (1972)
4.28 T.D. Plemons, J.A. Shooter, D. Middleton: J. Acoust. Soc. Am. *52*, 1503-1515 (1972)
4.29 T.D. Plemons: J. Acoust. Soc. Am. *56*, 1422-1425 (1974)
4.30 R.S. Andrews, L.F. Turner: J. Acoust. Soc. Am. *58*, 331-335 (1975)
4.31 K.J. Diercks, T.G. Goldsberry, C.W. Horton: J. Acoust. Soc. Am. *35*, 59-64 (1963)
4.32 T.G. Goldsberry: J. Acoust. Soc. Am. *42*, 1298-1305 (1967)
4.33 R.E. Bunney, R.R. Goodman, S.W. Marshall: J. Acoust. Soc. Am. *46*, 1223-1233 (1969)
4.34 S.W. Marshall, T.G. Olson: J. Acoust. Soc. Am. *47*, 949-950 (1969)
4.35 W.G. Newbauer, L.R. Dragonette: J. Acoust. Soc. Am. *48*, 1135-1149 (1970)
4.36 C.W. Horton, M.V. Melcher: J. Acoust. Soc. Am. *51*, 295-303 (1972)
4.37 I.D. Groves Jr., V.P. Benedetti: J. Acoust. Soc. Am. *54*, 717-729 (1973)
4.38 G.R. Barnard, J.L. Bardin, J.W. Whiteley: J. Acoust. Soc. Am. *57*, 577-584 (1975)
4.39 R.H. Wallace, H.V. Hillery, G.R. Barnard, B.M. Marks, C.M. McKinney: J. Acoust. Soc. Am. *57*, 862-869 (1975)
4.40 J.C. Bertrand, J.W. Young: J. Acoust. Soc. Am. *60*, 1265-1269 (1976)
4.41 L.A. Mole, J.L. Hunter, J.M. Davenport: J. Acoust. Soc. Am. *52*, 837-842 (1970)
4.42 I.S. Kol'tsova, I.G. Mikhailov: Sov. Phys.-Acoust. *15*, 453-455 (1969)
4.43 S.L. Campanella, A.G. Favret: J. Acoust. Soc. Am. *46*, 1234-1245 (1969)

4.44 G.A. Sandness: "Experiment and Numerical Evaluation of Acoustic Scattering at a Rough Surface", University of Wisconsin, Geo. and Polar Res. Center, Res. Rpt. 73-2 (1973)
4.45 G.R. Garrison, S.R. Murphy, D.S. Potter: J. Acoust. Soc. Am. *32*, 104-111 (1960)
4.46 B. Kinsman: *Wind Waves* (Prentice Hall, Englewood Cliffs, NJ 1965)
4.47 R.C. Spindel, P.M. Schultheiss: J. Acoust. Soc. Am. *52*, 1065-1068 (1972)
4.48 A.H. Schooley: Trans. Am. Geophys. Union *36*, 273-278 (1955)
4.49 A.H. Schooley: Trans. Am. Geophys. Union *39*, 405-408 (1958)
4.50 R.C. Spindel: "An Experimental Investigation of a Statistically Distributed Scattering Surface and Acoustic Scattering", Ph.D. Thesis, Yale University, New Haven, CT (1971)
4.51 J.G. Zornig, J.F. McDonald: Rev. Sci. Inst. *44*, 1217-1222 (1973)
4.52 J.F. McDonald: IEEE Trans., COM-*22*, 138-145 (1974)
4.53 L.L. Scharf, R.L. Swarts: J. Acoust. Soc. Am. *55*, 247-253 (1974)
4.54 F.B. Tuteur, H. Tung: "Asymmetric Doppler Amplitudes in the Surface Scatter Channel", Techn. Rpt. CS-7, Dept. of Eng. and Appl. Sci., Yale University (1976)
4.55 F.M. Labianca, E.Y. Harper: J. Acoust. Soc. Am. *61*, 378-389 (1977)
4.56 D. Middleton: J. Acoust. Soc. Am. *61*, 1231-1250 (1977)
4.57 L.E. Estes, G. Fain: J. Acoust. Soc. Am. *62*, S17 (1977)

# 5. Oceanography in Underwater Acoustics

J. P. Dugan

With 15 Figures

The topic under discussion concerns that oceanography which presently appears to be relevant to underwater acoustics. The basic organization of the sections, after a discussion of the important physical parameters, follows the scales of variations that exist in the ocean. First, the oceanic "climate" that is represented by mean ocean basin scale conditions with its seasonal changes is presented. Then, the ocean "weather" that is represented by eddies and fronts, and dominates the fluctuations on scales of many tens of kilometers to several hundred kilometers, captures a good portion of our attention. Next, the subjects of internal waves and fine structure representing horizontal scales of tens of meters to order ten kilometers are presented. Finally, we include a short discussion of the scales of variability in the near-surface mixed layer.

As in any monograph, the author may exercise his prerogative and include only those subjects that interest him. Thus, the subjects that are included should be germane to the application, but some important phenomena have to be excluded. In this case, the subject receiving most of our attention is the interior of the ocean. Thus, we consider regions that largely are away from all boundaries — lateral as well as surface and bottom. In particular, important topics such as shallow water (or continental shelf) oceanography, the surface wind wave field, and the structure of the bottom are ignored.

Most of our attention will be focused upon the description of spatial and temporal distributions of the physical parameters, with only incomplete descriptions of the underlying processes that cause the distributions. Only passing consideration is given the subjects of instruments and methods and of mathematical models, and, in all cases, many details are left to references where more complete descriptions are available.

## 5.1 Properties of Seawater

The physical property of seawater of most concern in acoustic wave propagation obviously is the isentropic sound speed (Sect.2.1). This parameter can be measured directly with high accuracy and precision, but it can also be calculated from measurements of other physical parameters. The usual water property of interest to the physical oceanographer is density, since it is the density differences that contribute fundamentally to drive the motions in the ocean. Other properties of interest are temperature and salinity since these two, with local pressure, combine to give the density. Of course, all these properties are not independent, as there exist well-known empirical relations between several of them. For example, the frequently used Wilson's equation [5.1] gives sound speed in terms of temperature, salinity, and pressure.

Temperature is the most easily measured variable so it is a prime candidate in any measurement system. Also, as we will discuss later on, it is the temperature and the salinity that are fixed by air-sea interaction processes in the vicinity where the basic water masses in the ocean are generated, so that these properties are independently of interest, in addition to their particular combination that yields density. Thus, physical oceanographers actually are interested in these properties, and sound speed is a secondary quantity obtained from them.

However, there is a problem. The salinity, roughly defined as the weight of salts in grams per kilogram of water (parts per thousand or per mil and designated $^{o}/oo$), is not measured directly. A popular measurement for many decades has been to take advantage of the fact that the constituent salts for all practical purposes are fixed in their relative amounts. Thus, an accurate titration method that actually measures chlorinity (the amount of chloride ion) has been used. A simple relation exists between the chlorinity and salinity to yield the latter. More recently, another property, the electrical conductivity, has been used to determine salinity. This is more precise than the titration method [5.2], although practically by definition titration still is the standard. More important than the high precision is the fact that the measurement of conductivity can be done over the side by profiling instruments called CTDs (conductivity-temperature-depth) instead of bringing a water sample aboard for titration. The acronym CTD is a misnomer since the last measurement actually is pressure and not depth. This method yields nearly continuous profiles of the measured parameters and, by empirical relations, other profiles of concern.

From the point of view of the acoustician, this is rather round about and not particularly accurate, especially because: 1) sound speed can be measured directly by a profiler, and 2) salinity often is not measured but is inferred from historical data. It even has been suggested seriously [5.3] that a temperature and sound speed measurement might be a good way to obtain the salinity.

The sensitivity of sound speed on these other parameters roughly is 4.6 $m\,s^{-1}\,{}^{o}C^{-1}$, 1.3 $m\,s^{-1}\,{}^{o}/oo^{-1}$, and 0.02 $m\,s^{-1}$ $dbar^{-1}$ pressure (one dbar pressure is approximately equivalent to one meter depth). Since ocean water usually varies by only several parts per thousand in salinity but by tens of degrees Celsius in temperature, a frequent approximation for acoustics that is fairly accurate in all but polar and sub-

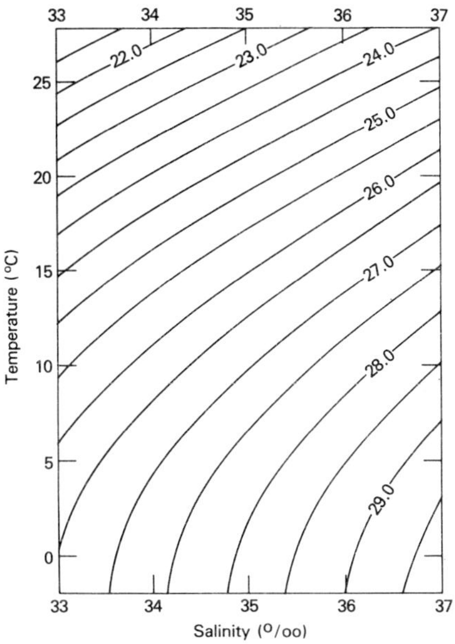


Fig. 5.1. $\sigma_t$ as a function of temperature and salinity. $\sigma_t$ is defined by (5.1)

polar waters is to assume that the salinity is constant, effectively removing this parameter from further consideration. However, salinity variations often have subtle, and in the several geographic regions mentioned above, have very important implications in the sound-speed profile. In the following, the point of view will be that of the physical oceanographer—salinity is a basic parameter and sound speed is a derived parameter.

Figure 5.1 shows curves of constant values of density on a uniform grid of the relevant ranges of temperature and salinity in the open ocean. The quantity $\sigma$ is defined by

$$\sigma = (\rho - 1)10^3 \quad , \tag{5.1}$$

so that $\sigma = 27.30$, for example, is shorthand for a density of $\rho = 1.0273$ g/cc. Note the curvature of the lines of constant density, especially apparent at the lower temperatures, which indicates the nonlinear relation between the variables. It is this curvature at low temperatures, as well as the generally colder waters at shallow depths at high latitudes, that causes the relatively more important effect of salinity in the Arctic and Antarctic.

This nomogram is accurate for a pressure of one atmosphere, and the quantity $\sigma_t$ stands for the density of a parcel of water at its in situ temperature and salinity but at a pressure of one atmosphere. The pressure affects the density through the compressibility of water, and the difference between the in situ and the surface value of density for this effect alone can amount to more than 2% for deep water.

There is a further complication, as the adiabatic expansion of water as it is brought up from one depth to another affects the temperature. This effect is accounted for by including the temperature change in referencing the density calculation to different depths. The parameter $\sigma_\theta$ is the density, defined as in (5.1), computed as if the water parcel were lifted adiabatically to the sea surface. The quantity $\sigma_\theta$ is called the potential density, and $\theta$ denotes the potential temperature, which is that temperature a water parcel cools to when it is lifted adiabatically to one atmosphere pressure. Other levels also are useful for comparison, such as $\sigma_4$, which is the density computed as if the water parcel were moved to 4000 dbar pressure [5.4].

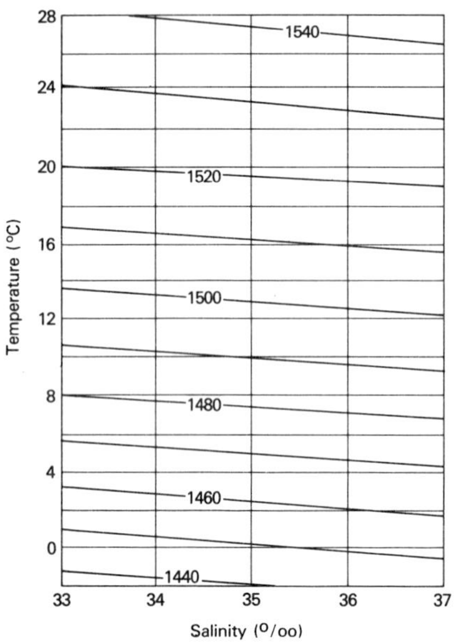


Fig. 5.2. Sound velocity as a function of temperature and salinity at the surface. Units are meters per second

Figure 5.2 shows curves of constant values of sound speed at atmospheric pressure on a uniform grid of temperature and salinity. Over the range of temperature and salinity of interest, the sound speed varies only in a relatively simple, almost linear manner in the parameters. In the density nomogram, $\sigma_t$ was plotted in order to avoid the pressure effect. That is, in a two-dimensional diagram, $\sigma_\theta$ could only have been plotted for one initial pressure. In the case of sound speed, no common nomenclature exists for such a three parameter dependence, so the absolute numbers are valid only for atmospheric pressure. However, the effect of pressure is relatively simple, with the sound speed increasing almost uniformly by about 0.018 $ms^{-1}$ per dbar increase in pressure.

## 5.2 Ocean Variability

Oceanic conditions can be broken down into mean and fluctuating components in several ways. This breakdown is not unique because, as is often quoted, one man's noise is another's signal. For the purpose of present day acoustics it appears most useful to use the distinction used by DESANTO, see (2.55), wherein there is assumed to be: 1) a mean vertical profile that represents local climatology, 2) a mesoscale component that is deterministic on the acoustic time scale, and 3) a statistical component that represents smaller scale fluctuations due to internal waves and fine structure that must be considered random in some sense on the acoustic time scale.

The mean vertical profile is representative of long-term mean conditions quite analogous to the climatology in the atmosphere. This structure is depth dependent, but it only has horizontal scales defined by the general circulation in the ocean basin of interest. It is also seasonally variable in the upper few hundred meters. In the deeper parts of the ocean, the variables when viewed in the appropriate parameter space have not varied measurably through this century of accurate measurements that are on record. This mean structure represents a long-term balance between locations and times where water of particular characteristics is formed and those where it is destroyed (Sect.5.2.1).

The mesoscale structure by definition is comprised of spatial scales from the order of tens of kilometers to several hundred kilometers in the horizontal, up to the ocean depth in the vertical, and temporal scales of days to months (Sect.5.2.2). The readily classified components of this class commonly are called eddies and fronts. These structures represent coherent departures from the climatology and are analogous to high and low pressure cells and fronts in the atmosphere. Thus, this component of ocean variability often is referred to as the ocean weather system, and eddies have been referred to as ocean storms. Generally, these features can be tracked for tens of days or even months at a time so that they can be considered to be deterministic perturbations (large ones at that) from the mean structure.

Smaller scale fluctuations (Sects.5.2.3 and 5.2.4) occur on time and space scales more comparable to the acoustic wave scales or observation scales of interest. That is, these fluctuations occur on time scales up to many hours and space scales up to kilometers in the horizontal and tens of meters in the vertical. The present vogue is to represent these structures in some statistical manner. The components making up this class generally include internal/inertial waves, fine and microstructure, and turbulence.

Finally, a class of oceanic motions of interest in acoustics, but not as easily classified in terms of scales as the above components, is comprised of near-surface effects (Sect.5.2.5). This class is made up of the dynamics that cause variability in the structure of the near-surface mixed layer, most notably in the parameters commonly called mixed layer temperature and mixed layer depth. All the previously discussed phenomena affect the mixed layer, but the strong influence of wind stress and solar heating make this part of the ocean unique enough to warrant separate discussion.

### 5.2.1 Ocean Climatology

The largest scale of variability in the ocean is limited by the size of the basin, and this scale has commonly been called the general circulation by oceanographers. The structure of this general circulation, with its seasonal fluctuations, has more recently been termed the ocean climatology. The motions are determined by balances of heat input and efflux, wind stress forcing, and nonequilibrium with connecting basins. The sources of the great water masses largely are fixed by air-sea interactions that determine the water density through its temperature and salinity. Gravity acts on the density differences, and motions due to gravity or the wind stress are influenced by basin shape, bottom topography, and rotation of the earth.

The basic principles of water mass production are well understood, as the origins of most of the distinctive water types have been identified and studied in some detail. The monograph of SVERDRUP et al. [5.5] remains a comprehensive, readable, and popular reference on this topic as attested to by being in its 22$^{nd}$ printing. In most water mass generation regions, heat and mass fluxes across the air-sea interface at certain times of the year result in the formation of large quantities of water of rather repeatable temperature and salinity characteristics. This water acquires a different density than its surroundings, and the resulting gravitational instability forces it to move. For example, the bottom waters are comprised of water of the greatest density that is formed in high latitudes. The Antarctic bottom water, which covers the bottom in the southern oceans and intrudes to about 30$^\circ$N in one portion of the North Atlantic, is formed in the Weddell Sea during expansion in the ice cover in the southern winter. Because of its high density, this water sinks, spreads out northward, and mixes vertically with overlying water types. Initially it has a temperature near -1$^\circ$C and a salinity near 34.6 $^o/_{oo}$, and, as with all water types, it can be traced by this T-S combination (T-S stands for a particular temperature-salinity couple). The overlying water masses, each having its own source, basically slide along lines of constant density as they intermingle. As the water moves along, it mixes with the adjacent water types, and its characteristics diffuse to some mean of the components.

This relatively simple picture of the ocean interior is complicated by smaller scale effects (Sects. 5.2.2-4), by near-surface seasonal effects (Sect.5.2.5), but also by other components of the general circulation. The vector curl of the mean wind stress over large areas of a basin subject the surface waters to lateral motions that in the end force the great boundary currents like the Gulf Stream and others like the Antarctic Circumpolar Current. The boundary currents are not locally in equilibrium with the wind stress since they are the result of larger scale forcing.

Instead, they are balanced by lateral density gradients in accord with the geostrophic relation. This balance is between the lowest-order terms in the hydrodynamic equations, the horizontal pressure gradient and the Coriolis acceleration, and it causes significant lateral gradients in the water density and other parameters in the region of the current.

The depth dependence of a single parameter like temperature changes drastically across a boundary current, with the depths of some isotherms regularly changing by 500 m or more. In terms of water mass types, though, the depth of a parcel of water is not as important as its T-S characteristics. A plot of temperature versus salinity at an ocean station, called a T-S diagram, often is more illuminating than a plot of either parameter versus depth at that station. If the ocean had two distinct water types, one overlying the other, the T-S diagram would consist of two points. Since there would be a mixing zone between the two types, there actually would be a straight line connecting the two points. All water parcels must lie along this line, and the position of a particular parcel along the line is determined uniquely by the percentage of the two original components in the parcel.

Figure 5.3 is an example of the power of the T-S diagram in ordering the variability in historical upper ocean data. The data presented here represents all available historical bottle samples in a small area of the subtropical Northwest Pacific for the months of October and November. The data was accumulated over many years from research vessels and ocean weather station vessels that took bottle stations; that is, a string of a finite number of water samples from the surface downward. The temperature-salinity-depth points are represented several ways. The temperature-depth and salinity-depth plots show the scatter in temperature and salinity that exists in the region. Some of the scatter is due to spatial variability, but since many years of data are represented, much of the scatter near the surface is due to temporal variability of the atmospheric weather from year to year. If a scatter plot were made of the sound velocity, it would vary by as much as 30 $ms^{-1}$ even at 500 m depth, mostly due to the temperature variability.

The third plot in Fig.5.3 shows the T-S scatter diagram for the same data, and it obviously orders the data considerably. The decrease in scatter only occurs for temperatures less than about 14°C and salinities less than about 34.6 ‰, and a line has been drawn through the T-S data at this point. A second line is drawn through the T-S data at 4°C, which roughly corresponds to the temperature at 800 m depth in the region. Returning to the temperature-depth plot, the 14°C line is drawn at the appropriate point, and it separates the depth regimes in which the T-S information is or is not useful in ordering the scatter. The points to the right probably do not correlated well with a tight T-S relation, while points to the left do. Similarly, the salinity-depth plot has a line near 34.6 ‰, and the scatter to the left and below this line correlates well with temperature variability through the T-S relation.

Figure 5.3 is a good illustration of the use of the T-S diagram to identify water types. The salinity minimum near 6°C and 34.0 ‰ occurs near 600 m depth in this

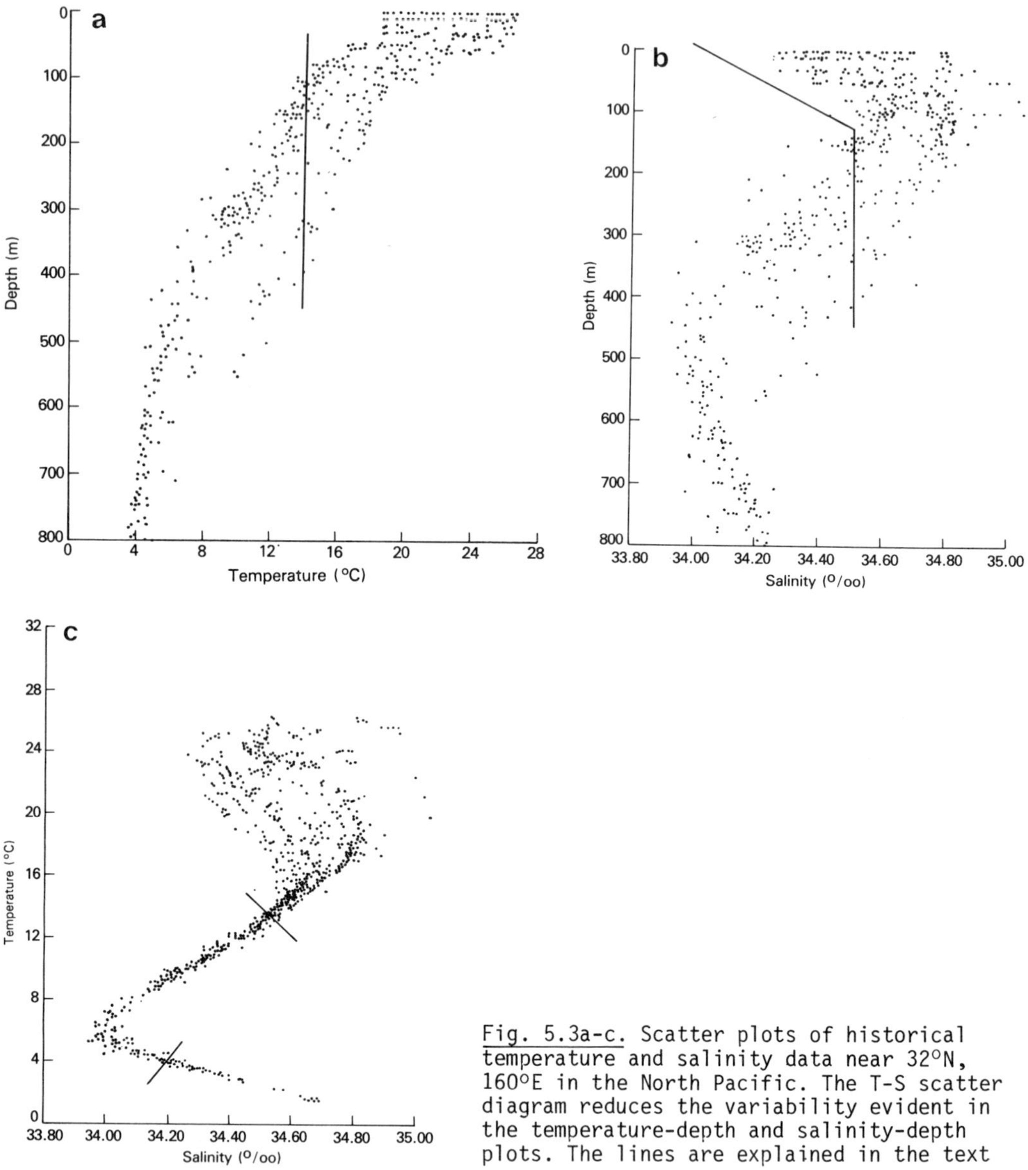


Fig. 5.3a-c. Scatter plots of historical temperature and salinity data near 32°N, 160°E in the North Pacific. The T-S scatter diagram reduces the variability evident in the temperature-depth and salinity-depth plots. The lines are explained in the text

region, and it is characteristic of North Pacific intermediate water. This relatively fresh water would appear to be a southward intrusion from north of the subarctic front between the lighter Western North Pacific central water and the heavier North Pacific deep water. The fact that straight lines fit the data very well above and below the salinity minimum indicates simple mixing of the three water types. The central water is formed mostly during the winter season southeast of Japan near 18°C and 34.9 ‰, and the deep water characteristically is near 1°C and 34.7 ‰.

There are several additional points worth mentioning before leaving the discussion of this particular figure. It can be concluded that in this region of the ocean the T-S relation can be used with temperature information for all depths below about 150 m to calculate the salinity associated with any particular temperature measurement with a certain degree of accuracy. Thus, a temperature measurement alone is adequate to estimate the sound velocity accurately. Also, this nice ordering of the data in the main thermocline restricts the number of physical processes that can occur and still cause the large variability in the individual parameters at a given depth. For example, the depth of any particular T-S point is not unique because certain mesoscale processes can move water parcels up or down several hundred meters while preserving their identity. This would cause the scatter in the temperature-depth and the salinity-depth data while preserving the tight relation in the T-S diagram.

Another interesting case of an apparent bulge on a T-S diagram is due to the Mediterranean outflow in the North Atlantic. Deep Mediterranean water is warmer and saltier than North Atlantic deep water because evaporation exceeds precipitation and runoff in the nearly enclosed Mediterranean Sea. Sea level is preserved by inflow of North Atlantic surface water, but, in addition, a considerable volume of warmer, salty Mediterranean water flows outward over the sill in the Strait of Gibraltar. This water, initially at about 13°C and 38.4 $^{0}/_{00}$, is mixed rapidly as it flows down the continental slope of the Iberian Peninsula, and out into the North Atlantic at a depth determined by its own density. By the time it flows out into deep water, its temperature is reduced to about 11°C and its salinity to about 36.0 $^{0}/_{00}$. Figure 5.4 gives T-S data for several individual stations across the North Atlantic

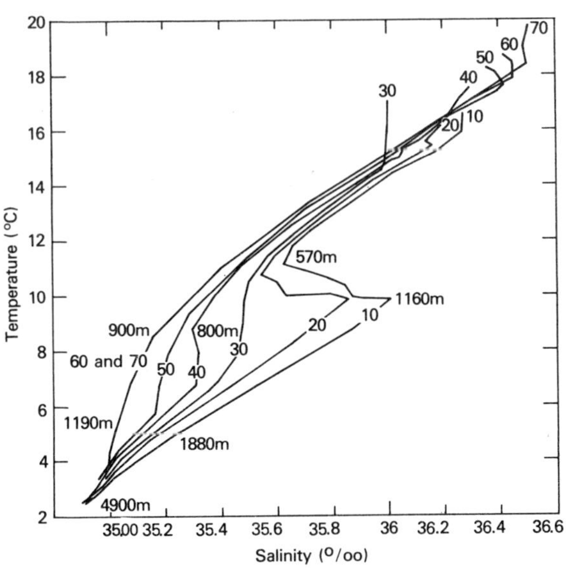


Fig. 5.4. T-S diagrams for stations across the North Atlantic near 36°N from 10°W to 70°W. Data obtained from Chain Cruise Number 7

near 36°N. There is one curve for each ten degrees of longitude from 10°W to 70°W. The topmost point on each line is the surface value, and straight lines have been used to connect the actual discrete points obtained in the bottle casts. Several depth values give an indication of the approximate depth range spanned by parts of the data.

Several interesting points can be gleaned from the diagram. The deep and bottom waters of Arctic origin are uniform across the basin. This water is very weakly stratified, as the T-S lines lie almost along constant density lines over several kilometers of depth (Fig.5.1 is misleading in this respect because the actual isopycnal lines at great depth are not as steep as those of $\sigma_t$). Also, the coincidence of T-S lines above 10°C is the North Atlantic central water which is characterized by nearly straight lines between 19°C, 36.7 °/oo and 8°C, 35.1 °/oo. The western end of this data set is in the formation region of this upper layer water, as the surface values lie close to the straight lines of the central water. The prominent bulge between 8°C and 11°C in the east is the influence of the Mediterranean outflow. This influence decreases with increased mixing with the upper and lower layers with increasing distance from Gibraltar. The bulge remains roughly along the $\sigma_t = 27.7$ isopycnal.

Figure 5.5 shows temperature-depth and sound velocity-depth plots of the same data as in Fig.5.4. The permanent thermocline separating the surface from the deep water occurs in the zone of 500 to 1500 m, and its structure changes markedly across the basin. The Mediterranean water causes the temperature inversions at depth in the 10°W and 20°W profiles, and, as mentioned previously, the effects get weaker in the west because of mixing. Also, a prominent feature in the temperature plot is the large amount of almost isothermal water above the permanent thermocline in the west. This is always present in the Sargasso Sea, having been formed in the winter season, and it commonly is referred to as "18°C Water". The corresponding sound velocity profiles are strongly affected by these prominent features. In the east, the Mediterranean water causes a local sound velocity maximum, so that the sound channel has two distinct minima. This feature gradually gives way to a single stronger sound velocity minimum in the west. However, the 18°C water causes a second acoustic duct in the top several hundred meters of water. This duct can reach a depth of 600 m in the Sargasso Sea northwest of Bermuda in a severe winter season.

### 5.2.2 Ocean Weather

Ocean fluctuations on spatial scales of fifty to five hundred kilometers generally are classified as being mesoscale variability, since this is between the large scale general circulation and the smaller scale represented by internal waves and the like. The time scale represented by this variability usually is the order of many days to several months. The features that cause the variability are fronts and eddies, and,

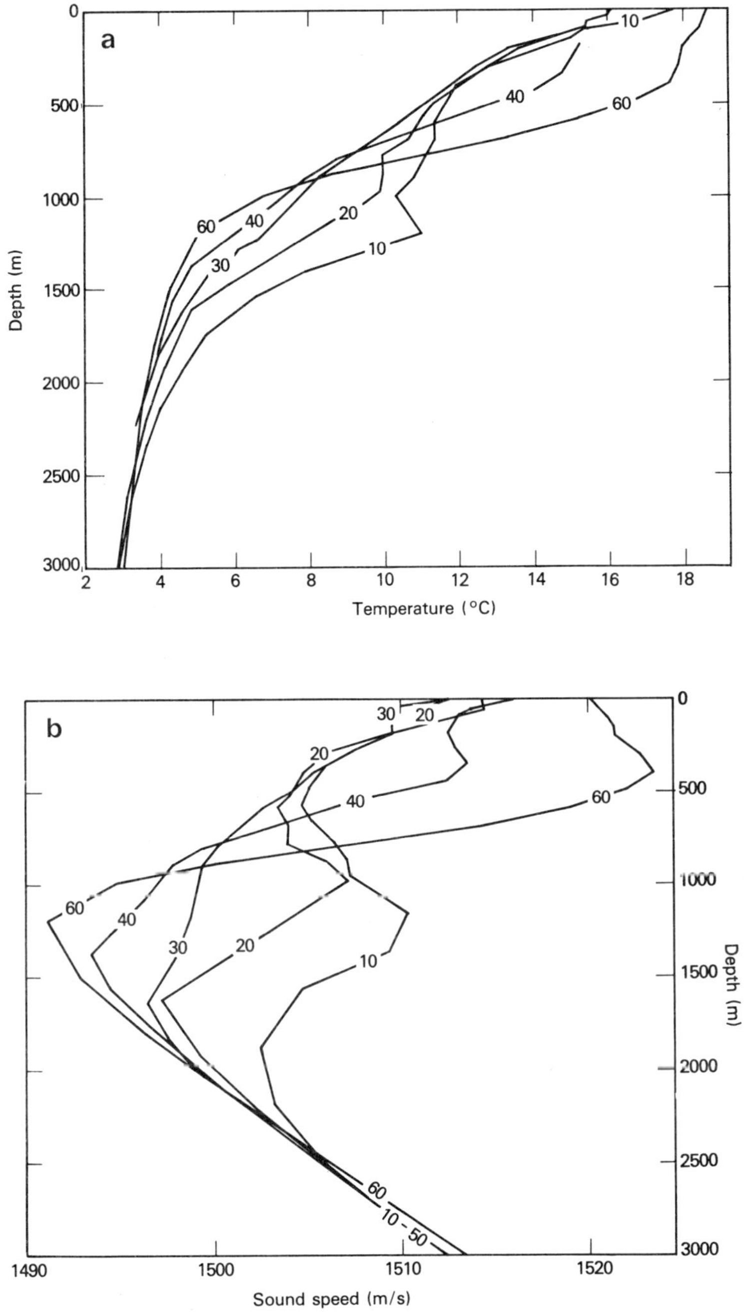


Fig. 5.5a and b. Temperature-depth and sound velocity-depth profiles for data shown in Fig.5.4. Numbers on plots indicate approximate longitude (°W)

because of analogies with atmospheric fronts and highs and lows, this scale of motions often has been referred to as the ocean weather system.

An oceanographic front is a zone separating two regions having different water characteristics. The characteristics that are different can be any one or more of the physical parameters previously discussed, so that the actual differences might or might not affect water density (this definition is different from that in meteorology). A density gradient has to be balanced by motions of some kind, while a constant density front can be practically stationary.

Figure 5.6 is a cross section of a much studied density front—the Gulf Stream north wall. As discussed in Sect.5.2.1, this density front is a necessary adjunct of the Gulf Stream as, to lowest order in the equations of motion, there is a geostrophic

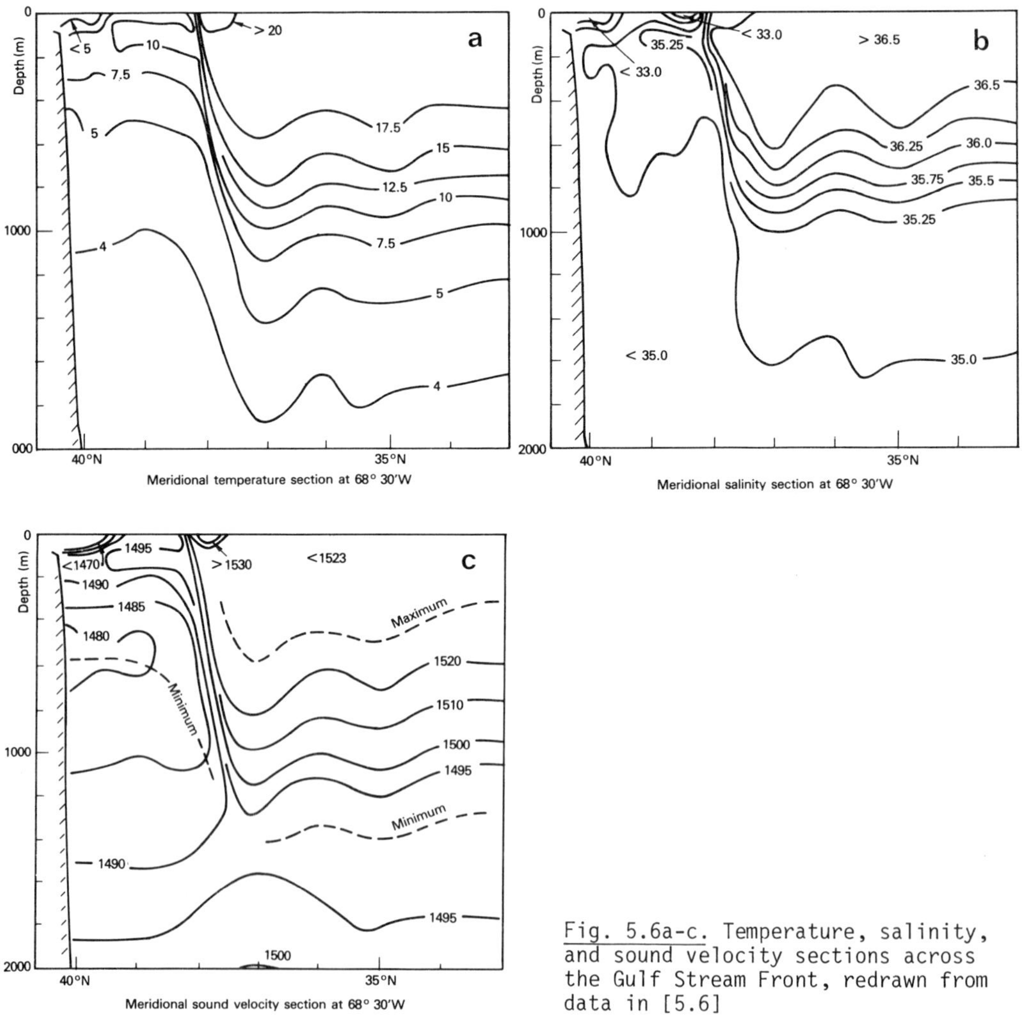


Fig. 5.6a-c. Temperature, salinity, and sound velocity sections across the Gulf Stream Front, redrawn from data in [5.6]

balance between the pressure gradient due to the density gradient and the fluid velocity. In fact, a good estimate of the water velocity can be obtained by measurement of the two-dimensional density structure across the front.

The three cross sections in Fig.5.6 illustrate several interesting aspects of some fronts. The data is redrawn from a north-south section across the Gulf Stream acquired in late winter, and reported by FUGLISTER [5.6]. The Gulf Stream is more like the jet stream than a front in the meteorological context, but it nevertheless serves as a useful example of an acoustic boundary zone between two distinctly different acoustic regimes.

The temperature section shows a mixed layer almost 400 m deep in the southern part. This often happens in this region in later winter. It is the same 18° water that is seen, for instance, in the temperature profile near 60°W in Fig.5.5. The front is in the region 37°N to 38°N, so all the interesting behavior occurs in a region less than 100 km wide. The Gulf Stream flows into the page, and, although there is significant current to depths of almost one kilometer, it is most conspicuous to surface observations by the region of warm water in the vicinity of the surface near 37.5°N. This water is warmer than its surroundings because it has been advected northward from its warmer origin. Also, as is usual, it is only slightly warmer than the Sargasso Sea water to its right, but it is considerably warmer than the slope water to the north. There is a temperature inversion just to the north of the surface front.

The salinity section is similar to the temperature section in gross structure, but it differs in several respects. The two sections look especially alike in the cold water below one kilometer depth and in the thermocline water in the Sargasso Sea because of the tight T-S relation there. As one would expect, T-S points in these regions plot well on the North Atlantic central water T-S diagram at 70°W in Fig.5.4. However, the relation breaks down in the region of the front and especially at shallow depths in the north. The region of high salinity near 150 m depth between 38°N and 39°N stabilizes the temperature inversion there.

Finally, the temperature and salinity sections can be combined to give the sound velocity section. The dominance of temperature in determining the sound velocity is obvious, as most features on the temperature section are mirrored in the sound velocity section. The local maximum sound velocity near 400 m depth in the Sargasso Sea causes a significant near-surface duct there. Also apparent is the variability across the front in the deep sound channel axis. The axis depth changes by almost 700 m across the front, but just as important, the channel is more diffuse in the vicinity of the front.

As mentioned before, the presence of this front is a feature of the general circulation. However, the structure of the front and its variability are more properly mesoscale features. Descriptions of other persistent fronts are given by RODEN [5.7-9] and KAWAII [5.10] in the North Pacific, WOOSTER [5.11] in the equatorial East

Pacific, VOORHIS [5.12], VOORHIS et al. [5.13], and BARANOV [5.14] in the North Atlantic, EMERY [5.15] and GORDON et al. [5.16] in the Antarctic, and BRISCOE et al. [5.17] in the Mediterranean.

There has not been a significant amount of work in the spatial and temporal variability of the structure within fronts, except for observations in the Gulf Stream reported by VON ARX et al. [5.18] and WEBSTER [5.19]. These observations indicate that there are sizeable fluctuations on horizontal space scales of several kilometers and time scales of a day or so, at least in the top 300 m where the data were dense in space and time. The corresponding changes in sound velocity locally were more than 20 $ms^{-1}$.

Fronts are quite variable on the large scale also, as their positions can fluctuate several hundred kilometers unless a particular front is associated with a topographic feature. A timely review of the large scale variability of ocean currents with experimental and observational results and theoretical and numerical models was given by ROBINSON [5.20]. Boundary currents and their associated fronts appear to be unstable in the sense that perturbations from the mean location in the form of travelling waves can amplify with time until a catastrophic event occurs. The front can bend around to form a large loop commonly called a meander. This meander can extend into either the warm or the cold side of the front, and it can be pinched off, forming an eddy in the process. The front then appears to reform somewhere near its mean position. The eddy retains characteristics of the water on the other side of the front in its interior. Eddies appear to retain their coherent form with a lifetime of months to several years before dissipating. The generating process by meandering of a front has been observed in some detail on several occasions, but the dissipation process is not well understood except in the case where the eddy enters shallow water on the continental shelf. In the open ocean, eddies move at speeds of one to eight kilometers per day, and are from 50 to 250 km in diameter (they usually are nearly round except when first generated). Apparently, they do not simply drift along in some local mean current, but instead they seem to propagate through the ocean as an independent entity. For example, eddies shed by the Gulf Stream travel westward, which is not obviously the direction of the mean currents on either side of the Stream. The appropriate dynamics appear to be associated with Rossby waves, though the features are more like solitons than waves in character. The horizontal density gradients are balanced by geostrophic velocity shear, so the eddies rotate in conformity with the lowest-order terms in the equations of motion. Thus, cold eddies rotate cyclonically (counterclockwise in the northern hemisphere) and warm ones anticyclonically.

The amplitude, defined as the depth perturbation of an isotherm or isopycnal in the permanent thermocline above or below the ambient depth in the vicinity, and the diameter are parameters often used to describe these features. Cold eddies are ones with elevated isotherms and warm ones have depressed isotherms. If the eddies are produced by meanders in a front, cold ones occur only on the warm side of the front and warm ones on the cold side. The limiting amplitude of an eddy shed from the front appears to be the maximum isotherm elevation across the front.

Some eddies like those in the vicinity of the Gulf Stream are called rings because they initially have a complicated ring-like structure. Near the surface in a cold ring, cold water from north of the front is surrounded by Gulf Stream water which in turn is surrounded by Sargasso Sea water. As mentioned earlier, the Gulf Stream water is slightly warmer than the local Sargasso Sea water. Just how the different water types intermingle as the feature propagates away and dissipates is not known.

Finally, eddies appear to be ubiquitous. They have been observed in the Eastern North Atlantic [5.21], Western North Atlantic [5.22], mid North Atlantic [5.23], North Pacific [5.10,24,25], and off Eastern Australia [5.26] to name only a few observations. Also, large parts of coordinated observation programs called POLYGON [5.21,27], the Mid Ocean Dynamics Experiment [5.28,29], POLYMODE, and the Ring Dynamics Experiment have been designed to intensively study eddy motions in parts of the North Atlantic. In each case, as many observations as possible have been made by a large number of investigators to quantify the variability and to determine the energy balances and transfer rates. One important finding of the many worth reporting is that the departures from geostrophic flow are at most the order of either the accuracy of the instruments or the resolving power of the observation arrays. A second result is that most of the variance on this scale can be accounted for by the barotropic and the first baroclinic mode of oscillation [5.30].

To explain this last statement, it is necessary to depart from the descriptive discussion in order to present a short summary of Rossby wave theory, as Rossby waves are the basic building blocks of mesoscale fluctuations. If it is assumed that the departures from rest are quasi-geostrophic, the governing equation for small amplitude, low frequency motions is equivalent to the conservation of potential vorticity [5.31]

$$\nabla_{\perp}^2 \psi_t + f^2/N^2 \psi_{zzt} + \beta\psi_x = 0 \quad , \tag{5.2}$$

where $\psi$ is the geostrophic stream function. This is linearly related to the other dependent variables which are the velocity, pressure, and density perturbation. The quantity

$$\nabla_{\perp}^2 = \partial^2/\partial x^2 + \partial^2/\partial y^2 \quad ,$$

is the horizontal Laplacian operator where x denotes the zonal direction (East), y the meridional direction (North), and z the direction of gravity. The parameter

$$f = 2\Omega \sin\varphi \quad , \tag{5.3}$$

is the Coriolis parameter or the inertial frequency (Sect.2.2), where $\Omega$ is the earth's rotation rate ($\sim 7.3\times10^{-5}$ s$^{-1}$), and $\varphi$ is latitude. The parameter $\beta$ is the local meridional gradient of the rotation rate ($\sim 2\times10^{-3}$ s$^{-1}$ km$^{-1}$). The term in (5.2) that includes $\beta$ is the restoring force which acts on parcels of water that have generated relative vorticity (compared to the planetary gradient) by excursions to different latitudes, driving them toward their equilibrium latitude. The quantity N is the buoyancy or Brunt-Väisälä frequency defined by (Sect.2.2)

$$N^2 = g/\rho \; d\rho/dz \quad , \tag{5.4}$$

in the absence of compressibility effects, and this, of course, is greater than or equal to zero for stable stratification. The buoyancy frequency is a strong function of depth but varies much more weakly horizontally. It normally has a maximum in the thermocline of order $10^{-3}$ s$^{-1}$, even rising to order $10^{-2}$ s$^{-1}$ in a strong seasonal thermocline, and it decreases to order $10^{-4}$ s$^{-1}$ in the deep ocean. It is nearly zero in the near-surface and near-bottom mixed layers. Actual profiles are quite variable on all vertical scales because the buoyancy frequency is a derivative of the density profile which has variability on each scale discussed in Sect.5.3. For the purpose here, the profile can be assumed to be a smooth curve obtained by appropriate averaging of oceanographic data.

Equation (5.2) is the $\beta$-plane model, and it admits solutions of the form

$$\psi \sim \tilde{\psi}(z)\, \exp[i(kx+ly-\omega t)] \quad , \tag{5.5}$$

so that (5.2) is reduced to

$$\tilde{\psi}_{zz} - N^2(z) f^{-2}[(k^2+l^2)+\beta k\omega^{-1}]\tilde{\psi} = 0 \quad . \tag{5.6}$$

In the case of constant depth, this equation has a discrete spectrum of solutions in which the greatest activity is in the thermocline where $N^2$ is largest. For

$$N^2(z) = N_0^2[1+\mu(z)] \quad ,$$

(5.6) has solutions for

$$N_0^2 f^{-2}[(k^2+l^2)+\beta k\omega^{-1}] = -\lambda_n \quad , \quad n = 0,1,2,\ldots, \tag{5.7}$$

where $\lambda_n \geq 0$. The characteristic frequencies are

$$\omega_n = -\beta k\left(k^2+l^2+f^2\lambda_n N_0^{-2}\right)^{-1} \quad , \tag{5.8}$$

and the characteristic functions are called Rossby waves. The exact values of $\lambda_n$ depend upon the buoyancy frequency profile, the depth, and the form of the free surface boundary condition. The mode $n=0$ has no vertical structure, so all fluid velocities are purely horizontal. It is called the barotropic mode. All higher modes have vertical shear, and are called baroclinic modes. As this is a Sturm-Liouville system, each higher mode has one more null representing a zero value of horizontal velocity. Thus, the first baroclinic mode essentially consists of water above the permanent thermocline sliding back and forth over the deep water. Finally, it is noteworthy from the form of the dispersion relation (5.8) that the waves are dispersive and anisotropic. The direction of all phase velocities is westward, whereas the direction of the group velocity depends upon the wavenumber.

The Rossby waves can be considered to be the building blocks of mesoscale features. Eddies cannot always be considered to have small amplitude, and large ones do not appear to disperse as a combination of linear Rossby waves would. Nevertheless, the theory is useful in characterizing the motions and in understanding nonlinear effects [5.30].

Returning to the descriptive aspects of eddies, there have been a few studies directed toward discovering regions in the oceans of high and low eddy activity by using historical data files. These data mix space and time scales so that there is no distinguishing between variability due to eddies from that due to either the meandering or the intermittency of fronts. Nevertheless, DANTZLER [5.32] has prepared a map of what is termed the eddy potential energy in the tropical and subtropical North Atlantic from historical XBT (expendable bathythermograph) data, and WYRTKI et al. [5.33] have used historical ship set and drift data to map what they call eddy kinetic energy over the world ocean. The latter data set suffers the additional disadvantage of averaging over relatively large space scales (the distance a ship travels in 24 hours), but the results in both cases are quite interesting. The regions of the western boundary currents and their extensions out into the interior of the basins are found to be quite high both in kinetic and potential energy. Poor historical data coverage in the southern oceans and in the Arctic prevent the computation of estimates of the variability in those places.

Various attempts have been made over the years to obtain nearly synoptic (i.e., all at the same time) pictures of fronts and eddies. Gulf Stream '60 [5.6] is an example of a multi-ship effort to map the Gulf Stream region with special attention to the location and orientation of the fronts. WILSON and DUGAN [5.25], and DUGAN et al. [5.23] have conducted a series of multi-ship surveys whose purpose was to quantify mesoscale variability in midocean regions. Fig.5.7 illustrates data acquired by five ships steaming at high speed in a line abreast from east to west. The time period was May so there was no significant seasonal thermocline, and though the region is not especially close to the Gulf Stream, it is in the Gulf Stream circulation system. The temperature section from the surface to 800 m depth shows several pro-

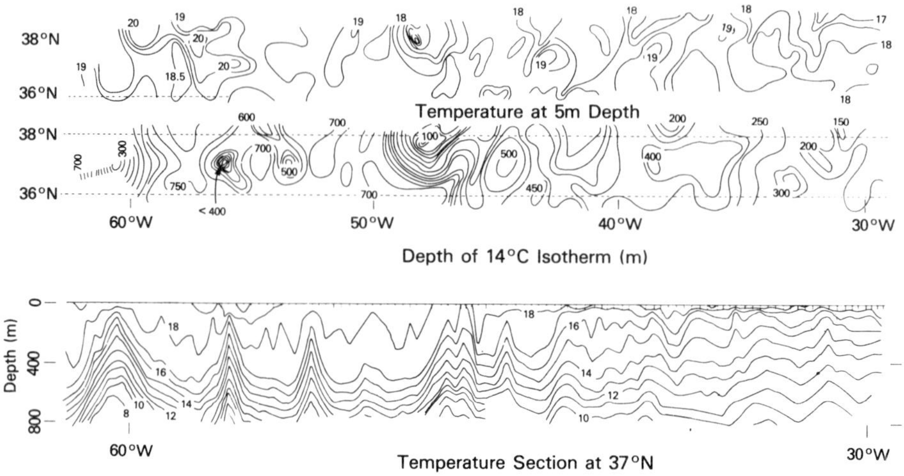


Fig. 5.7. Top) Map of the near-surface temperature; Middle) Map of depth of the 14°C isotherm surface; and Bottom) Temperature section at 37°N. Data acquired by five ships in May 1977 in North Atlantic, and discussed by DUGAN et al. [5.23]

minent features. There is a front in the region between 42°W and 48°W that separates the 18° water of the Sargasso Sea from the more uniform thermocline structure to the east. Putting the information in this section together with the topographic map of the 14°C isotherm surface, it appears that the cold feature near 48°W is a cut off eddy that could have been formed by a westward (or southward) meander of the front. Since there are no apparent historical T-S anomalies in this region, this thermal front also is a density and a sound velocity front; by geostrophy, it must support a significant southward going current. Also, the features near 54°W, 56°W, and 61°W appear to be isolated eddies. These quite likely could have come from north of this survey, especially, if the front curves around to the northwest.

The near-surface water is isothermal to about the depth of the 17° isotherm in much of the region west of 48°W, and a deep surface duct is interrupted only by the several large eddies there. These large changes in the duct occur over a scale of order 100 km.

Finally, the near-surface temperature has been plotted to indicate the relation (or lack of one) between the near surface temperature and the variability of the structure in the thermocline. The feature at 48°W has a cold spot over it that is 2°C cooler than surrounding water. This could be caused by advection of cold surface water from the other side of the front or by entrainment of colder subsurface water by vertical mixing due to wind stress. In comparison, though, the other eddies do not exhibit such a surface signature.

Two more figures are useful at this point to illustrate further the interesting problem posed by the above observation—that is, some mesoscale features have thermal surface expressions that could be sensed remotely but others apparently do not

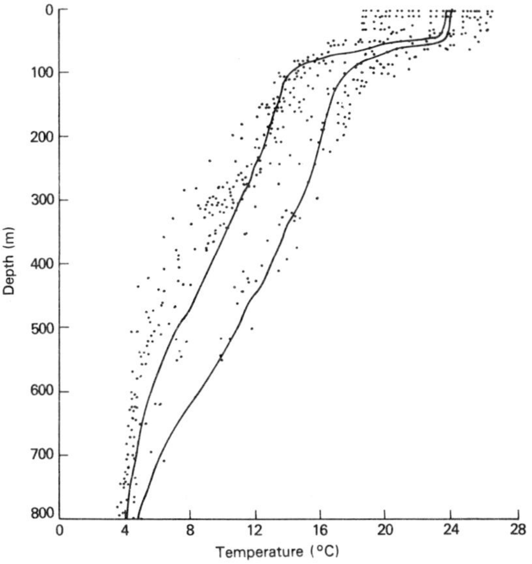


Fig. 5.8. Temperature data near 32°N, 160°E in the North Pacific. Continuous profiles are 100 km and 4 hours apart, while dots are historical data, as in Fig.5.3

have them. It currently is thought that eddies lose their surface signatures with increasing time from their formation, but this has not been documented well, and the time scales and appropriate parameters have not been identified.

Figure 5.8 is another plot of the North Pacific temperature data exhibited in Fig.5.3. The dots are historical data while the lines are two continuous temperature profiles that were acquired in the same region. The profiles were chosen from a large collection gathered in a short period of time; they were separated by less than 100 km in distance and 4 hours in time. The colder of the two profiles was in the middle of a cold eddy. The temperature change in the permanent thermocline is significantly greater than that in the near-surface mixed layer, and it also is a much larger fraction of the historical variability. This leads to the possibility that a large fraction of the historical variability in the thermocline in this region is due to mesoscale eddies, while that in the upper layer probably is due to year to year differences in seasonal heat flux. The identification of the large temperature profile change as being due to a mesoscale feature is entirely consistent with the tight T-S relation in the thermocline. A large amplitude eddy like this one can raise isotherms several hundred meters, and the raised parcels of water retain their salinity characteristics as long as the eddy was not recently shed from a front separating distinctly different water masses. Thus, while the T-S point is fixed, the temperature and salinity change significantly at a given depth. The only change that is apparent on the T-S diagram is that a given T-S point could have been collected at depths varying by several hundred meters.

It has been suggested that infrared imagery from weather satellites will revolutionize oceanography by identifying all mesoscale and larger features that are present

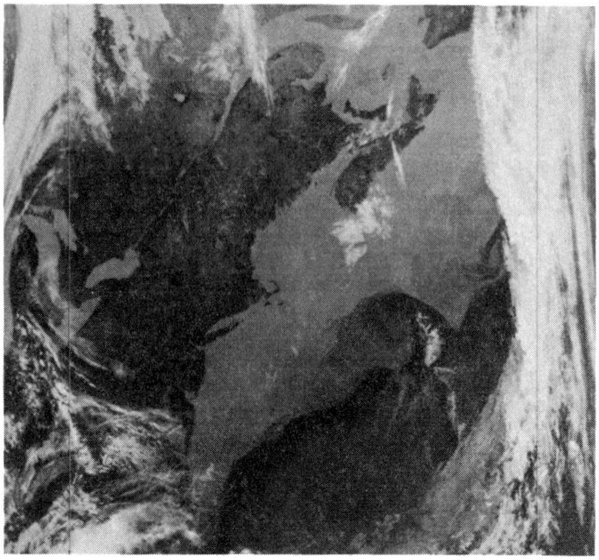

Fig. 5.9. Infrared image showing surface temperature patterns in the western North Atlantic. Data kindly supplied by Dr. R. Legeckis of NOAA/NESS

by their surface temperature signatures. Such identification would be exceedingly useful to the oceanographer and the underwater acoustician, but it depends upon the availability of a reliable model to transform two-dimensional surface temperature patterns into the three-dimensional mixed layer and thermocline structure. Figures 5.7 and 5.8 suggest that such a model would require much more information about the dynamic interaction of the surface mixed layer and the mesoscale variability in the thermocline than now is available.

Figure 5.9 is a suggestive example of the large amount of information that is available from remote sensing instruments. It is an infrared image of the western North Atlantic, with cold temperature being shaded light and warm regions shaded dark. The Great Lakes appear to the left, Long Island and Cape Cod in the center, and Nova Scotia at top center. Clouds appear as cold areas on the right and the extreme left. The Gulf Stream, or at least its surface expression, is the dark ribbon that emerges from the clouds in the vicinity of Cape Hatteras. The slightly cooler Sargasso Sea water is to the southeast, and the much colder Slope and Shelf waters to the north and west. The variability due to meanders in the Stream is apparent, and, in particular, there is a large northward meander south of Nova Scotia.

Before leaving this discussion, we mention that there presently is a lively debate over whether or not other surface signatures like dynamic topography or surface currents possibly could be remotely sensed and the information utilized for tentative identification of mesoscale features in the manner infrared images now are used. Dynamic topography is the surface of deflection of sea level from the geoid due to features in the general circulation or the mesoscale. This deflection of the surface is due to the difference in density structure of the water in a feature, and it can amount to 20 cm or more in large mesoscale features and to a meter across strong boundary currents. Similarly, the surface currents are the near-surface part of the geostrophic current associated with the feature. These currents are the order of 20 to 50 cm $s^{-1}$ over large eddies and several meters per second in boundary currents. At the time of this writing, it is not clear that the mesoscale signal could be pulled out of the many other sources of noise (most of which are not discussed herein) in these quantities, even if satellites could measure sea level and surface currents accurately.

### 5.2.3 Internal Waves

There are several classes of internal wave motions in the ocean. The internal tide is a long-wave motion that appears to be significant at least in the vicinity of continental shelves. This wave is generated by flow that is induced by the surface tide over bottom topography. These waves are order 150 km in length in deep water and have an amplitude of several meters in the open sea and often somewhat larger on the continental shelf. These motions have been observed on the shelf by LAFOND [5.34] and APEL et al. [5.35] and in deep water by REGAL and WUNSCH [5.36] among others. A comprehensive review of the theory and observations of the internal tide was given by WUNSCH [5.37].

Our interest here is on shorter scale, more random internal wave motions. The horizontal scale is order 100 m to 10 km, the vertical scale is order 1 to 100 m, and the time scale is order 10 minutes to about 1 day. These waves exist only because of increasing density of the ocean with increasing depth and because of rotation of the earth, as both of these effects tend to restore a water parcel to an equilibrium position once it has been moved from that position. As a simple explanation of the mechanism driving an internal gravity wave, consider the vertical displacement $\eta$ of a fluid parcel from its equilibrium depth in a density stratified fluid. The force tending to restore the parcel toward its equilibrium depth is the buoyancy difference it experiences in its new surroundings. To first order in a Taylor's series expansion of the density profile about the equilibrium depth, this force is equal to

$$-Vg\eta\, d\rho/dz \qquad (5.9)$$

where V is the volume of the fluid parcel. This force is balance by acceleration according to Newton's law

$$V\rho\eta_{tt} + Vg\eta\, d\rho/dz = 0 \quad , \qquad (5.10)$$

so that the displacement of the fluid parcel obeys the equation

$$\eta_{tt} + N^2\eta = 0 \quad , \tag{5.11}$$

where N is the buoyancy frequency. Thus, the buoyancy frequency is the natural frequency of oscillation of the fluid parcel. Similarly, the mechanism for the existence of an inertial wave can be explained simply by consideration of a fluid parcel in a rotating frame of reference. In a frame of reference fixed to the earth, fluid accelerations are balanced by a Coriolis acceleration

$$\underline{v}_t + f\underline{i}_\perp \times \underline{v} = 0 \quad , \tag{5.12}$$

where f is the rotation frequency, $\underline{v}$ is the fluid parcel velocity in the plane of rotation, and $\underline{i}_\perp$ is a unit vector along the axis of rotation. Reducing this to scalar form, the equations for the scalar velocity components (u,v) are

$$u_{tt} + f^2u = 0 \quad , \quad v_{tt} + f^2v = 0 \quad . \tag{5.13}$$

Thus, the velocity components (and parcel displacement) obey the harmonic oscillator equation with the natural frequency of oscillation being the inertial frequency. Of course, these models are overly simplistic, as they ignore all interactions of the fluid parcels with their surroundings.

Including these interactions, the equation governing small amplitude oscillatory motions in a stratified, rotating fluid is given by (2.53). Periodic solutions of this equation are admitted only for frequencies between the inertial and buoyancy frequencies. The dispersion relation for a plane wave in a locally uniformly stratified region is

$$\omega^2 = (f^2\ell^2+N^2k^2)(\ell^2+k^2)^{-1} \quad , \tag{5.14}$$

where $\ell$ is the vertical component of the wavenumber and k is its horizontal component. Thus, as in the case of Rossby waves, these waves are anisotropic and dispersive. As the wave frequency approaches the inertial frequency, the wavenumber vector becomes nearly vertical, and all fluid motions lie in nearly horizontal circles. Since the buoyancy frequency locally is a strong function of depth and only a weak function of lateral coordinates, the simplest analyses are one-dimensional. In regions of large buoyancy frequency like the seasonal thermocline, the governing equation admits a discrete spectrum of trapped waves that propagate horizontally along the thermocline. The frequency of these waves is bound above by the maximum buoyancy frequency and below by either the inertial frequency or the buoyancy frequency in the underlying ocean, whichever is greater. The dispersion relation for this discrete spectrum of waves is a function of the buoyancy frequency profile. For long wavelengths and high frequencies $\omega \gg f$, the lowest-order mode behaves as if the seasonal thermocline were an interface between the mixed layer and the deeper ocean. As the wave passes, the whole thermocline simply heaves up and down. For the second mode, there is a level of no vertical motion within the thermocline, and upper and lower parts oscillate out of phase. The phase speeds of these trapped waves are of order

$$c^2 = f_n(k) g \, \Delta\rho/\rho \quad , \tag{5.15}$$

where $\Delta\rho$ is the total density change across the seasonal layer and n is the mode number. The exact form of the function $f_n(k)$ depends upon the buoyancy frequency profile and the depth of the mixed layer, and it always decreases with increasing mode number.

For lower wave frequencies near the inertial frequency, the deep ocean is weakly stratified so that a continuous spectrum of waves are free to propagate through the water column according to the dispersion relation (5.14). If N(z) varies only slowly in comparison to the vertical wavelength $2\pi\ell^{-1}$, the WKB approximation yields (5.14) as the relation for $\ell$ in terms of the other parameters, so that

$$\ell = k\left[N^2(z)-\omega^2\right]^{\frac{1}{2}}(\omega^2-f^2)^{-\frac{1}{2}} \quad . \tag{5.16}$$

In this expression, the frequency $\omega$ and the horizontal wavenumber k are fixed, so the vertical wavenumber varies in response to vertical changes in N(z). The wave experiences a turning point at depths where $N(z) = \omega$, and it is evanescent in depth regions where $N(z) < \omega$.

The theory given above is overly simplified because the ocean has larger scale currents and current shear that affect these wave motions. These currents are due to mesoscale features and energetic low frequency waves, and they affect the higher-frequency waves. Including a unidirectional current in the linear theory, but ignoring rotation and wave motions out of the plane of the current for simplicity, the governing equation is

$$[\partial/\partial t+U(z)\partial/\partial x]^2\nabla^2 w + N^2(z)w_{xx} - [\partial/\partial t+U(z)\partial/\partial x]U''(z)w_x = 0 \quad , \tag{5.17}$$

where U(z) is the steady, unidirectional current profile and w is the vertical wave velocity component. Periodic solutions of the form

$$w(x,z,t) = W(z)\exp[i(kx-\omega t)] \quad , \tag{5.18}$$

are admitted for solutions of the equation

$$W_{zz} + \left\{N^2(z)\left[U(z)-\omega/k\right]^{-2} - U''(z)\left[U(z)-\omega/k\right]^{-1} - k^2\right\}W = 0 \quad . \tag{5.19}$$

The additional effect of the background current has complicated the equation considerably, but the most important addition is due to the denominator of the terms in brackets. This denominator goes to zero at depths for which $\omega/k \rightarrow U(z)$, resulting in a singularity in the equation. Regions where this occurs are called critical layers, and they are thought to be regions of strong interaction between the wave and the current [5.38]. The frequency of a plane wave having vertical wavenumber $\ell$ is given by

$$\omega = U(z)k \pm N(z)k(k^2+\ell^2)^{-\frac{1}{2}} \quad , \tag{5.20}$$

in the case in which the current profile has weak curvature. In the vicinity of a critical layer, the phase speed approaches the fluid velocity and the vertical wavenumber becomes unboundedly large. Both components of the group velocity go to zero relative to the fluid, so a wave packet of finite extent never reaches the level, and the energy density increases unboundedly also. The inclusion of the inevitable time dependent, diffusive, and nonlinear effects will change the picture somewhat, but the conclusions that critical layers are important and that they occur for realistic conditions in the ocean are immutable.

Another complication in the theory, even in the absence of background currents, is the occurrence of instabilities in the waves [5.39]. There is a debate over which of the several types of instability is predominantly responsible for the disintegration of a uniform wave train. Self-induced instabilities excited through nonlinear interactions have wavelengths longer and shorter than the basic wavelength in the wave train. The instabilities grow exponentially with time, and the theory indicates that the shorter waves can abstract energy from the primary wave at fast enough rates to destroy the wave train in a matter of the wave travelling the order of ten wavelengths. This nonlinear transfer among the different wavelengths tends to shift energy to higher wavenumbers where it eventually contributes to mixing and is dissipated [5.40].

As an illustration of the variability in the ocean in the internal wave band, Fig.5.10 shows several months of current meter data acquired in the permanent thermocline at one kilometer depth in the Sargasso Sea. The four time series represent the north-south component of velocity, the east-west component of velocity, and the current direction and speed. The near-daily fluctuations mainly are inertial waves that apparently are propagating past the meter, since the velocity vector has a tendency to rotate monotonically. The mooring was near $30^\circ$N, so the lunar tidal frequency is relatively near the inertial frequency, and the separation of the internal tidal motions from the inertial wave motions is not trivially accomplished. The mean current averaged over several days is of order 5 cm $s^{-1}$, and it shifts regularly over the several month period of observation, which indicates the presence of mesoscale motions. This mesoscale current is most easily seen in the figure by taking a running

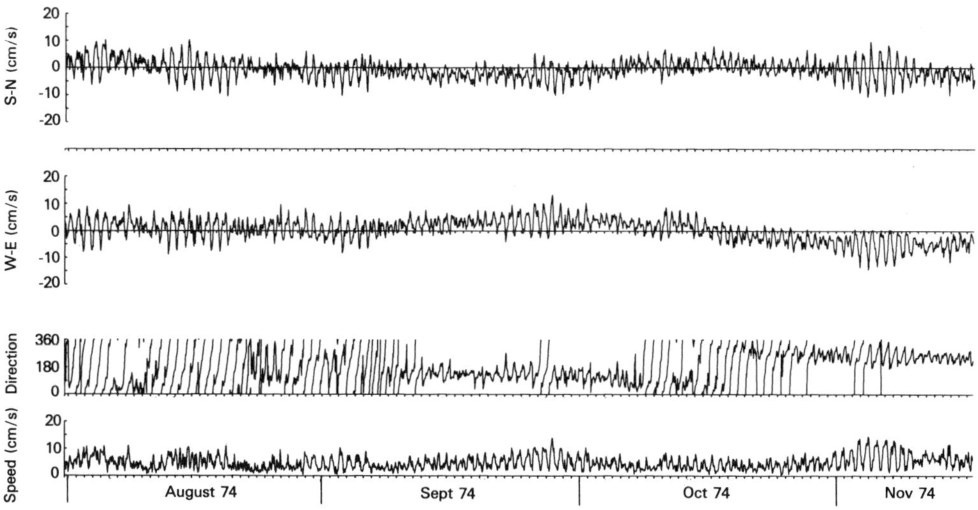


Fig. 5.10. Water velocity from moored current meter near 1000 m depth in the Sargasso Sea. Data kindly provided by the Buoy Group of the Woods Hole Oceanographic Institution

mean through the north-south and the east-west time series. The temperature variability at the current meter has a mesoscale trend having a range between 6.2°C and 7.0°C, and an inertial frequency component of range 0.2°C.

The wave component in this current meter data is quite intermittent, as there are times when the inertial motions are dominant, and other times when they appear to be weak. The cause of the intermittency is not clear, but the possible causes for it have to include an intermittent generating mechanism and the simple kinematic property of interference due to finite bandwidth of the process. The intermittency smears frequency spectra, and it contributes to the difficulty in resolving the inertial wave motions from the internal tide because the length of stationary data required for the resolution may be longer than the lifetimes of the motions. WUNSCH [5.37] has a good example of a frequency spectrum that resolves the two peaks. Also, this spectrum, as well as that in the velocity data like that in Fig.5.10, exhibits a definite spectral gap between the inertial frequency and the lower frequencies that represent the mesoscale. The kinetic and potential energies at the lower frequencies are much greater because the spectral amplitudes are as high as those in the inertial wave regime and the bandwidth is much greater.

Observations of variability in these scales and frequencies have been brought together on a common framework built around the linear internal wave theory by GARRETT and MUNK [5.41]. They synthesized measured spatial and temporal spectra and spatial coherences with the linear dispersion relation (5.14) to generate a model of a multi-dimensional spectrum. The model for the energy spectrum they propose in frequency/vertical wavenumber space is

$$E(\omega,\ell) = EA(\ell/\ell_*)\ell_*^{-1}\, 2\pi^{-1} f\omega^{-2}(1-f^2/\omega^2)^{-\frac{1}{2}} \quad , \quad f < \omega < N \quad , \tag{5.21}$$

where E is a constant and $\ell_* = 6\pi N$. The vertical wavenumber influence on the shape of the spectrum is given by

$$A(\lambda) = (\lambda-1)(\lambda+1)^{-t} \quad , \tag{5.22}$$

where in this case $\lambda = \ell/\ell_*$, and t was chosen to be 2.5. Using the dispersion relation for internal waves, this can be transformed into the horizontal/vertical wavenumber spectrum

$$E(k,\ell) = 2\pi^{-1} EfN\ell_*^{-1}\ell A(\ell/\ell_*)(N^2k^2+f^2\ell^2)^{-1} \quad , \quad 0 \leq k \leq \ell(1-f^2/N^2)^{\frac{1}{2}} \quad . \tag{5.23}$$

This exhibits the simple scaling ratio of f/N between horizontal and vertical wavenumbers that is in the dispersion relation. The various constants were chosen to give a reasonable fit to the various data sets that were available at the time. One-dimensional spectra of horizontal velocity or vertical displacement are obtained by integrals over one variable in either (5.21) or (5.23) after being multiplied by $N\omega^{-2}(\omega^2+f^2)$ and $N^{-1}\omega^{-2}(\omega^2-f^2)$, respectively. The spectra are assumed to be isotropic in the horizontal plane.

This model has been compared favorably with spectral estimates and coherences from moorings [5.42], horizontal tows [5.43], vertical profiles of temperature [5.44] and of horizontal velocity [5.45], and thermistor chain tows [5.46,47]. All these spectra exhibit red behavior in which most of the energy is contributed by the longest scales measured. In frequency space, the spectrum increases like $\omega^{-2}$ from the buoyancy to the inertial frequency. At higher frequencies, the spectrum rolls off precipitously.

The simple fact that a kinematic internal wave model with universal constants can be fitted well to data acquired in different ways in many places and times implies that some saturation limit has been reached. That is, no matter where in the deep ocean one is, the internal wave scales and amplitudes can be scaled by the local buoyancy and inertial frequencies. For this to occur, there must be a relatively uniform energy source and reasonably strong nonlinear coupling that redistributes the energy over the spectrum, no matter where it is put in. Such coupling is implied by the stability calculations of MIED [5.39] and the transport calculations of MÜLLER and OLBERS [5.40].

While this all appears to be well in hand, problems still abound [5.48,49]. The logarithmic scale upon which spectra usually are plotted can hide spectral amplitude differences of three or more between various spectra. Also, there is an indication at least in the seasonal thermocline that a considerable portion of the variance occurs in packets [5.50,51]. Typical cross sections of the ocean acquired by a ther-

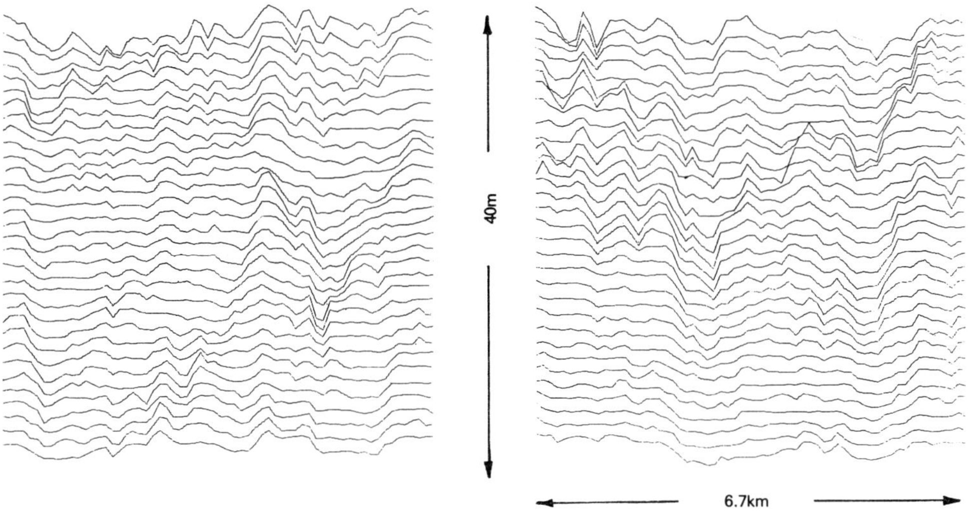


Fig. 5.11. Vertical cross sections of isotherm displacements obtained by thermistor chain tows [5.46]

mistor chain in the seasonal thermocline and analyzed by BELL et al. [5.46] are not homogeneous, and spectra are not always an appropriate statistical representation of the data. Fig.5.11 shows two examples of the cross sections. The lines are estimates of the isotherm displacements, as the data actually is the temperature deviation divided by the mean temperature gradient, and they represent the actual isotherm displacements only in the absence of fine structure contamination (Sect.5.2.4). The fluctuations are wavelike and they appear to be of several types. There are many small scale disturbances, some of which appear to be propagating at an angle to the horizontal, but there also is considerable energy in long vertical modes that are not completely resolved by the data sets. These latter waves could be low mode number trapped waves propagating along the seasonal layer. If so, the GARRETT and MUNK spectrum is not exactly the appropriate representation of them. The sections are distorted somewhat by larger scale fluctuations, and the analysis exhibits no apparent spectral gap (in wavenumber) between these internal wave scales and larger scales. There is an apparent spectral gap between the internal wave and the mesoscale fluctuations in the frequency domain, so it is doubtful that energy is being transferred down into the internal wave band from the mesoscale by nonlinear interactions. However, as stated before, a universal spectrum requires a relatively uniform source of energy. It is not known for sure where most of the internal wave energy comes from, though the likely sources have been identified [5.40,52]. The source of most of the wave energy near the inertial frequency apparently is the wind stress, but the details of the transfer mechanism from the wind to waves below the mixed layer have not been completely worked out.

The effect of internal waves on acoustic wave propagation has been estimated under certain restrictive deterministic circumstances in many instances. These scales are expected to be more important in calculations of coherence in space and time than they are to the overall propagation loss, but the details of random scattering are still being worked out [5.53,54]. Estimates of sound speed fluctuations due to internal waves have led most past work to be more concerned with the temperature fluctuations than with the fluid velocity fluctuations themselves. However, most of the energy in this frequency band is near the inertial frequency where the fluid motions essentially are horizontal. The vertical shear due to these waves is of order 1 cm $s^{-1}$ $m^{-1}$ in the thermocline, and it is coherent over several kilometers in the horizontal, so it is expected that the influence is most significant on nearly horizontal ray paths. The actual effect of the velocity shear has not been thoroughly studied, but several initial efforts are available [5.55,56]. One immediate impact is the loss of reciprocity, if that is important on these scales.

There are many useful general references on internal waves, including major chapters in the monographs by PHILLIPS [5.57] and TURNER [5.58]. A review [5.59] and a collection of papers [5.60], most of which were referenced in the foregoing, attest to the high present level of research activity in this area.

### 5.2.4 Fine Structure

Going down only slightly in spatial scale, the region of fine structure occupies vertical lengths of one to several tens of meters and horizontal lengths of tens to many hundreds of meters. The symptom of the presence of fine structure is apparent "sheets" and "layers" in a vertical density or temperature profile. The layers are regions of relatively uniform density separated by sheets of high density gradient. This step-like structure is dominant in some profiles like those of TAIT and HOWE [5.61], where the layers take up the order of 80% of the profile while the remaining 20% contains nearly all of the density change with depth. These layers in the base of the Mediterranean outflow are especially large at thicknesses between 25 and 40 meters. The change in $\sigma_t$ across the sheets is about 0.005, while that of temperature and salinity is about 0.3°C and 0.06 °/oo, respectively. The corresponding sound velocity across the sheets is about 1,5 m $s^{-1}$, so the layers are significant features on the sound velocity profile. These layers in the Mediterranean outflow have been traced for tens of kilometers.

WOODS [5.62] has observed smaller scale, distinct layering in the seasonal thermocline near Malta, but the distinction between sheets and layers is not always as clear in all observations [5.63]. The fluctuations are theorized to have several mechanisms for their origins and maintenance, and the different mechanisms apparently account for the differences in observed intensity. The sources include internal waves, lateral intrusions, and diffusive mechanisms. A good review of work prior to 1975 is available in [5.64].

The first mechanism thought to be responsible for fine structure is internal waves. Large amplitude waves having a small vertical wavelength can distort the mean density profile into a step-like profile, as explained by ORLANSKI and BRYAN [5.65]. This step-like structure would propagate along with the wave packet, and this explanation has been lent support by observational evidence that the steps can propagate vertically [5.66]. In addition to this kinematic effect, it is possible that the breaking of internal waves can maintain the layers. Once the sheets are present, they provide selective sites for enhanced shear during the passing of waves. This shear contributes to the generation of Kelvin-Helmholtz instabilities on the sheet which, in turn, cause enhanced mixing. Thus, the internal waves can supply energy almost directly to mixing in the vicinity of fine structure. However, whatever the effects of internal waves on the layers, it is clear that the presence of fine structure in turn contaminates the measurement of the larger scales and thereby causes problems in the interpretation of oceanographic data [5.67,68].

The presence of layers in regions of lateral intrusions like the Mediterranean outflow is common. A foreign water mass intrudes along its own density surface, and, as it progresses in a layer, it mixes with the water above and below. Most of the velocity shear due to the advection of fluid in the intruding layer occurs in a mixing region in the vicinity of the sheets, and the sheets remain relatively thin compared with the vertical scale of the layer. The interleaving pancakes of two water masses yield a vertical density profile of sheets and layers and a velocity profile having a series of weak, alternating jets. CAIRNS [5.69] has observed small thermal inversions that, if stabilized by salinity difference, can be interpreted as good examples of intrusions, and GARGETT [5.70] has observed T-S changes in layers that definitely point to intrusions. The mixing in the layers can be maintained by salt fingering and differential diffusion in certain combinations of opposing temperature and salinity gradients [5.58,71]. Even though the density gradient may be stable on the large scale, the difference in molecular diffusivities of heat and salt can drive convective instabilities that maintain well-mixed layers.

Because of the overlap in spatial scales with internal waves and the indistinct difference between the two phenomena, the statistical representation of data generally has been similar to that of waves, i.e., the calculation of spectra. Special fine scale profiling instruments have detected a small but significant bump in the vertical temperature gradient spectra in the range of one to ten centimeters [5.63]. McKEAN [5.68] and HAYES et al. [5.72] have described temperature gradient data by stochastic models. Also, a parameter called the Cox number has been defined by the ratio of the average variance of the temperature gradient to the square of the mean vertical gradient and is used to quantify the amount of small scale fluctuations in a profile.

The scales of fluctuations go on down to centimeters in the vertical and many meters in the horizontal, a parameter range called the microstructure. These scales are difficult to observe, but the mechanisms for generation and maintenance of the microstructure apparently are similar to those for fine structure. The picture is one of small scale layering, with some balance between the occasional catastrophic events of wave breaking and the ultimate smoothing by diffusive effects. There is little indication of any regular turbulence structure except during and after the occasional catastrophic events.

### 5.2.5 Near-Surface Structure

The scales of variability in the near-surface layers of the ocean are similar to those in the water column below, but the environment is special enough to merit separate attention. Near the surface, there exists a layer of well mixed water at most times of the year and locations in the ocean. This layer, usually between ten and several hundred meters deep, is sustained by mixing processes that are not fully understood. Presumably, wind stress maintains an intermittent turbulent state that exists in concert with surface waves, Langmuir cells, and thermal convection. Some of the impressed wind stress is carried off by wind waves which in turn give up the momentum at other places, and some is transferred to the layers below in the local region by turbulent stresses.

This near-surface layer, usually called the mixed layer or the sonic layer, has nearly constant temperature and salinity. The layer often is characterized by two parameters, the mixed layer depth (MLD) and the mixed layer temperature (MLT). The layer's importance to acoustics stems from its ability to behave as a duct. The sound speed increases weakly with depth below the surface because of the pressure effect, and then it decreases sharply in the thermocline.

As in most of the ocean, the significant horizontal scales of variability are orders of magnitude greater than the MLD, and until recently, most modeling efforts have attempted purely one-dimensional characterizations of the layer. The seasonal variability has been studied in great detail, especially at the locations of the ocean weather stations. The temperature varies markedly because of changes in solar heating and in wind stress, and more weakly because of precipitation or evaporation. Fig.5.12 is a graphical representation of the gross structure of temperature profiles for subtropical regions for the four oceanographic seasons of the year. The mixed layer is relatively shallow, of order ten to thirty meters, and warm in the summer. A seasonal thermocline having a strong temperature gradient separates it from the permanent thermocline. Strong fall storms deepen the layer, and through turbulent entrainment of the cooler water in the layers below, the mixed layer cools significantly. In addition, solar heating is less, and the layer cools further. Near the end of winter, the layer can be quite deep, of order 100 to 500 m, and it may have very little gradient demarking its lower boundary from the thermocline. The deeper layer can warm slightly because of vertical transport of heat. Spring brings strong

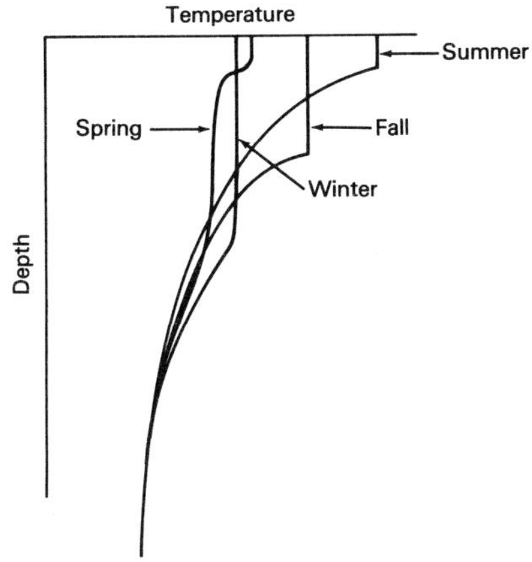


Fig. 5.12. Typical temperature profiles for the four oceanographic seasons in the subtropics

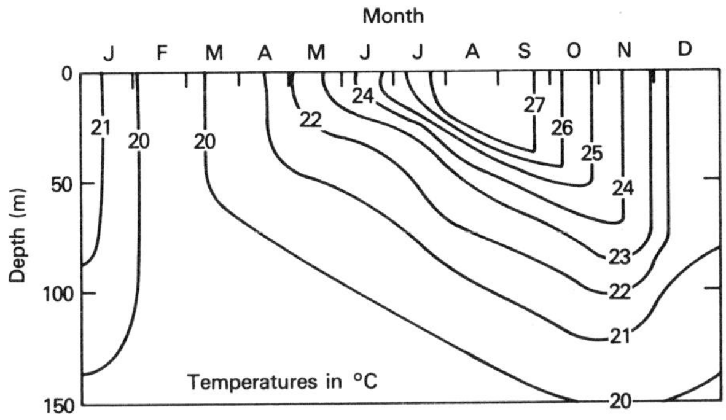


Fig. 5.13. Typical surface layer isotherm structure through the oceanographic year in the subtropics

insolation and calmer weather, and a shallow surface layer again is formed, repeating the cycle.

Figure 5.13 is a typical composite plot of the isotherms through the months of the year for the subtropics. The figure illustrates the annual cycle discussed above. The seasonal layer is evident from May through November, but the mixed layer is deeper than the measurements in winter. It is worth noting here that this description is somewhat misleading because these seasonal effects actually can be quite irregular. Not only does spatial variability distort this annual trend at any particular place, but the weather itself is quite variable. Even averaging out the spatial variability, the warmest summer temperature, for instance, can vary by more than one degree Celsius from one year to another.

The MLD is a strong function of the recent history of wind stress. TABATA et al. [5.73] calculated a positive correlation between deepening of the layer and average wind speed in the previous twelve hour period at Ocean Weather Station Papa in the Northeast Pacific. Thus, the recent passage of a series of storms can deepen the layer significantly, with changes of 50% not uncommon. Strong events such as storms are not homogeneous over one thousand kilometers, so there have to be two-dimensional effects; these have not been studied in detail. Even though these temporal and spatial effects occur regularly, monthly averages of historical layer depth and temperature remain of interest. They have been computed and are available in the form of maps that, in addition, illustrate large scale geographic variability. BATHEN [5.74] gave maps of MLD for the North Pacific. Similar maps of MLT are widely available in atlas form.

Besides the temporal and large-scale geographic variability in the mixed layer parameters, there is significant variability on smaller spatial scales due to forcing other than storms. Fronts and eddies in the mesoscale can cause MLT changes of several degrees Celsius and MLD changes of a factor of order two, and internal waves propagating in the thermocline cause fluctuations in the MLD of many meters. For both MLD and MLT, the magnitude of the variability decreases with scale.

An example of near-synoptic data for these parameters is Fig.5.7. The surface temperature map and the temperature section indicate the mesoscale variability in the parameters that is connected in some sense with the variability in the thermocline. The data were acquired in late winter, and the mixed layer in the west is over 400 m deep between the large eddies. In a sense, the eddies poke up through the layer, so that the effective layer depth is considerably smaller over the eddies.

Figure 5.14 shows successive temperature profiles at approximately twelve kilometer separation at a time when there was a strong seasonal thermocline. The tic marks at the top of the figure indicate a reference point of 25°C for each profile. Each tic mark is to the right of the respective profile, so the last profile indicates a surface temperature that is over 2°C cooler than the one second from the left. The reason this particular sequence of profiles was chosen for exhibition is the large though not especially unusual variability in the layer. Both the depth to the seasonal thermocline and the actual vertical structure of the layer are quite variable. The cause of the MLD variability in this instance has to be considered a mesoscale feature, as the wavelength of the fluctuation is order 100 km. On the other hand, the same data set from which this sample was chosen shows many instances

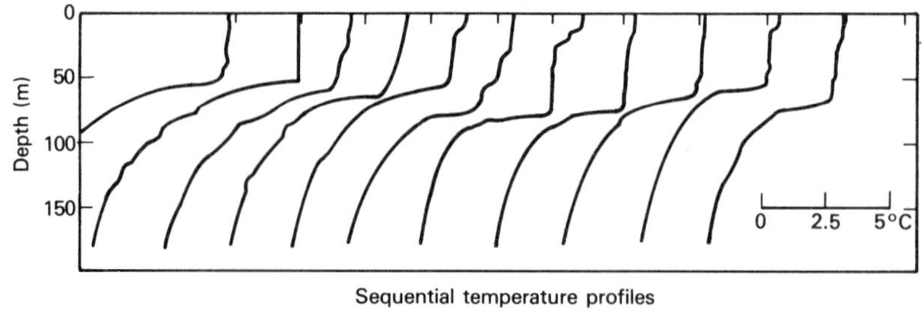


Fig. 5.14. Sequential temperature profiles 12 km apart that illustrate mixed layer variability

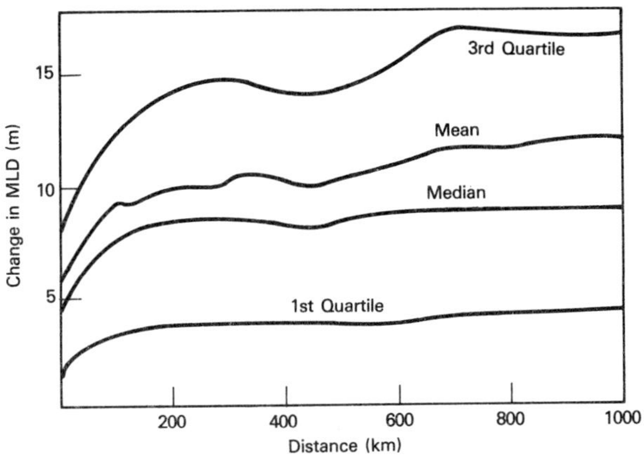


Fig. 5.15. Statistics of the change in mixed layer depth per unit distance between observation stations in a North Pacific oceanographic survey

of as large variability in MLD that is not as well sampled at this station spacing as this feature was. The change in the vertical structure in the mixed layer also is interesting. This structure is seriously undersampled, as the features are not coherent from profile to profile. The cause of the layering is not known, but several mechanisms could be responsible for it. One explanation is the occurrence of strong insolation during a period of little stirring followed by a period of mixing of this warmer water downward. Because of its intermittent occurrence in the data set, however, it more likely is due to fronts in the layer. VOORHIS et al. [5.75] have observed a simple advective relation between the layer temperature and mesoscale features in the permanent thermocline, and such interactions could cause fronts in the layer.

Figure 5.15 is a statistical synopsis of the spatial structure of the MLD variability in the complete data set from which Fig.5.14 is a sample. The curves indicate the probabilities of changes in MLD between any two points. Thus, at a distance of 200 km, the expected change in MLD is 10 m. There is a 50% probability of having a change greater than 8 m, and a 25% probability of having a change greater than 14 m. The structure of the curves indicates that a plateau is reached in the range 100-200 km, with only a slow increase at longer distances. This indicates the strong tendency of the significant variability to occur on scales less than about 200 km, while almost half of it occurs in less than 10 km. This also implies that the fluctuations essentially are uncorrelated for distances greater than 100-200 km.

These results are for the case of a strong seasonal thermocline, so large fluctuations are resisted by large buoyancy forces. If the same calculations were made for the data in Fig.5.7, the magnitude of the changes would be much larger. It is expected, however, that the 100-200 km range would again be the limit of significant correlation of MLD fluctuations.

An additional effect that selectively occurs in and near the bottom of the mixed layer is significant vertical velocity shear. The large density gradient there and in any (nearly horizontal) front in the interior of the mixed layer can support large shear. The Richardson number defined by

$$Ri = N^2/(dU/dz)^2 \quad ,$$

is a measure of the relative importance of buoyancy and shear effects [5.58], and voluminous experimental and theoretical results on the stability of stratified shear flows indicates that horizontal flows can sustain vertical shear without instability for $Ri \gtrsim 1/4$. Thus, regions of high stability can support large shears, and the mixed layer has a tendency to slide over the deeper layers as a slab. Because of reduced turbulent stresses in the high density gradient, the stress retarding the layer from accelerating with an implied wind stress is reduced. Thus, a shallow, buoyant layer which has less inertia than a deeper one can react easier to a given wind stress. In theory, the layer can accelerate until retarded by turbulent stresses in the thermocline or by loss of momentum to internal waves that are generated by nonuniformities in the layer depth and that radiate down into the ocean.

The scenario discussed above is partially conjecture, as it is extremely difficult to measure the velocity and velocity shear near the ocean surface. For instance, surface wave motions and mooring motions bias current meter measurements [5.28], and drogues are subject to uncertainties due to wind loading. There are, however, accepted observations and theories that the surface layer experiences energetic oscillations due to inertial waves generated by wind stress variations [5.76], so the conjecture of large and variable shears at the base of shallow layers is difficult to reject.

There have been a considerable number of attempts to model the physics governing the gross mixed layer dynamics. The models primarily have been one-dimensional, so that the effects of temporal (not spatial) variability in wind stress, solar heating, and buoyancy magnitude (or temperature gradient) in the seasonal thermocline have been studied. Most layer models are really energy balances in which a flux of turbulence energy works against gravity at the bottom of the mixed layer. The models differ mostly in the forms of the flux boundary conditions at the sea surface and at the bottom of the mixed layer, and in the form of dissipation and wave radiation [5.77]. For several reasons, including the evident importance of two-dimensional motions, the model results have not been highly satisfactory except in special circumstances. Two-dimensional numerical models have been applied to coastal upwelling phenomena, but the development and application of a realistic multi-dimensional model to the open ocean mesoscale problem has not been seriously attempted. The general subject of mixed layer modeling has been the subject of a major conference and workshop referred to above.

One final near-surface phenomenon of interest in underwater acoustics is the "afternoon effect". This is the diurnal temperature variation due to solar heating that occurs in the top few meters of the sea. Some 90% of the solar energy is absorbed in the top 10 m of average seawater, and, since the water warmed by the radiant heat is buoyant, the top layer tends to float even under the mixing conditions of a light to moderate wind. A sharp diurnal thermocline forms between this layer and the remainder of the mixed layer. This diurnal layer grows stronger and deeper in the afternoon after the sun passes its zenith, and it sometimes reaches a depth of 20 m and a temperature of up to several degrees Celsius warmer than the deeper mixed layer. The maximum temperature increase is proportional to the total daily heat input and inversely proportional to the diurnal layer depth, so these extremes are not reached simultaneously. At night, the heat is radiated from the surface to the relatively cold sky. The cooler water at the surface is denser than that below, and the resulting thermal convection can quickly destroy the diurnal layer as an entity. In the spring, more heat comes in than is radiated away, and this diurnal layer contributes to the warming trend in the mixed layer. In the fall, the net heat flux is upward, and the stronger convection can penetrate through the mixed layer and contribute to its deepening trend. Except for the scales and magnitude, the temporal-depth characteristics of the diurnal layer are similar to those in Fig.5.13.

The diurnal layer has received considerable attention in recent years by STOMMEL et al. [5.78], BOWDEN et al. [5.79], DELNORE [5.80], and KAISER [5.81]. The recent motivation for this work has been heat budget calculations that are important to air-sea interaction problems that affect meteorological processes and that affect the remote sensing problem discussed in Sect.5.2.2.

## References


5.1 W.D. Wilson: J. Acoust. Soc. Am. *32*, 1357 (1960)
5.2 R.A. Cox: "The Salinity Problem", in *Progress in Oceanography*, Vol.1, ed. by M. Sears (Macmillan Press, New York 1963)
5.3 W.J. Emery: J. Phys. Oceanogr. *7*, 293-297 (1977)
5.4 R.J. Lynn, J.L. Reid: Deep-Sea Res. *15*, 577-598 (1968)
5.5 H.U. Sverdrup, M.W. Johnson, R.H. Fleming: *The Oceans, Their Physics, Chemistry, and General Biology* (Prentice-Hall, Englewood Cliffs, NJ 1942)
5.6 F.G. Fuglister: "Gulf Stream '60", in *Progress in Oceanography*, Vol.1, ed. by M. Sears (Macmillan, New York 1963) pp.265-321
5.7 G.I. Roden: J. Geophys. Res. *77*, 7175-7187 (1972)
5.8 G.I. Roden: J. Phys. Oceanogr. *5*, 557-571 (1975)
5.9 G.I. Roden: J. Phys. Oceanogr. *7*, 761-778 (1977)
5.10 H. Kawaii: "Hydrography of the Kuroshio Extension", in *Kuroshio: Physical Aspects of the Japan Current*, ed. by H. Stommel, K. Yoshida (University of Washington Press, Seattle, WA 1972) pp.235-341
5.11 W.S. Wooster: Deep-Sea Res. *16* (Suppl.), 407-419 (1969)
5.12 A.D. Voorhis: Deep-Sea Res. *16* (Suppl.), 331-337 (1969)
5.13 A.D. Voorhis, D.C. Webb, R.C. Millard: J. Geophys. Res. *81* , 3695-3708 (1976)
5.14 Y.I. Baranov: Oceanology *12*, 187-192 (1972)
5.15 W.J. Emery: J. Phys. Oceanogr. *7*, 811-822 (1977)
5.16 A.L. Gordon, D.T. Georgi, H.W. Taylor: J. Phys. Oceanogr. *7*, 309-328 (1977)

5.17 M.G. Briscoe, O.M. Johanessen, S. Vincenzi: Deep-Sea Res. *21* , 247-262 (1974)
5.18 W.S. Von Arx, D.F. Bumpus, W.S. Richardson: Deep-Sea Res. *3*, 46-65 (1955)
5.19 F. Webster: Deep-Sea Res. *7*, 130-143 (1961)
5.20 A.R. Robinson: Rev. of Geophys. Space Phys. *13*, 598-601 (1975)
5.21 L.M. Brekhovskikh, K.N. Fedorov, L.M. Fomin, M.N. Koshlyakov, A.D. Yampolsky: Deep-Sea Res. *18*, 1189-1206 (1971)
5.22 F.C. Fuglister, L.V. Worthington: Tellus *3*, 1-14 (1951)
5.23 J.P. Dugan, A.F. Schuetz, W.S. Wilson: POLYMODE News *30*, 1977 (unpublished manuscript)
5.24 R.L. Bernstein, W.B. White: J. Phys. Oceanogr. *4*, 613-624 (1974)
5.25 W.S. Wilson, J.P. Dugan: J. Phys. Oceanogr. *8*, 537-540 (1978)
5.26 C.S. Nilsson, J.C. Andrews, P. Scully-Power: J. Phys. Oceanogr. *7*, 659-669 (1977)
5.27 M.N. Koshlyakov, Y.M. Grachev: Deep-Sea Res. *20*, 507-526 (1973)
5.28 W.J. Gould, W.J. Schmitz, C. Wunsch: Deep-Sea Res. *21*, 911-931 (1974)
5.29 W.F. Simmons, for the MODE Group: Deep-Sea Res. (submitted)
5.30 J.C. McWilliams, G.R. Flierl: Deep-Sea Res. *23*, 285-300 (1976)
5.31 J. Pedlosky: "Geophysical Fluid Dynamics", in *Mathematical Problems in the Geophysical Sciences, 1. Geophysical Fluid Dynamics*, ed. by W.H. Reid, Lectures in Applied Mathematics, Vol.13 (Amer. Math. Soc., Providence, RI 1971) pp.1-60
5.32 H.L. Dantzler, Jr.: J. Phys. Oceanogr. *7*, 512-519 (1977)
5.33 K. Wyrtki, L. Magaard, J. Hager: J. Geophys. Res. *81*, 2641-2646 (1976)
5.34 E.C. LaFond: "Internal Waves", in *The Sea*, Vol.1, ed. by M.N. Hill (Interscience, New York 1962) pp.731-763
5.35 J.R. Apel, H.M. Byrne, J.R. Proni, R.L. Charnell: J. Geophys. Res. *80*, 865-881 (1975)
5.36 R. Regal, C. Wunsch: Deep-Sea Res. *20*, 493-502 (1973)
5.37 C. Wunsch: Rev. Geophys. Space Phys. *13*, 167-182 (1975)
5.38 J.R. Booker, F.P. Bretherton: J. Fluid Mech. *27*, 513-539 (1967)
5.39 R.P. Mied: J. Fluid Mech. *78*, 763-784 (1976)
5.40 P. Müller, D.J. Olbers: J. Geophys. Res. *80*, 3848-3860 (1975)
5.41 C. Garrett, W. Munk: J. Geophys. Res. *80*, 291-297 (1975)
5.42 M.G. Briscoe: J. Geophys. Res. *80*, 3872-3884 (1975)
5.43 E.J. Katz: J. Geophys. Res. *80*, 1163-1167 (1975)
5.44 S.P. Hayes: J. Geophys. Res. *80*, 307-311 (1975)
5.45 T.B. Sanford: J. Geophys. Res. *80*, 3861-3871 (1975)
5.46 T.H. Bell, J.M. Bergin, J.P. Dugan, Z.C.B. Hamilton, W.D. Morris, B.S. Okawa, E.E. Rudd: Science *189*, 632-634 (1975)
5.47 T.H. Bell, Jr.: J. Geophys. Res. *81*, 3709-3714 (1976)
5.48 C. Wunsch: J. Geophys. Res. *80*, 339-343 (1975)
5.49 C. Wunsch: J. Phys. Oceanogr. *6*, 471-485 (1976)
5.50 L.M. Brekhovskikh, K.V. Konjaev, K.D. Sabinin, A.N. Serikov: J. Geophys. Res. *80*, 856-864 (1975)
5.51 W. Zenk, E.J. Katz: J. Geophys. Res. *80*, 3885-3891 (1975)
5.52 S.A. Thorpe: J. Geophys. Res. *80*, 328-338 (1975)
5.53 W.H. Munk, F. Zachariasen: J. Acoust. Soc. Am. *59*, 818-838 (1976)
5.54 S.M. Flatté, F.D. Tappert: J. Acoust. Soc. Am. *58*, 1151-1159 (1975)
5.55 T.B. Sanford: J. Acoust. Soc. Am. *56*, 1118-1121 (1974)
5.56 P.F. Worcester: J. Acoust. Soc. Am. *62*, 895-905 (1977)
5.57 O.M. Phillips: *Dynamics of the Upper Ocean* (Cambridge University Press, London 1966)
5.58 J.S. Turner: *Buoyancy Effects in Fluids* (Cambridge University Press, London 1973)
5.59 M.G. Briscoe: Rev. Geophys. Space Phys. *13*, 591-598 (1975)
5.60 M.G. Briscoe: J. Geophys. Res. *80*, 289-290 (1975)
5.61 R.I. Tait, M.R. Howe: Deep-Sea Res. *15*, 275-280 (1968)
5.62 J.D. Woods: J. Fluid Mech. *32*, 791-800 (1968); see also P. Bradshaw, J.D. Woods: In *Turbulence*, ed. by P. Bradshaw, Topics in Appl. Phys., Vol.12 (Springer, Berlin, Heidelberg, New York 1978)
5.63 T.R. Osborn, C.S. Cox: Geophys. Fluid Dyn. *3*, 321-345 (1972)
5.64 M.C. Gregg: Rev. Geophys. Space Phys. *13*, 586-591 (1975)
5.65 I. Orlanski, K. Bryan: J. Geophys. Res. *74*, 6975-6983 (1969)
5.66 J.R.N. Lazier: J. Phys. Oceanogr. *3*, 226-229 (1973)

5.67 C. Garrett, W. Munk: J. Phys. Oceanogr. *1*, 296-302 (1971)
5.68 R.S. McKean: J. Phys. Oceanogr. *4*, 200-213 (1974)
5.69 J.L. Cairns: J. Geophys. Res. *80*, 299-306 (1975)
5.70 A.E. Gargett: Deep-Sea Res. *22*, 767-776 (1975)
5.71 J.S. Turner: Deep-Sea Res. *14*, 599-611 (1967)
5.72 S.P. Hayes, T.M. Joyce, R.C. Millard, Jr.: J. Geophys. Res. *80*, 314-319 (1975)
5.73 S. Tabata, N.E.J. Boston, F.M. Boyce: J. Geophys. Res. *70*, 3867-3878 (1965)
5.74 K. Bathen: J. Geophys. Res. *77*, 7138-7150 (1972)
5.75 A.D. Voorhis, E.H. Schroeder, A. Leetmaa: J. Phys. Oceanogr. *6*, 953-961 (1976)
5.76 R.T. Pollard: "Observations and Models of the Structure of the Upper Ocean", in *Modeling and Prediction of the Upper Layers of the Ocean*, ed. by E.B. Kraus, NATO Advanced Study Institute, Urbino, Italy, 1975 (Pergamon Press, New York 1977) pp.102-117
5.77 P.P. Niiler, E.B. Kraus: "One-Dimensional Models of the Upper Ocean", in *Modeling and Prediction of the Upper Layers of the Ocean*, ed. by E.B. Kraus, NATO Advanced Study Institute, Urbino, Italy, 1975 (Pergamon Press, New York 1977) pp.143-172
5.78 H. Stommel, K. Saunders, W. Simmons, J. Cooper: Deep-Sea Res. *16* (Suppl.), 269-284 (1969)
5.79 K.F. Bowden, M.R. Howe, R.I. Tait: Deep-Sea Res. *17*, 401-411 (1970)
5.80 V.E. Delnore: J. Phys. Oceanogr. *2*, 239-247 (1972)
5.81 J.A.C. Kaiser: J. Phys. Oceanogr. *8* (in press)

# 6. Inverse Methods for Reflector Mapping and Sound Speed Profiling

N. Bleistein and J. K. Cohen

With 9 Figures

Over the past few years, the present authors have been engaged in a coordinated research program on inverse problems.

This chapter describes the two main segments in this research program:

I) physical optics far-field inverse scattering (denoted by the acronym, POFFIS);

II) seismic or subsurface profiling in media with small variations in propagation speed.

The first of these has application to the mapping of "high contrast" interfaces in the seabed while the second can be applied to small inhomogeneity problems.

## 6.1 The POFFIS Identity

The POFFIS identity was originally derived by BOJARSKI [6.1]. Analysis of the basic identity and its extensions can be found in [6.2-10]. The problem of interest here is reconstruction of the image of a convex scatterer from observations of the high frequency far-field scattering from the object in response to a known probing signal.

The POFFIS identity relates the scattered field to the Fourier transform of the characteristic function of the scatterer. The characteristic function is equal to unity in the region occupied by the scatterer and zero elsewhere. Thus, knowledge of this function makes possible the reconstruction of an image of the scatterer. The Fourier transform of the characteristic function is itself a function of a vector transform variable $\underline{k}$. The magnitude of the vector is proportional to frequency; the direction of the vector is determined by the source-receiver directions. Consequently, in any practical situation, with limited bandwidth and directions of observation, the Fourier transform is known only in some aperture limited — band limited and aspect angle limited — domain in $\underline{k}$ space. The problem of extracting information about the characteristic function from a high frequency aperture limited Fourier transform is solely a question in the Fourier analysis of piecewise constant functions. The problem is discussed from this point of view in Sect.6.1.2 and will prove useful in the discussions in Sect.6.2.1-3 below.

### 6.1.1 Derivation of the POFFIS Identity

A point source located at $\underline{x}_0$ in Fig.6.1, gives rise to a signal $p_I(\underline{x},\underline{x}_0,\omega)$ which is a solution of the reduced wave equation

$$\nabla^2 p_I + \omega^2 p_I/c^2 = -\delta(\underline{x}-\underline{x}_0) \quad ; \quad \underline{x} = (x_1,x_2,x_3) \quad . \tag{6.1}$$

Here, $\nabla^2$ is the three-dimensional Laplacian, $\delta(\underline{x})$ is a three-dimensional Dirac delta function and

$$p_I(\underline{x},\underline{x}_0,\omega) = \exp(i\omega|\underline{x}-\underline{x}_0|/c)/(4\pi|\underline{x}-\underline{x}_0|) \quad . \tag{6.2}$$

The signal $p_I$ is incident upon a convex scatterer (B in Fig.6.1), giving rise to a scattered signal $p_S$. The scattered field has the following integral representation in terms of its values on $\partial B$, the boundary of B (see, for example, [6.11]):

$$p_S(\underline{x},\underline{x}_0,\omega) = \int\limits_{\partial B} \left[p_S(\underline{x}',\underline{x}_0,\omega)\frac{\partial g}{\partial n}(\underline{x},\underline{x}',\omega) - g(\underline{x},\underline{x}',\omega)\frac{\partial p_S}{\partial n}(\underline{x}',\underline{x}_0,\omega)\right]dA \quad . \tag{6.3}$$

Here the integral is over the scattering surface in prime variables, g is the Green's function, given by (6.2) with $\underline{x}_0$ replaced by $\underline{x}'$, and $\partial/\partial n$ denotes the outward normal derivative to $\partial B$.

At this point, the *physical optics* approximations are introduced [6.11]. To do so, attention is first restricted to the backscatter case, $\underline{x} = \underline{x}_0$ in (6.3)[1]. Then, it is assumed that $\underline{x}_0$ is very far from the scatterer. Thus, in a coordinate system with origin "near" or inside the scatterer, $x_0 = |\underline{x}_0| \gg 1$. The scatterer is now di-

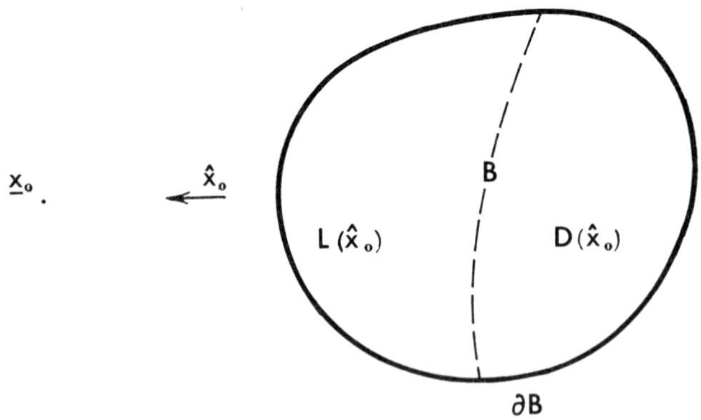


Fig. 6.1. A scattering object B divided into lit side (L) and dark side (D) relative to a point source at $\underline{x}_0$ subject to physical optics laws

[1]The physical optics approximations remain valid for a "small" separation angle between $\underline{x}$ and $\underline{x}_0$.

vided into two parts, a lit side $L(\hat{x}_0)$ and a dark side $D(\hat{x}_0)$ with respect to "physical optics" illumination by a plane wave from the direction of $\underline{x}_0$, that direction denoted by the unit vector $\hat{x}_0$ as shown in Fig.6.1.

Under the assumption that the scatterer is acoustically hard (so that the exact boundary condition is that the normal derivative of the total field is zero on $\partial B$), the physical optics approximations are as follows

$$p_S = \frac{\partial p_S}{\partial n} = 0 \quad , \quad \underline{x}' \text{ on } D(\hat{x}_0) \quad ; \tag{6.4}$$

$$p_S = p_I \quad , \quad \frac{\partial p_S}{\partial n} = -\frac{\partial p_I}{\partial n} \quad , \quad \underline{x}' \text{ on } L(\hat{x}_0) \quad . \tag{6.5}$$

For the acoustically soft case (where the exact boundary condition is that the total field is zero on $\partial B$), the minus sign in the second equation in (6.5) is deleted and a minus sign is inserted on the right side of the first equation in (6.5). This will only result in an overall minus sign in the results to be derived and, hence, only the case (6.4,5) will be discussed below.

Applying all of these results to (6.3) yields

$$p_S(\underline{x}_0,\underline{x}_0,\omega) = \int_{L(\hat{x}_0)} \left\{\frac{\partial}{\partial n}\left[p_I^2(\underline{x}',\underline{x}_0,\omega)\right]\right\} dA \quad . \tag{6.6}$$

The far field assumption is now made, namely that

$$x_0 \gg x \quad , \quad x = |\underline{x}| \quad , \quad \underline{x} \text{ on } \partial B \quad . \tag{6.7}$$

Then, for $\underline{x}$ on $\partial B$, $|\underline{x}-\underline{x}_0|$ may be expanded as follows:

$$|\underline{x}-\underline{x}_0| = x_0 - \underline{x} \cdot \hat{x}_0 + O(x^2/x_0) \quad . \tag{6.8}$$

The result (6.2) is now inserted in (6.6) and then from (6.8), a two-term expansion is used in the phase and a one-term expansion is used in the amplitude to yield

$$p_S(\underline{x}_0,\underline{x}_0,\omega) = \frac{\exp(2i\omega x_0/c)\rho(\omega\hat{x}_0/c)}{(4\pi x_0)^2} \quad . \tag{6.9}$$

Here $\rho(\omega\hat{x}_0/c)$ is a *phase and range normalized far field scattering amplitude* given by

$$\rho(\omega\hat{x}_0/c) = \int_{L(\hat{x}_0)} \left\{\frac{\partial}{\partial n}\left[\exp(-2i\omega\underline{x}'\cdot\hat{x}_0/c)\right]\right\} dA \quad . \tag{6.10}$$

If a similar experiment is performed from the opposite direction, the result is

$$\rho(-\omega\hat{x}_0/c) = \int_{D(\hat{x}_0)} \left\{\frac{\partial}{\partial n}\left[\exp(2i\omega\underline{x}'\cdot\hat{x}_0/c)\right]\right\} dA \quad . \tag{6.11}$$

From (6.10) and (6.11), it follows that

$$\rho(\omega\hat{x}_0/c) + \rho^*(-\omega\hat{x}_0/c) = \int_{\partial B}\left\{\frac{\partial}{\partial n}\left[\exp(-2i\omega\underline{x}'\cdot\hat{x}_0/c)\right]\right\}dA \quad . \tag{6.12}$$

Here (*) denotes complex conjugate.

Applying the divergence theorem here yields

$$\rho(\omega\hat{x}_0/c) + \rho^*(-\omega\hat{x}_0/c) = \int_B \nabla^2\left[\exp(-2i\omega\underline{x}'\cdot\hat{x}_0/c)\right] dV = -(4\omega^2/c^2)\tilde{\gamma}(2\omega\hat{x}_0/c) \ . \tag{6.13}$$

Here,

$$\tilde{\gamma}(\underline{k}) = \int_B \exp(-i\underline{k}\cdot\underline{x}')dV \quad , \quad \underline{k} = (k_1,k_2,k_3) \quad . \tag{6.14}$$

The function $\tilde{\gamma}(\underline{k})$ is the Fourier transform of the function

$$\gamma(\underline{x}) = \begin{cases} 1, & \underline{x} \text{ in B} \\ 0, & \underline{x} \text{ not in B} \end{cases} \quad . \tag{6.15}$$

This function is the characteristic function of the region B. If $\gamma(\underline{x})$ is known, the region B is known. [Actually, if the discontinuity of $\gamma(\underline{x})$ is known, B is known.] The result (6.13) relates the Fourier transform of $\gamma(\underline{x})$ to the phase and range normalized scattering amplitude. The value of the transform variable is

$$\underline{k} = 2\omega\hat{x}_0/c \quad . \tag{6.16}$$

Thus, (6.13) is the POFFIS identity.

Alternative forms of the POFFIS identity will now be derived. Returning to (6.9), the inverse time transforms of $p_S$ and $\rho$ are defined to be $P_S$ and V, respectively. From (6.9) it follows that

$$V(\hat{x}_0,t) = (4\pi x_0)^2 \, P_S(\underline{x}_0,t+2x_0/c) \quad . \tag{6.17}$$

That is, the Fourier transform of the phase and range normalized scattering amplitude is a range normalized and time-delayed scattering amplitude in the time domain and thus, is as easily observable as $\rho(\omega\hat{x}_0/c)$. The objective now is to show how to use this result and the basic POFFIS identity to derive a POFFIS identity in the time domain.

The Fourier inversion in (6.13) is first expressed as an integral in polar coordinates. The result is

$$\gamma(\underline{x}) = -\frac{1}{4\pi^3 c}\int_0^\infty d\omega \int_\Omega d\Omega\left[\rho(\omega\hat{x}_0/c)+\rho^*(-\omega\hat{x}_0/c)\right] \cdot \exp(2i\omega\hat{x}_0\cdot\underline{x}/c) \quad . \tag{6.18}$$

Here, $\Omega$ denotes the unit sphere with variable $\hat{x}_0$ and differential solid angle element on it, $d\Omega$. The volume element in polar coordinates has a factor of $\omega^2$; however, the integrand also has a factor $\omega^{-2}$ arising from solving for $\tilde{\gamma}$.

This result may be rewritten as an integral over the hemisphere centered at $\hat{x}_0$. The result is

$$\gamma(\underline{x}) = -\frac{1}{4\pi^3 c}\int_0^\infty d\omega \int_{\Omega/2} d\Omega\Big\{\big[\rho(\omega\hat{x}_0/c)+\rho^*(-\omega\hat{x}_0/c)\big]\cdot\exp(2i\omega\hat{x}_0\cdot\underline{x}/c)$$
$$+\big[\rho(-\omega\hat{x}_0/c)+\rho^*(\omega\hat{x}_0/c)\big]\cdot\exp(-2i\omega\hat{x}_0\cdot\underline{x}/c)\Big\} \quad . \qquad (6.19)$$

Here, the second pair of terms, with $\hat{x}_0$ replaced by $-\hat{x}_0$, provides the integral over the back hemisphere. From the definitions of $\rho$ in (6.10,11), it should be recalled that these terms are calculated by integrating over regions $L(-\hat{x}_0)$ and $D(-\hat{x}_0)$, respectively. These, in turn, are just $D(\hat{x}_0)$ and $L(\hat{x}_0)$, respectively. Furthermore, from (6.11)

$$\rho^*(\omega\hat{x}_0/c) = \rho(-\omega\hat{x}_0/c) \quad , \qquad (6.20)$$

where now, the change of sign is in $\omega$, with $\hat{x}_0$ and hence $L(\hat{x}_0)$ remaining unchanged. Thus, the fourth term provides the extension of the first term to negative frequencies; the second term provides the same extension of the third term. Thus,

$$\gamma(\underline{x}) = -\frac{1}{4\pi^3 c}\int_{-\infty}^{\infty} d\omega \int_{\Omega/2} d\Omega\big[\rho(\omega\hat{x}_0/c)\ \exp(2i\omega\hat{x}_0\cdot\underline{x}/c)+\rho(-\omega\hat{x}_0/c)\ \exp(-2i\omega\hat{x}_0\cdot\underline{x}/c)\big] \quad . \qquad (6.21)$$

Here, $\rho$ in the first line is defined by (6.10) and in the second line by (6.11). However, again using the fact that $D(\hat{x}_0) = L(-\hat{x}_0)$, the result can again be written as an integral over a sphere:

$$\gamma(\underline{x}) = -\frac{1}{4\pi^3 c}\int_{-\infty}^{\infty} d\omega \int_{\Omega} d\Omega\ \rho(\omega\hat{x}_0/c)\ \exp(2i\omega\hat{x}_0\cdot\underline{x}/c) \quad . \qquad (6.22)$$

The inverse time transform of $\rho$ was defined to be V. Therefore, (6.22) can be rewritten as

$$\gamma(\underline{x}) = -\frac{1}{4\pi^3 c}\int_{\Omega} d\Omega\ V(\hat{x}_0,-2\hat{x}_0\cdot\underline{x}/c) \quad . \qquad (6.23)$$

This is the POFFIS identity in the time domain.

The result (6.22) also provides an alternative to the POFFIS identity (6.13). In the form (6.21), the requirement of combining "front side" and "back side" observations before Fourier synthesis has been dispensed with. There is still a double covering in $\underline{k}$ space in (6.22), but this may be overcome by using (6.19) to conclude that

$$\gamma(\underline{x}) = -\frac{\mathrm{Re}}{2\pi^3 c}\int_0^\infty d\omega \int_\Omega d\Omega\ \rho(\omega\hat{x}_0/c)\ \exp(2i\omega\hat{x}_0\cdot\underline{x}/c) \quad . \tag{6.24}$$

In this form, the integral is again a Fourier transform. When (6.9) is used and the integral is expressed again in terms of

$$\underline{k} = \omega\hat{x}_0/c \quad , \quad \hat{k} = \hat{x}_0 \quad , \tag{6.25}$$

the result becomes

$$\gamma(\underline{x}) = -8\pi^{-1}\ \mathrm{Re}\left\{\int k^{-2}d^3k x_0^2 u_S(\hat{k}x_0,\hat{k}x_0,ck)\cdot\exp(2i\underline{k}\cdot\underline{x}-2ikx_0)\right\} \quad . \tag{6.26}$$

In this equation $x_0$ may be a function of $\hat{x}_0$ (i.e., the observation distance may be different in different directions). Thus, in general, $x_0$ must remain under the integral sign.

It should be noted that the POFFIS identity is a result derived for high frequency only. Thus, the identity is suspect when used to determine $\tilde{\gamma}(\underline{k})$ for small values of $|\underline{k}|$. Often, as a practical matter, low frequency (and hence, small $|\underline{k}|$) information is simply not available. Furthermore, the obstacle often cannot be probed from all directions $\hat{x}_0$. Thus, one is faced with the limited aperture problem mentioned in the introduction and addressed in Sect.6.1.2.

The POFFIS identities derived here can be generalized to treat localized anomalies in an otherwise flat region [6.8].

### 6.1.2 The Limited Aperture Problem for the POFFIS Identity

As noted in earlier sections, complete information required for Fourier inversion of the POFFIS identity is not available in practice. That high frequency far field scattering data should suffice to reconstruct the image of the scatterer has been rigorously proven by MAJDA [6.12].

In this section, a method for processing both band limited and aspect angle limited data will be described. The method was originally proposed by BOJARSKI [6.3] for high frequency band limited data and further analyzed for this case in [6.9]. Limited aspect angle is discussed in [6.10].

The method is based solely on the features of the Fourier transform of piecewise constant (in this case, one-zero) functions. The only additional assumption made here is that the region in which the function equals unity is convex, finite and has a smooth boundary surface. These are exactly the assumptions made for the scatterer B of Sect.6.1.1. Indeed, the function to be analyzed here is $\gamma(\underline{x})$ defined by (6.15) and its Fourier transform $\tilde{\gamma}(\underline{k})$ defined by (6.14).

The approach to be presented here is based on the following ideas from one-dimensional Fourier analysis. If $\gamma(x)$ is a one-dimensional characteristic function, then its first derivative consists of two Dirac delta functions at the boundaries of the region where $\gamma(x) = 1$. The function $\delta(x-a)$ has Fourier transform $\exp(-ika)$ and band limited inverse transform

$$\delta(x-a) \approx \pi^{-1}(x-a)^{-1} \sin[k(x-a)]\Big|_{k_-}^{k_+} \quad , \tag{6.27}$$

with $k_\pm$ the band limits. This function peaks at $x = a$, with peak value $(k_+ - k_-)/\pi$. Thus, the boundaries of the region of interest are readily recognized from this band limited processing. The purpose of this section is to show that this is true in two- and three-dimensions, as well, and also to show the effects of aspect angle limitations in these cases.

For the function $\gamma(\underline{x})$ defined by (6.15), the directional derivative is the direction defined by the unit vector $\hat{\eta}$ given by

$$\Delta(\underline{x},\hat{\eta}) = -\hat{\eta} \cdot \nabla\gamma(\underline{x}) \quad . \tag{6.28}$$

If $\partial B$ is defined by $\Phi(\underline{x}) = 0$, with $\Phi$ negative in B and positive outside of B, then

$$\Delta(\underline{x},\hat{\eta}) = \hat{\eta} \cdot \nabla\Phi\delta(\Phi) \quad . \tag{6.29}$$

The Fourier transform of $\Delta$ is

$$\tilde{\Delta}(\underline{k},\hat{\eta}) = \int \hat{\eta} \cdot \nabla\Phi(\underline{\xi})\delta[\Phi(\underline{\xi})] \exp(-i\underline{k}\cdot\underline{\xi})dV \quad . \tag{6.30}$$

Here, the domain of integration is all of $\xi$-space. The integration normal to the boundary is readily calculated by exploiting the properties of the delta function. The result is

$$\tilde{\Delta}(\underline{k},\hat{\eta}) = \int_{\partial B} \hat{\eta} \cdot \hat{n} \exp(-i\underline{k}\cdot\underline{\xi})dA \quad . \tag{6.31}$$

Here, $\hat{n}$ is the outward normal to $\partial B$.

The aperture limited inverse transform of $\tilde{\Delta}$ will be denoted by $\Delta_1$. If $A_\eta(\underline{k})$ denotes the aperture in $\underline{k}$ space, then

$$\Delta_1(\underline{x},\hat{\eta}) = \frac{1}{(2\pi)^3} \int_{A_\eta(\underline{k})} d^3k \int_{\partial B} dA\, \hat{\eta} \cdot \hat{n} \exp[i\underline{k}\cdot(\underline{x}-\underline{\xi})] \quad . \tag{6.32}$$

The apertures of interest are those for which $k = |\underline{k}|$ ranges between two values $k_-$ and $k_+$ and the angle of $\underline{k}$ is restricted to some region $\Omega$. Thus,

$$\Delta_1(\underline{x},\hat{\eta}) = \frac{1}{(2\pi)^3} \int_{k_-}^{k_+} k^2 dk \int_{\Omega} d\Omega \int_{\partial B} dA\, \hat{\eta} \cdot \hat{n} \exp[i\underline{k}\cdot(\underline{x}-\underline{\xi})] \quad . \tag{6.33}$$

Here, $d\Omega$ is the solid angle element in $\underline{k}$ space.

The four fold integral over $\Omega$ and $\partial B$ can be calculated by the method of four-dimensional stationary phase [6.13]. The integration over k can then be done explicitly. The method is described in [6.10], but the form of the result presented here is due to ARMSTRONG [6.14]. The conditions that the phase be stationary are as follows. Firstly, the point $\underline{\xi}$ on $\partial B$ must be such that $\underline{x} - \underline{\xi}$ is perpendicular to $\partial B$. Secondly, the angles of $\underline{k}$ are such that $\underline{x} - \underline{\xi}$ and $\underline{k}$ are colinear or anticolinear. Thus, the stationary points in $\underline{\xi}$ and $\underline{k}$ are such that $\underline{x} - \underline{\xi}$ and $\underline{k}$ lie along the normal to $\partial B$ at the stationary point.

For each choice of $\underline{x}$ there may be one or more stationary points (or none at all!). If the contribution from each stationary point is denoted by $\Delta_2(\underline{x},\hat{n})$, then one finds that asymptotically

$$\Delta_2(\underline{x},\hat{n}) = (2\pi)^{-1}\,\hat{n}\cdot\hat{n}(1-\mu K_1 D)^{-\frac{1}{2}}(1-\mu K_2 D)^{-\frac{1}{2}}\cdot D^{-1}\left\{\sin(kD)-i[\cos(kD)-1]\right\}\Bigg|_{k=k_-}^{k_+} \quad . \tag{6.34}$$

Here,

$$D = |\underline{x}-\underline{\xi}| \quad ; \tag{6.35}$$

$K_1$ and $K_2$ are the principal curvatures of $\partial B$ at the stationary point; $\mu = +1$ or $-1$ according to whether $\underline{x} - \underline{\xi}$ and $\hat{n}$ are anticolinear or colinear at the stationary point.

The real part $\Delta_2$ is seen to have the same qualitative behavior for D small as the one-dimensional result (6.27). Furthermore,

$$\lim_{D\to 0} \Delta_2(\underline{x},\hat{n}) = (2\pi)^{-1}\,\hat{n}\cdot\hat{n}(k_+-k_-) \quad . \tag{6.36}$$

For each point on $\partial B$ the contributions from stationary points with $D = 0$ will dominate all other contributions to $\Delta_1(\underline{x},\hat{n})$ if the bandwidth is sufficiently broad. There can be at most two such contributions. This will occur if the region $\Omega$ contains both the vectors $\hat{k} = \hat{n}$ and $\hat{k} = -\hat{n}$ for the given stationary point $\underline{\xi}$.

In [6.9], the following test of this method, as applied to the POFFIS identity, was carried out. The exact backscatter solution for a sphere of unit radius was used to generate $\rho(\omega\hat{x}_0/c)$ for the POFFIS identity (6.13). The angular aperture $\Omega$ was taken to be unrestricted; i.e., all angles were used. The vector $\hat{n} = (1,0,0)$ was used and $\Delta_1(\underline{x},\hat{n})$ was calculated for $x_2 = x_3 = 0$. The angular integrations were carried out analytically and the integration in k was done numerically with the trapezoid rule. An example of the output of this processing is shown in Fig.6.2. The circles represent the theoretical results based on the asymptotic analysis presented here. In this case $k_- = 15.75 \approx 5\pi$, $k_+ = 31.5 \approx 10\pi$. At the peak, the percentage error between theory and computation was 0.3%.

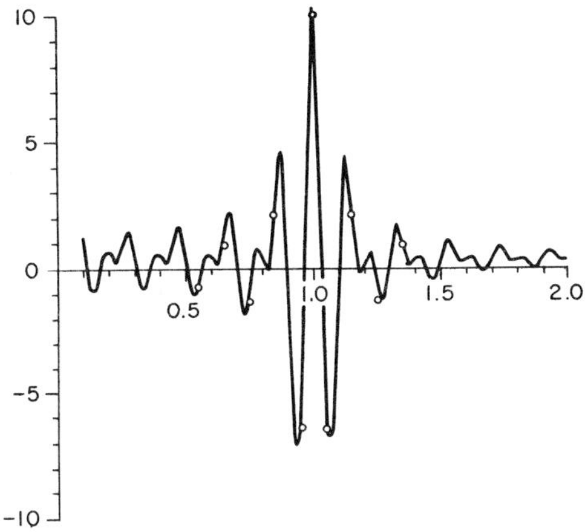


Fig. 6.2. The bandlimited output of the POFFIS formalism at fixed angle. The scatterer was a sphere of unit radius. The range in wave number was 15.75-31.5

Fig. 6.3. Computer output of function $\Delta_2$ for an ellipse with principle axes of length one and two and K range 9-27

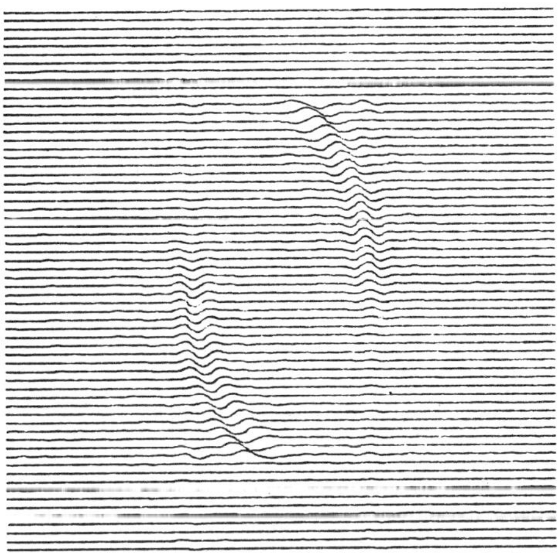

Fig. 6.4. Same configuration as in Fig.6.3, except that now the angular range is limited to 0-90° and 180-270°

The two-dimensional analog of the results presented here is carried out in exactly the same manner. In (6.33), $\partial B$ would now be a curve and $d\Omega$ is replaced by $d\theta$, the polar angle of $\underline{k} = (k_1, k_2)$. The factor $(2\pi)^{-3}$ is replaced by $(2\pi)^{-2}$. The result (6.34) is now replaced by

$$\Delta_2(\underline{x},\hat{n}) = (2\pi)^{-1}\ \hat{n} \cdot \hat{n}(1-\mu KD)^{-\frac{1}{2}} \cdot D^{-1}\left\{\sin(kD)-i[\cos(kD)-1]\right\}\Big|_{k_-}^{k_+} \quad . \tag{6.37}$$

Here $\mu$ and D are as defined below (6.34).

In [6.15], the two-dimensional result was tested for an ellipse with semi-axes of length one and two. The result for full angular aperture k range 9 to 27 and $\hat{n} = (1,0)$ is shown in Fig.6.3. The third dimension (the value of $\Delta_2$) is laid back down in the plane. On each line, the height is normalized with respect to maximum height. Furthermore, the maximum on each line is tested against the absolute maximum. When this relative maximum falls below a critical value, the entire line is zeroed out. This insures, for example, that "noise" will not be enhanced outside of the vertical extent of the ellipse. Nonetheless, near the top and bottom of the ellipse, where $\hat{\eta} \cdot \hat{n}$ is nearly zero, there is clearly some loss in resolution. This region could be resolved better if $\hat{\eta}$ were chosen to be (0,1).

In Fig.6.4, a limited aperture process is shown for $\Omega$ consisting of two quadrants. There is some "spill-over" outside of the angular aperture, but, for the most part, the ellipse is reproduced as in the previous diagram, but only for the region for which the normals to the ellipse lie in the angular aperture in $\underline{k}$ space.

In [6.10], further examples are carried out in which numerical data is generated for backscattering by a circular cylinder. The results confirm the POFFIS method of Sect.6.1 as well as the Fourier analysis of this section.

It should be noted that near the evolute of $\partial B$, (6.34) or (6.37) becomes invalid. In [6.14], ARMSTRONG shows that the contribution from that region is extremely small relative to the peak value of $\Delta_2$ at $D = 0$.

Based upon the results of this section, the practical implementation of any form of the POFFIS identity involves taking a formal derivative of the given formula for the characteristic function.

## 6.2 An Inverse Method for Determining Small Inhomogeneities in a Medium

When a medium is known to have small and perhaps slowly varying inhomogeneities, the method of Sect.6.1 is inapplicable to the problem of reconstructing the medium. The method presented here is a procedure for determining such variations. This method would be appropriate to the detection of velocity gradients in the seabed as well as to the general problems of subsurface mapping or geophysical exploration. Mathematically, the inverse problem considered here is to determine coefficients in a system of equations governing a wave propagation problem. It is assumed that the reference values of these coefficients (often constants) are available and that *variations from these reference values* are to be determined from observations of the field arising from known input sources. Physically, the coefficients to be determined usually characterize the medium velocities or the acoustic impedance or some similar quantity.

For clarity of exposition, the discussion here will be restricted to media in which propagation is governed by the three-dimensional wave equation. In [6.16] the inverse problem for electromagnetic waves and elastic waves was discussed, as well.

An essential feature of the inverse method presented here is that an integral equation is derived for a function which characterizes the velocity or impedance variation. This equation is a Fredholm integral equation of the first kind; it has parameters in it which characterize the source and receiver locations. In a number of cases of source-receiver configurations of practical interest, this integral equation is invertible. Thus, a velocity or impedance profile (depth section) is obtained by direct processing of the observed data (time section) itself; i.e., by performing weighted quadratures on the data. This direct inversion of time section to depth section involves only one theoretical assumption: the subsurface variation must be "small". Even this limitation is not overly restrictive as can be seen from one of the example below in which a 20% velocity variation was successfully treated. In fact, the real world data restrictions—noise, attenuation, discretization and finiteness of observations, etc.—are usually of greater concern than this theoretical linearization assumption.

### 6.2.1 An Integral Equation for Three-Dimensional Velocity Variation

It is assumed here that three-dimensional wave propagation is governed by the equation

$$\left[\nabla^2 - v^{-2}(\underline{x},z)\partial_t^{\,2}\right]P(\underline{x},z,t) = 0 \quad ; \quad \underline{x} = (x,y) \quad . \tag{6.38}$$

Waves propagate in the semi-infinite medium, $z > 0$, measured *downward*. The operator $\partial_t$ denotes a partial derivative with respect to time. It is the coefficient $v(\underline{x},z)$ which is to be determined from probes and observations made at the upper surface $z = 0$. For many cases of interest, the surface $z = 0$ separates two media of greatly differing impedances (e.g., air-water, air-earth). Hence, an appropriate boundary condition for (6.38) which introduces the probes is

$$\partial_z P = \Phi(t,\underline{x};\underline{\lambda}) \quad , \quad z = 0 \quad , \quad \underline{\lambda} = (\lambda,\mu) \quad . \tag{6.39}$$

The sources $\Phi$ characterized by the vector $\underline{\lambda}$ may be many types ([6.16] for a discussion of plane wave sources). For definiteness, here, the discussion will be restricted to impulsive point sources located at $\underline{x} = \underline{\lambda}$, $z = 0$. Thus, the boundary condition actually treated is

$$\partial_z P = -\delta(t)\delta(\underline{x}-\underline{\lambda}) \quad , \quad z = 0 \quad . \tag{6.40}$$

To complete the specification of the direct problem for P, it is assumed that

$$P \equiv 0 \quad , \quad t < 0 \quad . \tag{6.41}$$

The reference value of the velocity, which is assumed known, will be denoted by $c(\underset{\sim}{x},z)$. Then, the variation, denoted by $\alpha(\underset{\sim}{x},z)$, is defined by the equation

$$v^{-2}(\underset{\sim}{x},z) = c^{-2}(\underset{\sim}{x},z)[1+\alpha(\underset{\sim}{x},z)] \quad . \tag{6.42}$$

It is the variation $\alpha$ which is to be determined by means of observations of P made at surface points $\underset{\sim}{x} = \underset{\sim}{\xi}$, $z = 0$.

The total field P is decomposed into a primary field $P_I$, which is the solution in the absence of the variation $\alpha$, and a scattered field $P_S$, which is the response to $\alpha$. Thus,

$$P(t,\underset{\sim}{x},z;\underset{\sim}{\lambda}) = P_I(t,\underset{\sim}{x},z;\underset{\sim}{\lambda}) + P_S(t,\underset{\sim}{x},z;\underset{\sim}{\lambda}) \quad , \tag{6.43}$$

where $P_I$ satisfies the following

$$\left(\nabla^2 - c^{-2}\partial_t^{\,2}\right)P_I = 0 \quad , \quad z > 0 \quad ; \tag{6.44}$$

$$\partial_z P_I = -\delta(t)\delta(\underset{\sim}{x}-\underset{\sim}{\lambda}) \quad , \quad z = 0 \quad ; \tag{6.45}$$

$$P_I \equiv 0 \quad , \quad t < 0 \quad . \tag{6.46}$$

The argument $\underset{\sim}{\lambda}$ is introduced into P, $P_I$, and $P_S$ to emphasize their dependence on the source location. It then follows from the problems for P and $P_I$ that $P_S$ satisfies the following

$$\left(\nabla^2 - c^{-2}\partial_t^{\,2}\right)P_S = -\alpha c^{-2}\partial_t^{\,2}P \quad ; \tag{6.47}$$

$$\partial_z P_S = 0 \quad , \quad z = 0 \quad ; \tag{6.48}$$

$$P_S \equiv 0 \quad , \quad t < 0 \quad . \tag{6.49}$$

An adjoint function Q is now introduced, satisfying the following

$$\left(\nabla^2 - c^{-2}\partial_t^{\,2}\right)Q = 0 \quad , \quad z > 0 \quad ; \tag{6.50}$$

$$\partial_z Q = -(\tau-t)H(\tau-t)\delta(\underset{\sim}{x}-\underset{\sim}{\xi}) \quad , \quad z = 0 \quad ; \quad \underset{\sim}{\xi} = (\xi,\eta) \quad ; \tag{6.51}$$

$$Q \equiv 0 \quad , \quad t > \tau \quad . \tag{6.52}$$

It will be seen below that $\underset{\sim}{\xi}$, $\tau$ denote the locations and times of observations at the upper surface. Furthermore, by comparing the problems for $P_I$ and Q, it can be seen that

$$\partial_t^{\,2}Q(t,\underset{\sim}{x},z;\underset{\sim}{\xi}) = P(\tau-t,\underset{\sim}{\xi},z;\underset{\sim}{x}) \quad . \tag{6.53}$$

Upon applying Green's theorem (in space and time) to the quantity

$$P_S\left(\nabla^2 - c^{-2}\partial_t^{\,2}\right)Q - Q\left(\nabla^2 - c^{-2}\partial_t^{\,2}\right)P_S \quad ,$$

the following integral equation is obtained

$$\int_0^\infty dz \int_{-\infty}^\infty dx\, dy\; \alpha(\underline{x},z)c^{-2}(\underline{x},z) \int_0^\tau dt\; P(t,\underline{x},z;\underline{\lambda})P_I(\tau-t,\underline{\xi},z;\underline{x})$$

$$= \int_0^\tau dt\, P_S(t,\underline{\xi},0;\underline{\lambda})(\tau-t) \quad . \tag{6.54}$$

The right-hand side here is a function of the field observations at the upper surface and, hence, known. The left-hand side has two unknowns, namely $\alpha$ and $P_S$ [a nonlinear system for these two unknowns is supplied by (6.54) and (6.47,49)]. However, $P_S$ appears only in P and, therefore, only through the product $\alpha P_S$. However, from (6.47) it is seen that $P_S$ is itself of the order of $\alpha$. Thus, for $\alpha$ small, it is expected that $\alpha P$ can be reasonably approximated by $\alpha P_I$. In this case, (6.54) becomes an integral equation for $\alpha$ alone, namely

$$\int_0^\infty dz \int_{-\infty}^\infty dx\, dy\; \alpha(\underline{x},z)c^{-2}(\underline{x},z) \int_0^\tau dt\; P_I(t,\underline{x},z;\underline{\lambda})P_I(\tau-t,\underline{\xi},z;\underline{x})$$

$$= \int_0^\tau dt\, P_S(t,\underline{\xi},0;\underline{\lambda})(\tau-t) \quad . \tag{6.55}$$

This is the basic linear integral equation for $\alpha(\underline{x},z)$.

### 6.2.2 Direct Inversion for Backscatter Over a Medium with Two-Dimensional Velocity Variation

The integral equation (6.55) will now be specialized as follows. The reference velocity c will be taken to be a constant, so that $P_I$, defined by (6.44-46), is given by

$$P_I(t,\underline{x},z;\underline{\lambda}) = \delta(t-|\underline{x}-\underline{\lambda}|/c)/(2\pi|\underline{x}-\underline{\lambda}|) \quad . \tag{6.56}$$

The variation in $\alpha$ will be assumed to be independent of y, so that $\alpha(\underline{x},z)$ will be replaced by $\alpha(x,z)$. In this case, both the sources and receivers will be restricted to the x axis. Finally, only backscattered (or CDP stacked) observations will be made, so that

$$\underline{\lambda} = \underline{\xi} = (\xi,0) \quad . \tag{6.57}$$

For this case, it is possible to carry out the y and t integrations in (6.55) to obtain

$$\int_0^{\infty} dz \int_{-\infty}^{\infty} dx\ \alpha(x,z)H(c\tau/2-\rho)\left[(c\tau/2)^2-\rho^2\right]^{-\frac{1}{2}} = -2\tau(\pi c)^2 \int_0^{\tau} dt\ P_S(t,\underline{\xi},z;\underline{\xi})(\tau-t)\ . \quad (6.58)$$

Here,

$$\rho = \left[(x-\xi)^2+z^2\right]^{\frac{1}{2}}\ . \quad (6.59)$$

This integral equation can be solved by transform techniques. The result is

$$\begin{aligned}\alpha(x,z) &= 2ic^3\pi^{-1}\int_{-\infty}^{\infty} d\xi \int_{-\infty}^{\infty} dk_1 \int_{-\infty}^{\infty} dk_3 \int_0^{\infty} d\tau \int_0^{\tau} dt\Big\{k_3(\tau^2-\tau t) \\ &\quad \cdot P_S(t,\underline{\xi},z;\underline{\xi})\ \exp\left[2ik_1(x-\xi)-2ik_3z+i\omega\tau\right]\Big\}\ ; \\ \omega &= c\left[\mathrm{sgn}(k_3)\right]\left(k_1^2+k_3^2\right)^{\frac{1}{2}}\ . \end{aligned} \quad (6.60)$$

This is the direct inversion formula for the prescribed source receiver configuration for this section.

In [6.17] this result was implemented for high frequency data synthetically produced for a number of layered models. Because of the high frequency bandlimiting, the data was processed for $\partial\alpha/\partial z = \alpha'$ rather than for $\alpha(x,z)$ itself. This merely requires multiplication of the integrand in (6.60) by $2ik_3$. In accordance with the results of Sect.6.1.2, not only the *location* of discontinuities may be determined but the *magnitude* of the discontinuities may be determined, as well. This *direct inversion* procedure produces a mapping of the interfaces and an estimate of the velocity variations across them.

The model used embodies other real world restrictions band limiting, namely

I) the observations are made only at *discrete* points on the line;

II) the observations are made only over a *finite* length of the line.

In these examples, the spacing between $\xi$ points is 100 feet, the maximum array length is 8000 feet and the bandwidth is 6-24 Hz.

Synthetic data was generated for tilted planes at angles up to 75°. Direct inversion reproduced these planes with three-place accuracy both in tilt angle and velocity variation.

In an earlier paper [6.18] a direct inversion formula was presented for the parabolic approximation to the wave equation used by CLAERBOUT [6.19] for (15°) wave equation migration. Synthetic data for tilted planes is used and the direct inversion is carried out analytically. The results show an error in tilt angle and in velocity estimation on the order of the fourth power of the angle of inclination. Numerical tests confirm this result, as well. This kind of error is already known to users of

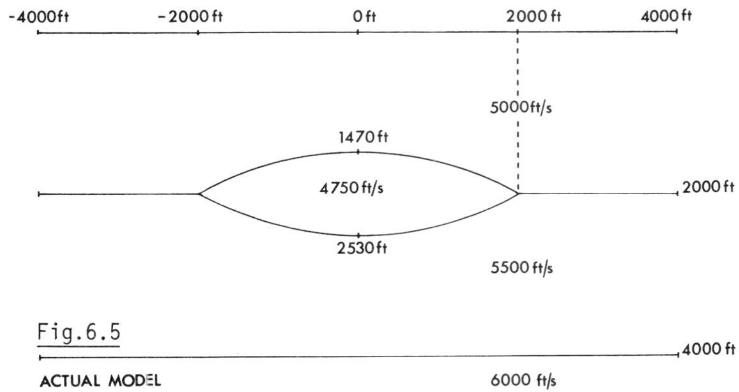


Fig.6.5

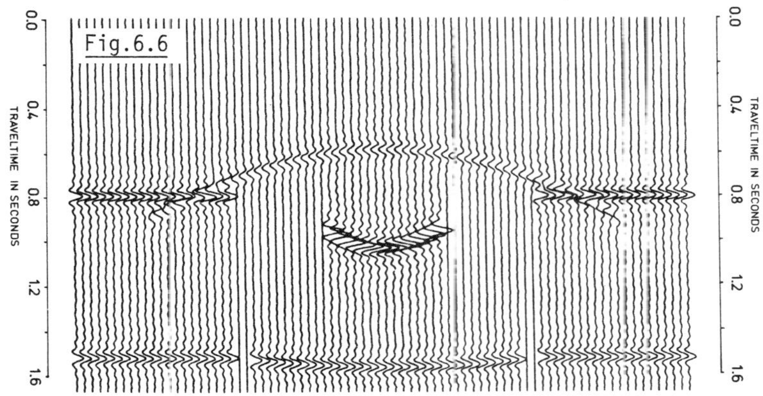


Fig.6.6

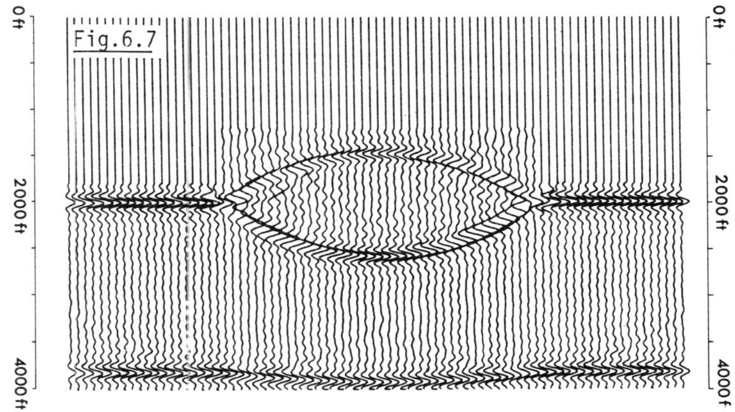


Fig.6.7

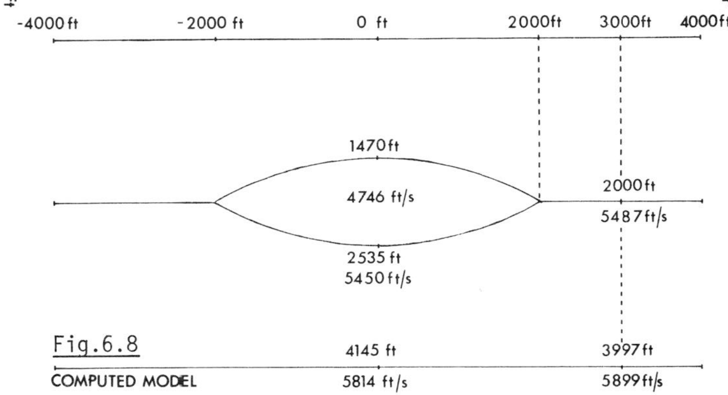


Fig.6.8

Fig. 6.5. Slice of a subsurface region with sound speeds as indicated in the diagram. (With permission of the Editor of Geophysics; this applies also to Figs. 6.6-8)

Fig. 6.6. Synthetic backscattered data for the configuration of Fig. 6.5, as generated by the research group at Marathon Oil Company

Fig. 6.7. Output of a computer algorithm to implement direct velocity inversion on the data of Fig. 6.6.

Fig. 6.8. Velocity and depth estimates as computed from the output which produced Fig. 6.7

$15^{\circ}$ wave equation migration. The error is due to the underlying parabolic approximation and not a flaw of the direct inversion procedure or the wave equation migration procedure. Since the computational requirements of direct inversion for the wave equation are no more difficult than those for the parabolic approximation, the former is preferred to the latter by the present authors.

Figure 6.5 shows a subsurface configuration for which synthetic data was graciously provided to us by the research group at Marathon Oil. The time section provided by them is depicted in Fig.6.6; Fig.6.7 is the result of our direct inversion procedure and Fig.6.8 shows our estimate from the output of various relevant quantities in the model. The lower two sets of numbers exhibit errors of less than 4%, while above the level the errors are less than 1%.

The result corresponding to (6.60) for fully three-dimensional variations is

$$\alpha(\underline{x},z) = 2ic^3\pi^{-2}\int_{-\infty}^{\infty} d\xi d\eta \int_{-\infty}^{\infty} dk_1 dk_2 \int_0^{\infty} dk_3 \int_0^{\infty} d\tau \int_0^{\tau} dt \Big\{ k_3(\tau^2-\tau t)$$

$$\cdot P_S(t,\underline{\xi},z;\underline{\xi})\ \exp\{2i[k_1(x-\xi)+k_2(y-\eta)-k_3 z]+i\omega\tau\}\Big\} \quad ;$$

$$\omega = c\left[\operatorname{sgn}(k_3)\right]\left(k_1^2+k_2^2+k_3^2\right)^{\frac{1}{2}} \quad . \tag{6.61}$$

## 6.2.3 Direct Inversion for a Case with Separated Source and Receiver

Here it is assumed that $\alpha = \alpha(z)$, i.e., that $\alpha$ varies with z alone. Intuitively, then one would expect that only one experiment would be necessary to determine $\alpha$. Again, it is assumed that c is constant so that $P_I$ is given by (6.56). In this case, $\underline{\lambda}$ and $\underline{\xi}$ are taken to be fixed with

$$\underline{\xi} = -\underline{\lambda} \quad . \tag{6.62}$$

In (6.55) the x, y and t integrations can all be performed yielding an elementary integral equation which can readily be inverted to yield

$$\alpha(z) = -4\pi c \int_0^{2(\lambda^2+z^2)^{\frac{1}{2}}/c} dt \left[\frac{2}{c}\cdot\frac{\lambda^2+2z^2}{(\lambda^2+z^2)^{\frac{1}{2}}} - t\right] P_S(t,\underline{\lambda},0;-\underline{\lambda}) \quad . \tag{6.63}$$

For backscatter, it is only necessary to set $\underline{\lambda} = \underline{0}$ here to obtain the result

$$\alpha(z) = -4\pi c \int_0^{2z/c} dt(4z/c-t)P_S(t,\underline{0},0;\underline{0}) \quad . \tag{6.64}$$

### 6.2.4 Direct Inversion for a One-Dimensional Problem

In one dimension, one can generate synthetic wide band data by straightforward and economic means. This was done by GRAY [6.20] as a prelude to the analysis of strongly depth dependent velocities in three dimensions. An example of the output of this analysis is shown in Fig.6.9. The solid wave is the assumed $\alpha(x)$ while the dots are the result of direct migration on synthetically generated wide band backscattered data.

In [6.16] the authors also treated a case in which $c(x)$ was not constant. The function $p_I$, the time transform of $P_I$, was expressible as a sum of exponentials and the integral equation was still invertible in closed form.

### 6.2.5 Direct Inversion in Free Space

Often in oceanographic and seismic exploration, the source is placed below the surface $z=0$. In this case, the basic problem (6.38,40,41) should be replaced by a *source* problem, with the delta function moved from the right side of (6.40) to the right side of (6.38) and the boundary condition (6.40) imposed on a new interface *above* $z=0$, say at $z=z_0<0$.

If the effects of the reflections from the interface at $z_0$ can be accounted for (by no means an easy task), then such an experiment can be modeled by a free space problem in which the medium is "known" for $z<0$ and assumed to vary only for $z>0$. In this case, one still obtains (6.55) as the basic integral equation, but with $P_I$ now a solution of the corresponding source problem in free space. For $c=\text{constant}$, the effect of this modification on $P_I$ is to replace the factor of 2 on the right side of (6.56) by a 4. Since two factors of $P_I$ appear in the kernel of the integral equation (6.55), the results (6.60,61,63) need only be modified by the introduction of a multiplier of 4 on the right side.

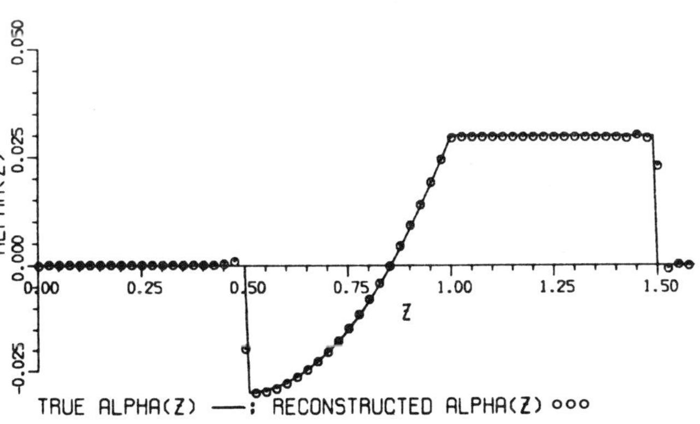


Fig. 6.9. Synthetic example processing wide band data

References


6.1 N.N. Bojarski: "Three-Dimensional Electromagnetic Short Pulse Inverse Scattering", Syracuse Univ. Res. Corp., Syracuse, N.Y., NTIS # AD-845 126 (1967)
6.2 R.M. Lewis: IEEE Trans. AP-*17*, 308-314 (1969)
6.3 N.N. Bojarski: Inverse Scattering Company Rpt. # N00019-73-C-0312/F, prepared for Naval Air Systems Command, AD-775 235/5 (1974)
6.4 W.L. Perry: IEEE Trans. AP-*22*, 826-829 (1974)
6.5 W. Tabbara: IEEE Trans. AP-*21*, 245-247 (1973)
6.6 W. Tabbara: IEEE Trans. AP-*23*, 446-448 (1975)
6.7 S. Rosenbaum-Raz: IEEE Trans. AP-*24*, 66-70 (1976)
6.8 N. Bleistein: J. Acoust. Soc. Am. *59*, 1259-1264 (1976)
6.9 N. Bleistein: J. Acoust. Soc. Am. *60*, 1249-1255 (1976)
6.10 R.D. Mager, N. Bleistein: IEEE Trans. AP-*26*(5), 695-699 (1978)
6.11 B.B. Baker, E.T. Copson: *Mathematical Theory of Hyugens Principle*, 2nd ed. (University Press, Oxford 1950)
6.12 A. Majda: Commun. Pure Appl. Math. *30*, 165-194 (1977)
6.13 N. Bleistein, R.A. Handelsman: *Asymptotic Expansions of Integrals* (Holt, Rinehart and Winston, New York 1975)
6.14 J. Armstrong: "An Analysis of the Aperture Limited Fourier Inversion of Characteristic Functions", Denver Res. Inst. Rpt. # MS-R-7812 (1977)
6.15 N. Bleistein, J.K. Cohen: "Application of a New Inverse Method to Non-Destructive Evaluation", Denver Res. Inst. Rpt. # MS-R-7716 (1977)
6.16 J.K. Cohen, N. Bleistein: SIAM J. Appl. Math. *32*(4), 784-799 (1977)
6.17 J.K. Cohen, N. Bleistein: "A Direct Inversion Procedure for Acoustic Waves", Denver Res. Inst. Rpt. # MS-R-7803 (1977)
6.18 N. Bleistein, J.K. Cohen: "A Note on a Direct Inversion Procedure for Claerbout's Equations", Denver Res. Inst. Rpt. # MS-R-7802 (1977)
6.19 J. Claerbout: *Fundamentals of Geophysical Data Processing* (McGraw-Hill, New York 1976)
6.20 S. Gray: "Direct Inversion for Strongly Depth Dependent Velocity Profiles", Denver Res. Inst. Rpt. # MS-R-7902 (1978)

# 7. Acoustic Probing of Space-Time Scales in the Ocean

R. P. Porter

With 23 Figures

The circulation of salt water about the earth's surface is a complex process that influences our lives. Boundary currents, such as the Gulf Stream, moderate our weather. The Gulf Stream changes shape and location in rough approximation to the seasons. The weather on the East Coast of the United States may worsen significantly if the stream moves further offshore. These circulating waters also carry nutrients essential to the maintenance of fishing grounds, such as the Grand Banks, or the anchovy fishery off Peru supported by the upwelling El Nino.

Boundary currents are not monolithic, the stream meanders, splits into multiple currents, and throws off circulating cells or eddies. One or more of these eddies may cool the local ocean surface enough to influence atmospheric circulation. Such a cell can also introduce nutrient-enriched water into a poorly endowed region, changing the distribution of fish.

We must learn the complexities of the circulation patterns before we can understand their influence on the ocean environment. Techniques are now available for tracking Gulf Stream eddies for many months. Satellite maps of the stream can track the surface water, but they do not reveal much about the underlying currents (Fig. 5.9). Ship surveys are useful but time consuming, and of short duration compared to the time scale of the phenomena. Only acoustic monitoring promises to sample the details of a sufficiently large volume of ocean for a long period of time.

The relationship between sound and the ocean is complex. The years during and immediately after World War II were spent trying to understand how the ocean alters sound passing through it. We now know that sound is reflected from the sea surface and that properties of surface waves can be extracted from the reflected sound. We know that sound passing through the Gulf Stream is altered by changes in the current and temperature. Can we sort out this relationship so that we may measure currents and temperature by observing variations in sound? Workers at several institutions seek answers to this question.

We now understand that the motions of waves in the ocean drastically affect sound propagation, and thus we can predict the characteristic behavior of these sound fluctuations. For example, cold Gulf Stream rings—those masses of counterclockwise circulating cold water thrown off by the stream—can cause the intensity of sound at a listening station to increase. As an eddy crosses a transmission path, the intensity of the sound wave can change by as much as a factor of ten.

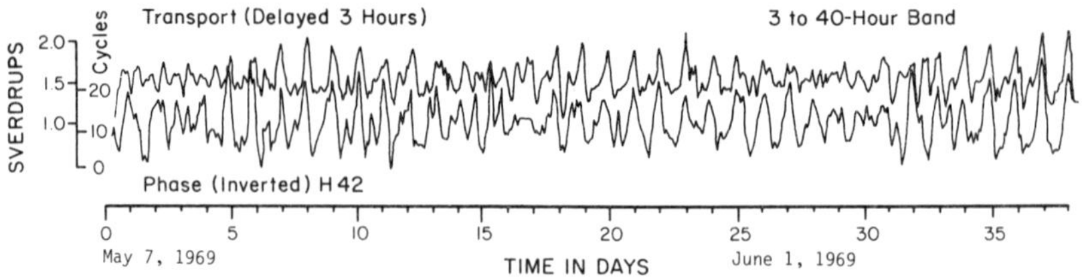


Fig. 7.1. Comparison of acoustic phase and mass transport across the Florida Straits. The transport has been delayed three hours to optimize the correlation. One Sverdrup is about 4 $km^3/h$. (Courtesy Palisades Geophysical Institute)

During the last decade, an acoustic range was established across the Florida Straits between Miami and Bimini [7.1]. Measurements taken over a long period of time disclosed a correlation between acoustic and oceanographic variables. The acoustic signal responds to changes in the Gulf Stream that are caused by tidal forces. Figure 7.1 shows a comparison between the phase which is a measure of the change in travel time and the transport of water by the Gulf Stream. We see that both vary with the daily tides.

As we have shifted from studying how sound travels through sea water to utilizing this knowledge to explore the dynamics of the ocean, it has become clear that sound is a sensitive indicator of average ocean conditions along a particular propagation path. It can, for example, sample large expanses of ocean, yielding measurements of the average temperature and current. Conventional measurement techniques, such as moored current meters and CTD probes (*C*onductivity *T*emperature and *D*epth), have disadvantages that make remote acoustic sensing techniques attractive.

CTD probes, for example, can yield a good snapshot description (typically one lowering per 25 square kilometers), but the time evolution of the dynamic structure requires a prohibitive number of lowerings. Moored current meters can record ocean currents for as long as a year but are too expensive to provide adequate coverage of large areas. In addition, the commonly used rotor meters sample only the water wetting the instrument and cannot measure currents less than 1 cm/s. Acoustic monitoring (remote sensing) has the promise of measuring the time and space evolution of the dynamic structure with only a few sensors. Acoustic monitoring is not limited to the immediate vicinity of the instrument; it can measure average ocean conditions over 1 to 100 kilometers acoustic paths.

## 7.1 Sound Probes of Ocean Currents

### 7.1.1 SOFAR Floats for Lagrangian Current Measurements

The SOFAR float program, developed by ROSSBY of the University of Rhode Island and WEBB of the Woods Hole Oceanographic Institution, is the most successful application of sound probing to date [7.2]. The floats are untethered aluminium tubes, containing batteries, electronics, and a sound emitter that is adjusted to be neutrally buoyant at a particular depth such as the axis of the sound channel. A picture of a typical float is shown in Fig.7.2. These devices emit bursts of sound picked up by sensitive listening equipment at widely scattered sites.

The acoustic travel times were used to calculate the location of each float and their continuous drift was taken as a measure of the local current. During September 1974, ten floats were placed in Gulf Stream rings by WEBB and tracked from Bermuda and other stations. Some showed coherent motions. Figure 7.3 shows a cartwheeling float track. Independent evidence reveals that this float was caught in a Gulf Stream ring that was moving slowly to the southwest. It remained in the ring and was tracked for more than two months [7.3].

During the *M*id-*O*cean *D*ynamics *E*xperiment (MODE), which began at sea in 1973, floats were implanted west of Bermuda and tracked by listening stations at Bermuda, Grand Turk, and Eleuthera. Their trajectories are a Lagrangian measure of the current. Some twenty floats were launched during MODE, with many continuing to transmit for more

Fig. 7.2. SOFAR float similar to those used during the Mid-Ocean Dynamics Experiment

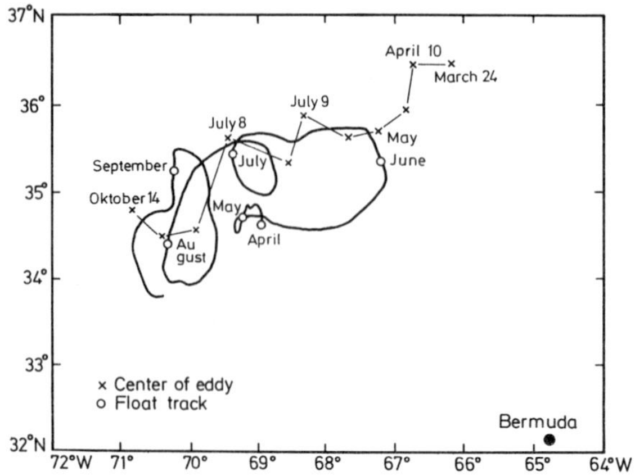


Fig. 7.3. Gulf Stream ring and entrained float. The location of the eddy center is shown for the data indicated by ×. The float track is the curve beginning in April and ending during September. (Courtesy Robert Cheney, Naval Oceanographic Office)

than two years. Because the floats constantly changed position, the acquired data is not always easy to interpret. The object of the float program is to measure current as a function of time and position.

### 7.1.2 Current Measurements by Reciprocal Transmissions

A set of more recent acoustic experiments by Munk, Worcester and Williams of the University of California, San Diego, CA., have been designed to measure Eulerian currents. Water movements are measured from nearly stationary sensing locations. These workers have attempted to monitor mid-ocean currents with scales large compared to the acoustic path length. In the preliminary experiment recently reported, identical transmitters and receivers were suspended from each of two surface ships [7.4] (in an experiment underway at this writing, the transmitter-receiver pairs are moored to satisfy the requirements of a Eulerian measurement). Acoustic pulses were transmitted simultaneously between the two ships separated by 25 km. The resulting receptions have been examined for differences in amplitude, structure and travel time. Here, we discuss the portion of the measurements relevant to the determination of current.

By placing the transmitters about 25 km apart and about 1 km deep, the pulses travel by only two wholly refracted paths: an upper path turning at a depth of 200 m and a lower path with a turning point at 1500 m. The propagation geometry is shown in Fig.7.4. The upper and lower paths can be separated from each other by differences in travel time (all the surface reflected and bottom reflected arrivals occur later). A series of arrivals at one ship is shown in Fig.7.5. The first arrival travels along the deep path while the second travels along the upper path.

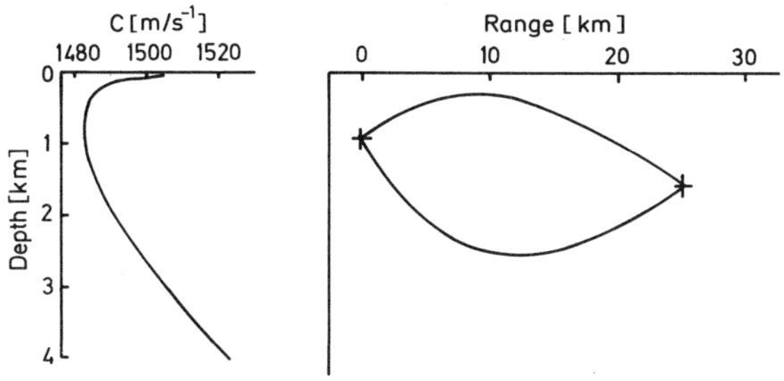


Fig. 7.4. Sound-speed profile and refracted ray paths off San Diego. Sound traveled by these paths during the reciprocal propagation experiment of WORCESTER

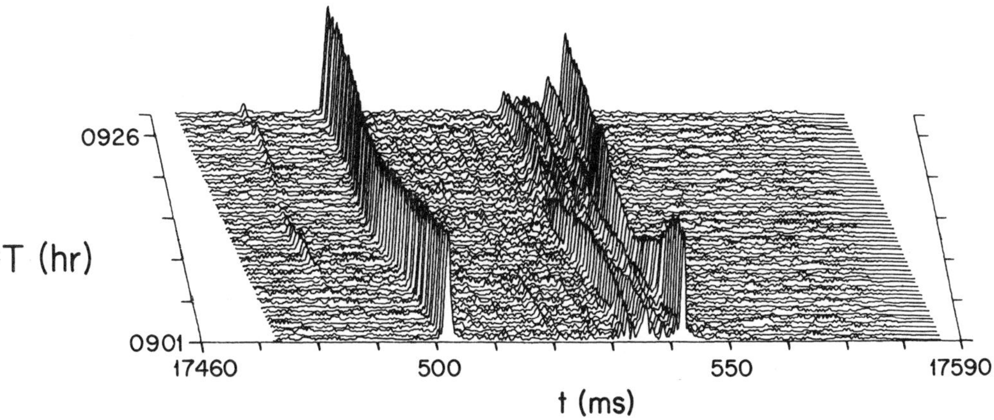


Fig. 7.5. Arrival amplitudes at 30 s intervals after high-resolution processing. The time marks are in ms from the start of the transmitted pulse. Later arrivals traveled along upper path in Fig.7.4. Note the relative complexity of the upper path

In the absence of a current the received pulse trains at the two ships will have the same travel time and the same pulse structure. The existence of current and current shear causes changes in the travel time and pulse shape for signals traveling up current and down current. With the simplest imaginable physics the travel time difference in the two way transmissions is proportional to the current between the stations. The pulse structure has been observed at the two vessels. Pulse splitting along the upper path (second arrival in Fig.7.5) has been observed from one ship but not always simultaneously from both. Actual use of non-reciprocal travel time differences to measure currents must await the results of a moored experiment.

## 7.2 Acoustic Fluctuations as a Measure of Ocean Dynamics

Various dynamic mechanisms have been investigated as causes for acoustic fluctuations (it is hoped that understanding these fluctuations will allow us to use acoustics to probe ocean dynamics). They include small scale turbulence and mixing, internal gravity waves, internal tides, and large scale eddies. Small scale turbulence is a significant source of fluctuations at high frequencies above 5 kHz. Early investigations have attempted to quantify fluctuations at high frequencies, short ranges, and single paths [7.5-7]. The greater part of these fluctuations appear to be caused by refraction and scattering from thermal microstructure. The fluctuation of lower frequency acoustic signals is of greater interest because these signals respond to dynamic events in the ocean such as internal waves and tides. Beginning in the 1960's experiments at low frequencies and long ranges began to reveal that ocean dynamics could be studied with acoustic probes [7.1,8,9]. Low frequency signals are characterized by relatively stable phase and random amplitude. Time series of one year duration reveal significant components of the phase fluctuations at tidal frequencies. Some spectra exhibiting a strong tidal component will be presented later.

### 7.2.1 CW Transmissions and Tides

Frequency spreading, induced by the ocean, of transmitted acoustic spectral lines has been studied intensively in acoustic ranges established across the Florida Straits and between Eleuthera and Bermuda by project MIMI [7.1,8,9]. In this work continuous modulated transmissions have been collected in time series lasting as long as a year. Both experiments have been conducted at frequencies near 400 Hz. The Florida Straits data is short range, 7-21 nm. The propagation is by surface reflected, refracted (SRR) paths primarily. The Eleuthera to Bermuda data is recorded at Bermuda and at another site at half the range to Bermuda. The source is located near the top of main thermocline; much of the propagation is by SRR paths.

Most of the amplitude fluctuation energy occurs at periods of 10 h or longer; the spectrum is nearly flat for periods as long as 100 h. Above 1 cycle per hour (cph) the energy drops off rapidly.

The phase data collected by project MIMI is more closely related to oceanic fluctuations. There is dominant energy at the tidal frequency (similar to Fig.7.19) that is discussed later. At frequencies above the tidal frequency the phase spectrum falls as $\omega^{-2}$. The role of internal waves as a source of phase fluctuations is unclear from these data.

DEFERRARI has looked at the influence of multipath interference on broad-band signals propagated across the Florida Straits [7.10]. He has observed both frequency-selective narrow-band fades and broad-band fading. Narrow-band fades occur when multipath arrivals are widely separated. Broad-band fades are possible when several paths arrive at nearly the same time. Figure 7.6 shows a cw fade with the associated $180^\circ$ phase jump across the fade. Discontinuities in phase are a consequence of loss of the signal and resultant inability to estimate phase. DEFERRARI has studied the behavior of fades in both time and frequency. Selective fades are observed as deep notches traveling, in time, across the frequency band.

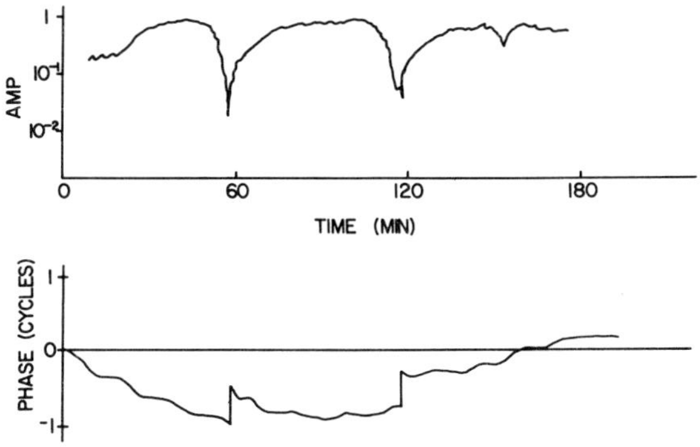


Fig. 7.6. Amplitude and phase time series illustrating that phase jumps are correlated with deep amplitude fades

Other continuous tone experiments have been conducted between Eleuthera and Bermuda by STANFORD [7.11]. The amplitude and phase spectral densities have been studied. His data agree well with the results collected by project MIMI. STANFORD has emphasized that his amplitude data exhibit a seasonal dependence at the higher frequencies. His phase data do not reveal the tidal line because his eight-hour time segments are too short. The slope of phase versus frequency is approximately the same as the MIMI data.

### 7.2.2 Short Pulse Transmissions Along Single Paths

EWART has conducted an experiment whose purpose was to compare the statistical behavior of amplitude and phase fluctuations with the results of acoustic scattering calculations based on existing models of oceanic fluctuations [7.12] (another single path experiment has been performed by KENNEDY [7.13]). Acoustic pulses were measured whose transmission paths between the source and three receivers were wholly refracted. Temporal and spatial scales appropriate to the tidal, internal gravity wave, and fine structure regimes of ocean dynamics were studied.

A schematic diagram of the experiment is shown in Fig.7.7. The receiver tower with the three receivers deployed horizontally was lowered onto the slope of Cobb seamount at a depth of 951 m. The transmitter tower was located 17 km away at a depth of 973 m. The data was radiated to an underwater data link and then telemetered to a nearby research vessel. Pulses were sent from the transmitter to the receiver at 4 and 8 kHz at regular intervals. A self-propelled, programmable research vehicle (SPURV) was used to examine horizontal and vertical temperature microstructure. The experiment was designed so that the first arrival travelled along a wholly refracted path as shown in Fig.7.7. For the transmission path shown here and a transmitter elevation of 15 m, the first 2 ms of the pulse arrival was free of reflected multipath interference.

Amplitude and phase time series were collected for 145 hours at 4 and 8 kHz. Short pulses were transmitted. The phase spectra at both 4 and 8 kHz exhibit significant

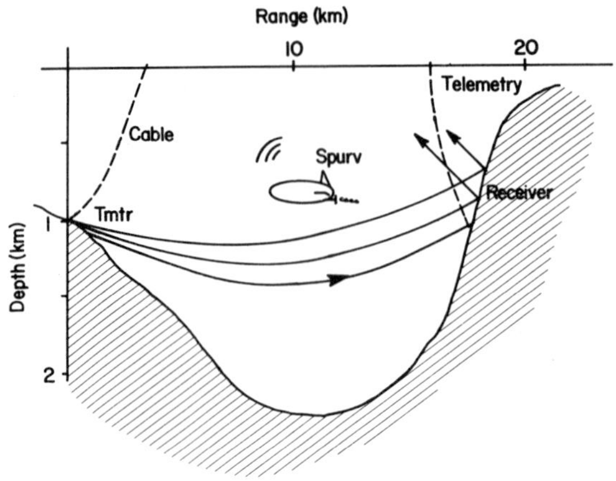


Fig. 7.7. Geometry for Cobb seamount experiment. Pulses discussed by EWART traveled along wholly refracted lower path

tidal peaks and are virtually identical out to the Brunt-Väisälä frequency N of internal gravity-wave energy which is discussed in the next section in more detail. The region between the inertial frequency f and N is characterized by an $\omega^{-3}$ falloff. The 4 kHz data exhibit a change in slope at N. For the amplitude spectra the tidal peaks are much less pronounced. Between f and N the spectra fall off at about $\omega^{-1}$. Above N the falloff is faster than $\omega^{-3}$. At frequencies less than N the amplitude spectra for the 4 and 8 kHz frequencies are nearly identical.

Much of the data shows amplitude fading that is characteristic of multipath interference in spite of the ray-trace of Fig.7.7 indicating a wholly refracted path. When actual sound speed profiles measured with a CTD are used, fluctuations in the sound speed are found that can cause the one refractive path to split into less well defined small-scale multipath.

The phase spectra do show distinct peaks at the diurnal and semi-diurnal tidal frequencies. These peaks have been compared with tidal data collected at Cobb seamount by LARSEN and IRISH [7.14]. The dominant energy is at the tidal frequencies. The energy in the internal wave and fine structure regions of the spectrum is small. EWART concluded that energy at tidal frequencies must be removed from the data if we expect to examine the effect of internal waves and fine structure.

### 7.2.3 Spatial and Temporal Fluctuations of CW Transmissions

Long-range experiments designed to study acoustic fluctuations have been conducted with moored and drifting hydrophones [7.15,16]. These experiments yield evidence that acoustic fluctuations with time scales less than a day and space scales less than a kilometer are caused by the interaction of sound and internal gravity waves. The magnitude of measured phase and amplitude fluctuations are consistent with predictions based on the internal-wave field.

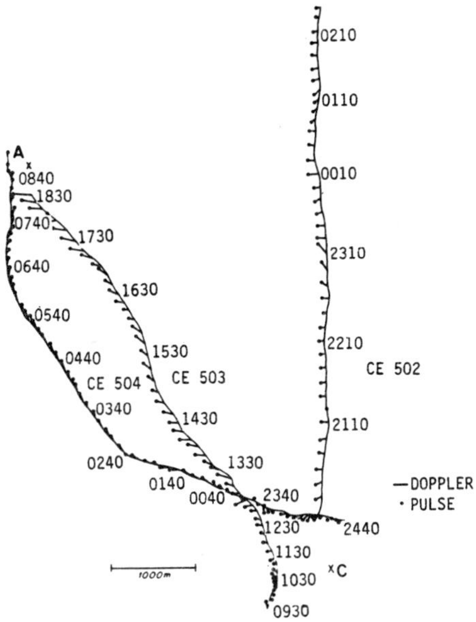


Fig. 7.8. Drifting hydrophone tracks obtained near Bermuda. Solid lines are Doppler tracks; dots are pulse positions every 10 min. Time of day is indicated at hourly intervals along the track. Positions of navigation beacon moorings are labeled A, B, C

A dual frequency experiment (220 and 406 Hz) was performed in September 1974 northeast of Eleuthera Island at a range of 250 km. The two-sound sources were located within 14 km of each other along a bearing nearly perpendicular to the transmission path. All data were recorded on sonobuoys (free-floating acoustic receivers equipped with a UHF transmitter for data communication) drifting at a depth of 305 m. These buoys were tracked by a Doppler navigation system using bottom-moored beacons. Typical drift tracks are shown in Fig.7.8. Hydrophone positions determined by a pulse navigation system are also shown for comparison with the tracks measured using the Doppler system. This experiment has permitted a study of frequency scaling of acoustic amplitude and phase fluctuations between 220 and 406 Hz.

A digital data acquisition system has been used to collect acoustic data from drifting hydrophones. The salient features of the analysis are summarized here; detailed discussions are available elsewhere [7.17,18]. Figure 7.8 shows three beacons labeled A, B, and C that are moored on the ocean floor. Given the location of the beacons and a hydrophone position at the beginning of a run (obtained with an acoustic pulse navigation subsystem) we can calculate a continuous hydrophone drift track. A continuous projection of this track along the transmission bearing is also calculated and made available as a continuous Doppler shift $f_D(t)$. The phase change due to hydrophone motion along the transmission path is

$$\theta_D(t) = \int_0^t f_D(t)dt \quad . \tag{7.1}$$

The incoming 220 and 406 Hz transmissions are passed through coherent phase detectors and then low-pass filtered. The digital processing includes a coherent phase detector (realized by software) to remove the Doppler shift and a 10 mHz low-pass filter for removing Doppler shift due to ocean-surface scatter as well as maximizing the signal-to-noise ratio. The output of the digital processor are the quadrature components

$$X(t) = B(t)\cos\theta(t) \tag{7.2}$$

$$Y(t) = B(t)\sin\theta(t) \quad . \tag{7.3}$$

Statistical estimates of the amplitude B(t) and the phase rate defined as

$$\dot{\theta}(t) = \frac{X(t)\dot{Y}(t)-\dot{X}(t)Y(t)}{X^2(t)+Y^2(t)} \quad , \tag{7.4}$$

have been formed. The phase rate is single valued and, for this reason, is a preferred estimator instead of the phase which is multivalued.

### 7.2.4 Spectra of Phase and Amplitude Fluctuations

Spectra for the sonobuoy data collected at 220 and 406 Hz are shown in Fig.7.9. Phase-rate spectra are shown on the left, spectra of the log amplitude in the middle, and amplitude spectra on the right. The sonobuoys drifted slowly averaging 100 m/h along the transmission path. The drift speed and direction were generally constant. The phase-rate spectra show nearly the same characteristics for both the 220 and 406 Hz carrier frequencies. The spectra are peaked at low frequencies, flat between 1 and 10 cph and then fall at a rate between $\omega^{-1}$ and $\omega^{-2}$. It is possible that the spectral shape below 0.5 cph is distorted by the motion of the sonobuoy. Difficulties in surveying the Doppler net were experienced permitting uncertainties in the phase rate of about 1 cph of drift. Amplitude spectra are shown on the right of Fig.7.9. These spectra fall as $\omega^{-1.5}$ to a frequency of 10 cph; the spectra then break and fall at a rate in excess of $\omega^{-3}$. Spectral contamination is possible especially above 10 cph.

It is of considerable practical importance to determine how acoustic fluctuations scale with carrier frequency. Also frequency scaling provides information useful for understanding the physics of the interaction between the acoustics and gravity-wave fields. The phase-rate spectrum for the 406 Hz signal has been divided by the spectrum for the 220 Hz signal (Fig.7.10). If the phase fluctuations vary linearly with frequency then the ratio of the power spectra should be 5 dB. We see that level at low frequencies; at high frequencies the ratio fluctuates with the average near 6 dB.

The failure of the linear scaling law at fluctuation frequencies above 2 cph is readily understood. These spectral fluctuations result from the interaction of sound with the high-frequency, large wave-number domain of the internal gravity-wave field. In contrast, the low-frequency acoustic fluctuations result from interaction with the

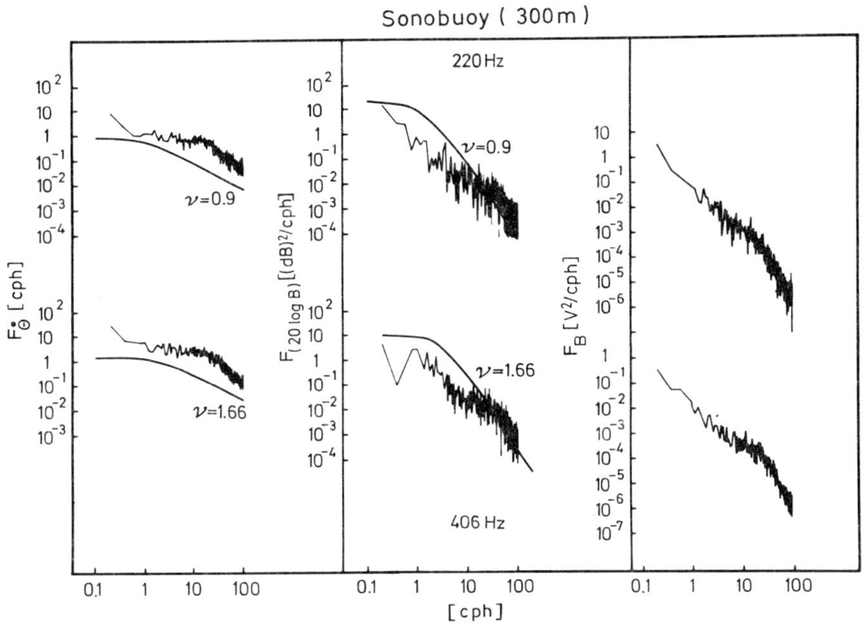


Fig. 7.9. Spectra of 300 m data at 220 and 406 Hz. Phase-rate spectra are shown on the left, log-amplitude spectra in the middle, and amplitude on the right. Solid lines are predictions based on the random multipath model of DYSON et al. [7.45]

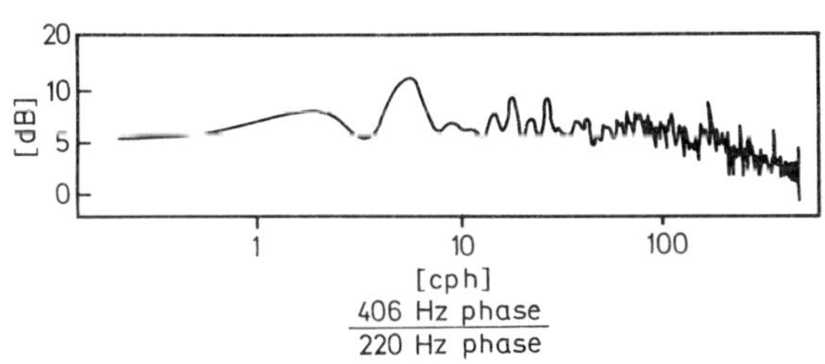


Fig. 7.10. Ratio of 406 and 220 Hz phase-rate spectra shown in Fig.7.9. Vertical scale is 10 log of the ratio of the power spectra

long-wave domain of the internal-wavefield. Since the two sources are separated by 14 km, the acoustic transmission paths interact with different internal-wave fields at sufficiently high fluctuation frequencies. At low frequencies (less than 1 cph) the two paths sample the same realization of the internal gravity-wave field.

A single frequency experiment at 220 Hz was performed southeast of Bermuda in September 1975. For this experiment, a 220 Hz sound source was moored at the axis of the deep sound channel, a depth of 1100 m. The data discussed here were collected on hydrophones suspended from a drifting ship at several depths. Ship-suspended hydrophones were tracked by the Doppler navigation system.

Figure 7.11 presents phase-rate and amplitude spectra for drifting hydrophones at 500, 1000, and 1500 m depths. The drift rate averaged 500 m/h along the propaga-

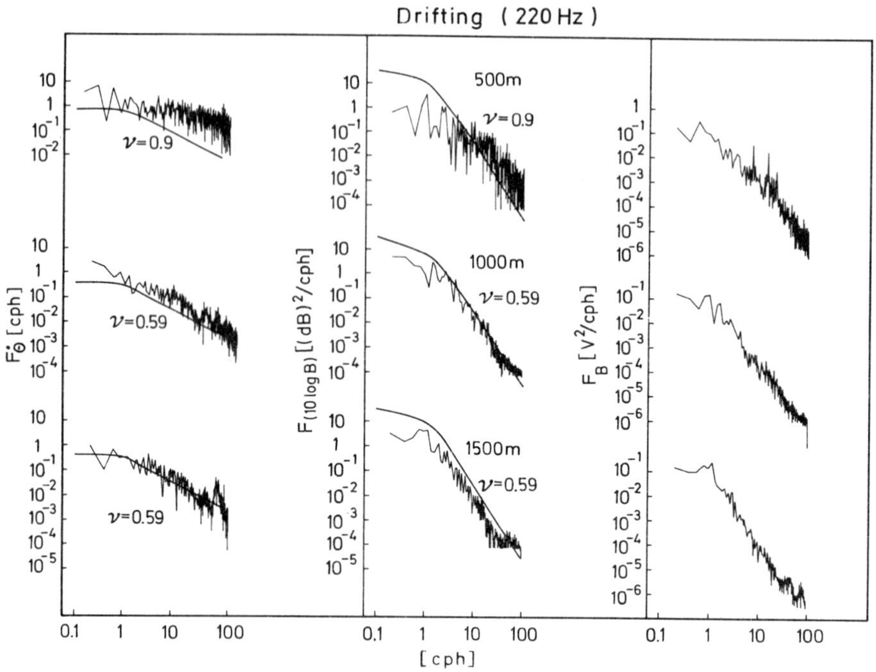


Fig. 7.11. Spectra of ship-suspended hydrophone data recorded at 220 Hz during an experiment near Bermuda. Phase-rate spectra are shown on the left, log-amplitude spectra in the middle, and amplitude spectra on the right. Solid lines are predictions based on the random multipath model of DYSON et al. [7.45]

tion path for these data. Typical drift paths are shown in Fig.7.8. Runs at different depths were taken in random order over a three-day interval. The phase-rate spectra fall monotonically at a rate between $\omega^{-0.5}$ and $\omega^{-1}$. There is some tendency for the spectra from the deeper tows to fall off at a faster rate. There is no hint of the break at 10 cph observed in the sonobuoy data shown in Fig.7.9. The amplitude spectra also fall monotonically at a rate similar to that of the sonobuoy spectra below 10 cph. Again, there is no high frequency roll off in the spectra shown in Fig.7.11.

Figure 7.12 shows the phase-rate time series for a 100 m drifting hydrophone. The phase-rate spectrum is shown in Fig.7.11. The phase-rate time series is dominated by a few very sharp spikes with phase rates as high as ±45 cph. The occurrence of these fades correlates with very deep amplitude fades as was observed by DEFERRARI [7.10]. These phase rates are too high to be caused by internal gravity waves. Rather, they occur because the signal has nearly vanished allowing the phase rate to vary rapidly and randomly. These spikes can obviously dominate the high-frequency region of the spectrum. We have formed estimates of the spectra with the spikes removed. This is accomplished by band limiting the phase rate to a maximum of ±10 cph as shown in Fig.7.12.

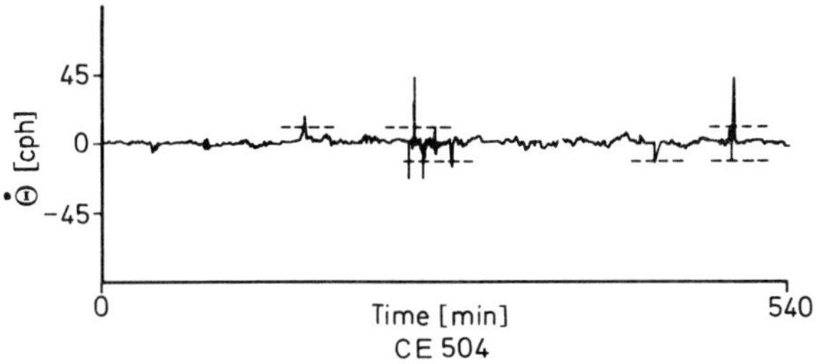


Fig. 7.12. Time series of phase-rate data for 1500 m drifting hydrophone. Spikes correlate with deep amplitude fades. Hard limiting of phase rate indicated by dashed lines at ±10 cph

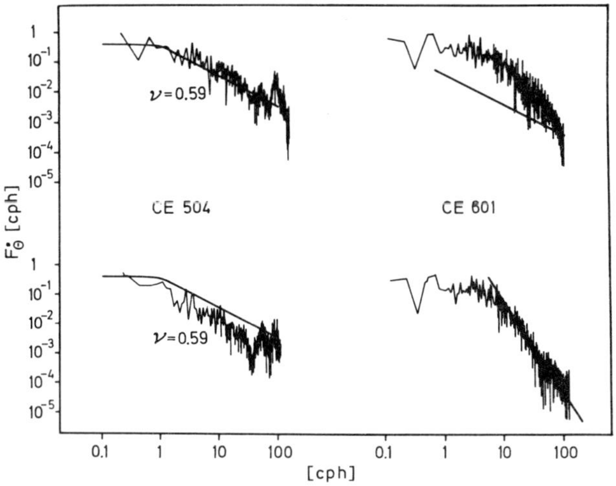


Fig. 7.13. Phase-rate spectra for drifting phone (left) and moored phone (right) at 1500 m depth, showing the effects of spike removal. Spikes have been clipped for time series used to calculate lower curves. Random-multipath-model predictions are shown by solid lines for left-hand curves

Figure 7.13 presents spectra before and after spike removal. The phase spectra on the left are for a drifting hydrophone; the spectra on the right are for a moored hydrophone. Consider the moored hydrophone. Before the spikes with phase rates greater than 10 cph were removed, the spectrum fell as $\omega^{-1.5}$ at high frequencies. Removal of the spikes reveals the underlying data whose spectra fall as $\omega^{-3}$. The low frequency spectra for the moored phone is not altered appreciably by removing the spikes. The drifting hydrophone data also falls slightly faster after removing the spikes. Some of the side lobe structure at high frequencies is a result of clipping the spikes. The same technique was applied to the sonobuoy data. Again, the spectral level is reduced and small changes occur in the decay rate at frequencies above 10 cph.

More recently, a thorough investigation of low frequency fluctuations as a function of range was performed. Sound sources at 220 and 100 Hz were moored near the axis of the sound channel at a depth of 1100 m. Digital recording buoys were moored near the sound channel axis and deep ($\sim$3000 m) at various ranges from the source.

The experiment was conducted on longitude line due south of Bermuda. The *T*rident *V*ertical *A*rray (TVA) was furthest from the 220 Hz source at a range of 550 km and a depth of 4200 m. An extensive series of environmental measurements was performed. Up to 30 continuous days of acoustic data at 220 Hz was collected on some receivers. The geometry of the experiment is shown in Fig.7.14.

We have performed spectral analysis of the intensity and phase rate. These spectra are estimated from two adjacent time series each 17 hours longer. The spectra of the two adjacent segments are averaged to produce each plotted estimate. In addition, the individual spectra are Hanned. Nine neighboring frequency bins ( ~0.3 cph averaging interval) have been averaged to smooth the spectra.

Figure 7.15 is a collection of consecutive spectra representing a total of 272 hours of data collected on mooring 40 which is 280 km from the 220 Hz source. Both intensity and phase rate spectra are shown. As expected, the high frequency regions of the spectra (1-100 cph) are repeatable throughout the data. Some non-stationarity does exist at low frequencies. At high frequencies the phase-rate spectra roll off as $\omega^{-2}$ to $\omega^{-3}$ while the intensity spectra fall very rapidly. Both spectra fall more rapidly than those of drifting hydrophones. Hydrophone motion appears to translate

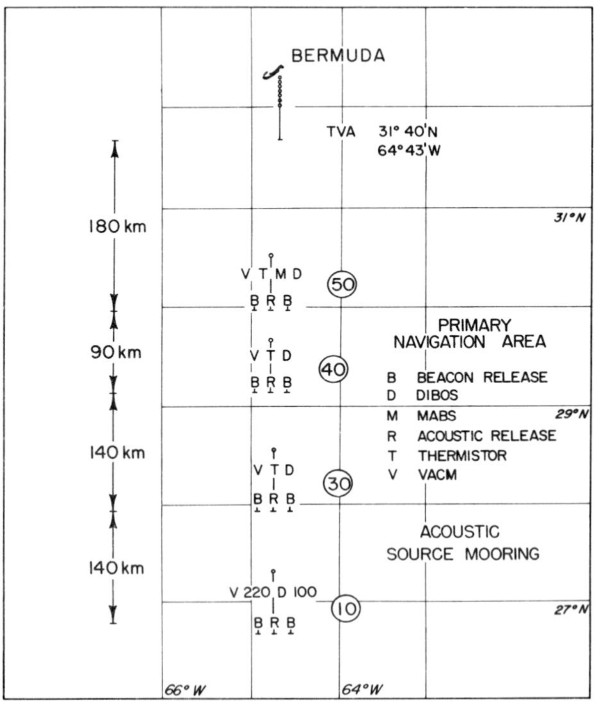


Fig. 7.14. Geometry of Bermuda experiment conducted during 1976. Five moorings, including the source mooring, were deployed along a line due south of the TVA array. Moored environmental sensors are indicated as well as the acoustic instruments

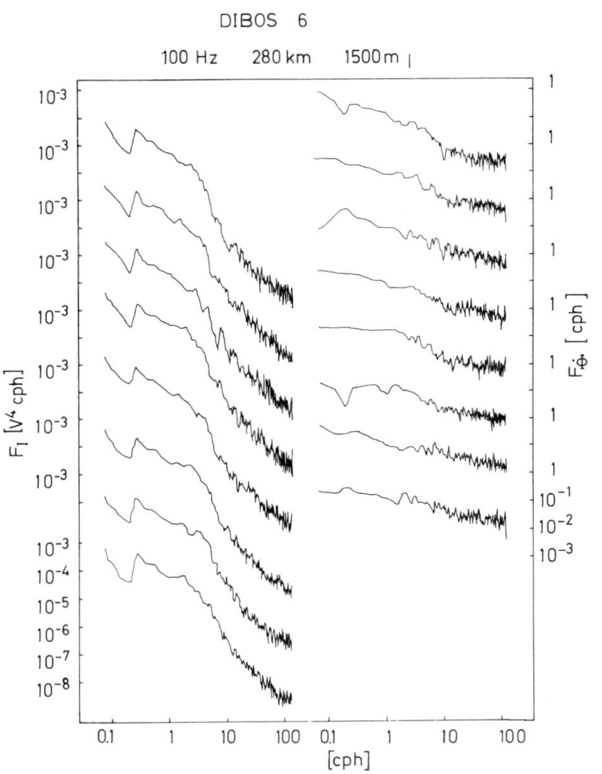


Fig. 7.15. Successive intensity and phase-rate spectra obtained from mooring 40. Each spectrum represents successive 34 hours of data with the earliest recorded spectra at the top of the figure

energy into the high frequency region of the spectrum. At low frequencies, both the intensity and phase-rate spectra are nearly flat. This again is contrasted with the spectra for drifting hydrophones which have a significant slope at low frequencies.

Spectra for one of the phase detector outputs (7.2,3) are shown in Fig.7.16 for data taken at various ranges from 9-550 km. Each spectrum shown is an ensemble average of two segments, each 17 hours long. All these data are flat at frequencies below 1 cph and then fall rapidly. There is no consistent trend with range. Intensity spectra for the same data are shown in Fig.7.17. The only noticeable trend with range is a flattening of the spectra at low frequencies.

The phase-rate spectra for the same data are shown in Fig.7.18. Note that we have included 100 Hz data collected at a range of 40 km. These spectra dramatically change character with range. At 9, 40, and 140 km we can fit an $\omega^{-1}$ slope at low frequencies. Above 3 cph the slopes fall dramatically to $\omega^{-2}$ or $\omega^{-3}$. It is intriguing to interpret these data in terms of internal-wave spectra but let us resist the temptation here. At long ranges, the low frequency regions of the spectra level off and are nearly flat. The spectrum at 280 km is in transition between the two spectral types. The

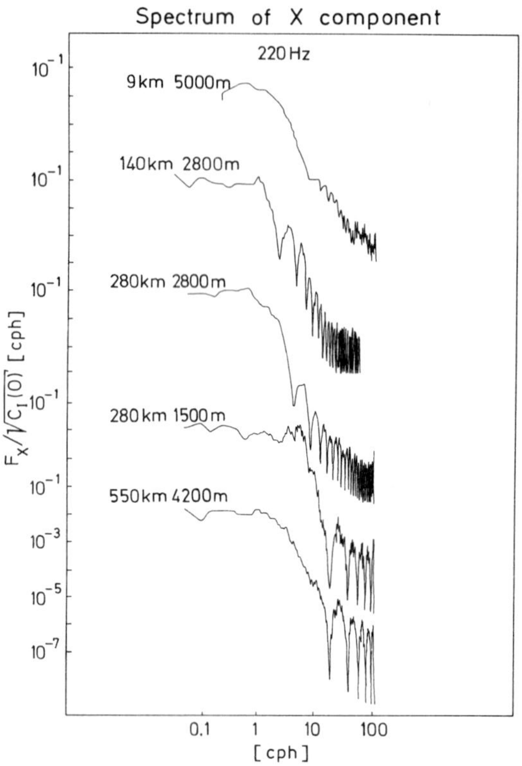


Fig. 7.16. Spectra of the X-component of the acoustic field recorded by several moored instruments. These data were recorded by buoys placed as shown in Fig.7.14

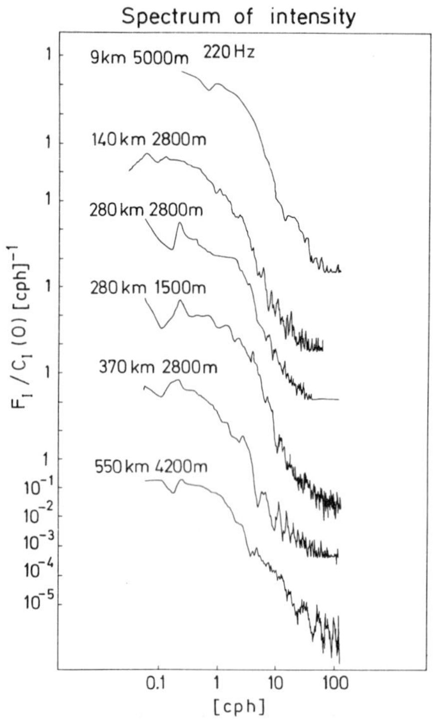


Fig. 7.17. Spectra of the intensity of the acoustic field recorded by several moored instruments. These data were recorded by buoys placed as shown in Fig.7.14

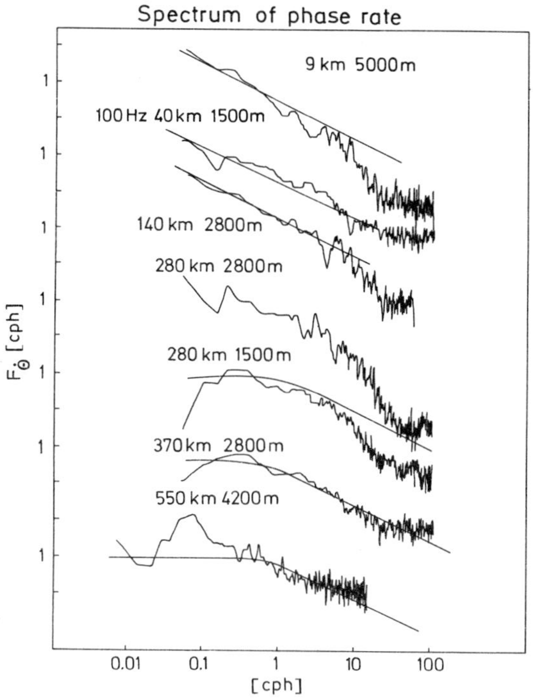


Fig. 7.18. Spectra of phase-rate of the acoustic field recorded by several moored instruments. Data fits on the upper three curves are based on the internal-wave spectra. The lower three spectra are fit to the random multipath model developed by DYSON et al. [7.45]

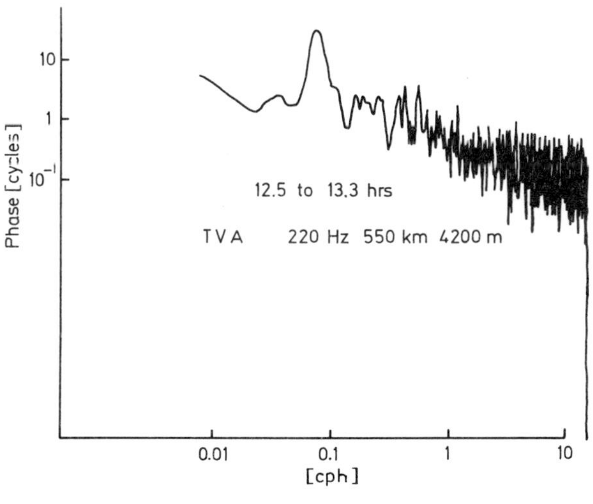


Fig. 7.19. A phase-rate spectra for acoustic data recorded with the TVA array located near Bermuda. These spectra have been averaged over only three neighboring frequencies to reveal the tidal peak

TVA data, at a range of 550 km, deviates from a flat spectrum at 0.08 cph. These data processed with less frequency smoothing reveal a sharp peak at the frequency of the semi-diurnal tide (Fig.7.19). Evidently, the tidal contribution is added to the phase data at the end of the transmission path near Bermuda. A dominant tidal line has been observed in data collected by EWART and project MIMI.

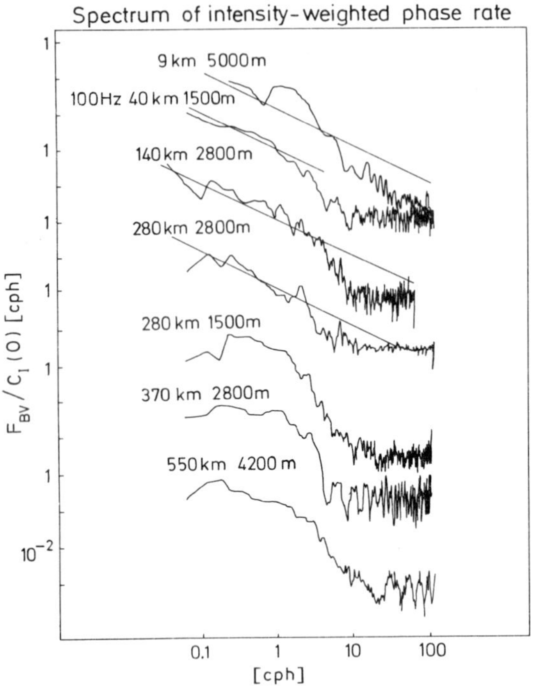


Fig. 7.20. Spectra of intensity-weighted phase-rate. The upper three curves have been fit to internal wave spectra. All spectra fall off sharply above the rms phase rate of 1.5 cph

Figure 7.20 presents spectra of the intensity-weighted phase rate. This quantity $I\dot{\theta} = BV$, where $V = B\dot{\theta}$, suppresses the phase-rate contributions during a fade where $I = 0$. This is a continuous mathematical procedure that eliminates the phase-rate spikes shown in Fig.7.12. The low-frequency regions of the spectra are not very different than those for the unweighted phase rate. As before we can fit an $\omega^{-1}$ line at ranges of 9, 40, and 140 km. Again the spectra are flat at low frequencies. The most striking change is at high frequencies; fade suppression yields a very sharp break at about 3 cph almost independent of range.

## 7.3 Sound Speed Variations and Internal Gravity Waves

Most quantitative studies of the dependence of acoustic signals on ocean dynamics concern internal-gravity waves and tides. The dynamic phenomena are sufficiently well understood to permit quantitative comparison of theory and experiment. It is this comparison that has motivated a great deal of the theoretical developments presented in a later section. Here we turn our attention to a description of the internal gravity-wave field in the ocean (other discussion of internal waves is available in Chaps.2 and 5).

### 7.3.1 Internal Gravity Wave Spectra

The density of the oceans varies with depth due to the stratification of pressure, temperature, and salinity. An element of water transferred adiabatically from one depth to another will experience a buoyancy force proportional to the potential-density difference. The resulting motion may or may not be stable depending on the density variation as a function of depth. For stable conditions a displaced element of water will tend to oscillate about its equilibrium depth with a characteristic buoyancy or Brunt-Väisälä frequency n(z) [7.19].

Internal gravity waves are pervasive throughout the world's oceans. They dominate the dynamic oceanic behavior for periods above the peak Brunt-Väisälä $n_0$ and below the inertial period, 12 h/sin$\xi$, where $\xi$ is the latitude (24 h at 30°N). Extensive current meter measurements made in the North Atlantic have revealed that the energy in the flow spectrum is greatest at the inertial frequency and decreases as $\omega_g^{-2}$ at higher frequencies [7.20-22]. Towed thermistor chain and towed STD (salinity, temperature, and depth measurements) have produced wavenumber spectra that decrease roughly as $\alpha^{-2}$ [7.23,24]. The measurements have been made in the Atlantic and Pacific Oceans as well as the Mediterranean Sea [7.25]. While there are differences from place to place, weighting of the data by the local Brunt-Väisälä frequency (see following discussion) puts all the spectra on nearly the same universal curve. Even though the evidence is not conclusive, GARRETT and MUNK have postulated a universal frequency-wavenumber spectrum that fits nearly all available data between the inertial and Brunt-Väisälä frequencies [7.26].

A linearized treatment of the dynamic equations produces a set of wave equations driven by the Coriolis force with a natural oscillation at the characteristic buoyancy frequency [7.27]. Thus, it is understandable that the ocean will have oscillations at both the inertial and Brunt-Väisälä frequency. A detailed analysis shows that waves at any frequency between the inertial frequency f and the Brunt-Väisälä frequency N can propagate horizontally with modes being established in depth. Because the basic equations are nonlinear, the driving force is able to produce gravity waves throughout the spectral range (f,N). It is interesting to note that internal waves are trapped in a deep ocean channel because of the behavior of the temperature and salinity gradients, and hence N, with depth (Fig.7.21).

Many of the important phenomena in sound propagation have analogues in internal-wave propagation. The analog to Snell's law is used to calculate turning points for either ray theory or WKB mode theory. For a given gravity-wave frequency $\omega_g$, the ray will turn at a depth $z_t$ such that $\omega_g = N(z_t)$, where N(z) is the local Brunt-Väisälä frequency. The measured buoyancy frequency profile of Fig.7.21 is a reasonably accurate representation between Eleuthera and Bermuda. Internal-wave frequencies $\omega_g < 1.8$ cph will propagate horizontally with depth distribution between the surface and about 1500 m. At higher frequencies two channels exist; the minimum-predicted wavelength of 100 m indicates that they are strongly coupled.

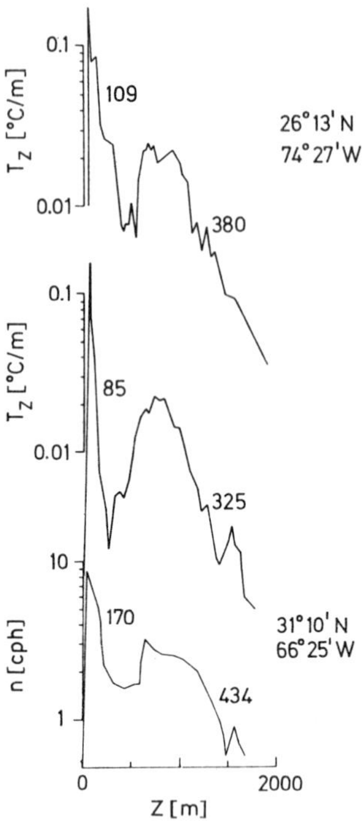


Fig. 7.21. Plots of temperature gradients and buoyancy frequency with depth. The upper plot is the temperature gradient measured northeast of Eleuthera; the lower plots are temperature gradient and buoyancy frequency measured southwest of Bermuda. The stratification depths in meters are indicated

In general, N(z) is maximum just below the mixed layer and near the top of the main thermocline. Near the surface, stability is quite variable. In the GARRETT and MUNK generalization of the stability of the world's oceans, the gravity-wave spectrum assumes an exponential distribution of the buoyancy frequency. The acoustic-internal wave experiments of interest here took place in the western Sargasso between Eleuthera and Bermuda [7.16]; the buoyancy-frequency depth distribution is shown at the bottom of Fig.7.21. The distribution of Fig.7.21 is poorly approximated by a single exponential. We have fit two exponentials, one for the seasonal and one for the main thermocline, with the indicated slopes. The stratification depth for the empirical buoyancy frequency profile is defined as

$$b = N_0^{-1} \int N(z)dz \quad , \tag{7.5}$$

where $N_0$ is the peak buoyancy frequency and where the integral is evaluated over the entire water column. Note that b is the 1/e depth (the stratification depth of GARRETT and MUNK) for an exponential profile. From the profile of Fig.7.21 we find that $N_0 = 9$ cph and $b = 340$ m. To simplify the following calculations, we specify all frequencies as cyclical, in units of Hertz (Hz) for acoustic signals and cycles per hour (cph) for internal waves. All wavenumbers are in units of 1/m.

As noted above, oscillations of internal waves are found from the linearized dynamical equations of motion which exclude tidal forces, but include the Coriolis force. GARRETT and MUNK provide a detailed derivation in their paper, while we present a brief summary of those aspects which are pertinent to the present analysis (note that we have made their scaling explicit in the following equations). The spectrum presented in their 1972 paper is discussed here. Further refinements are discussed in Chap.5. Phase fluctuations due to internal tides and geostrophic flows are specifically excluded from this internal-wave model. Assume that the internal-wave displacement is a progressive wave characterized by a wavenumber-frequency energy density $E(\alpha,\omega_g)$ under conditions of a stationary, homogeneous, and horizontally isotropic gravity-wave field. Note that $\int E(\alpha,\omega_g)d\alpha\, d\omega_g$ has units of $(\mathrm{m\ s^{-1}})^2$. GARRETT and MUNK have proposed the power spectrum

$$E(\alpha,\omega_g) = 4b^3N_0^3Ef^2\mu^{-1}\omega_g^{-1}\left(\omega_g^2-f^2\right)^{-\frac{1}{2}} \quad , \quad \alpha^{(1)} < \alpha < \alpha_c \tag{7.6}$$

$$= 0 \quad , \quad \text{otherwise,}$$

where

$$\mu(\omega_g) = j\pi\left(\omega_g^2-f^2\right)^{\frac{1}{2}} \tag{7.7}$$

and

$$\alpha_c = \mu(\omega_g)/2\pi bN_0 \quad . \tag{7.8}$$

Here $\alpha^{(1)}(\omega_g)$ is the dispersion relation for the lowest-order mode, and j is the number of propagating modes at the inertial frequency f. In fitting this spectrum to observation, GARRETT and MUNK have chosen

$$E = 2\pi \times 10^{-5} \quad , \quad j = 20 \quad .$$

The internal-wave displacement for a given member in $\alpha$, $\omega_g$ space can be written as

$$\delta = Z(z)N_0^{-1}q(\alpha,\omega_g)\ \exp[i2\pi(\alpha r-\omega_g t)] \quad , \tag{7.9}$$

where Z(z) is the normalized wave function and q is the wave-velocity density. They have argued that the $\alpha,\omega_g$ space is densely populated and thus can be approximated by a continuum. Since there are many closely grouped modes with random amplitude, it is further argued that the vertical wave function Z(z) may be replaced by its root-mean-square (rms) value at any depth where

$$\overline{Z^2(z)} = N_0\left(\omega_g^2-f^2\right)/N(z)\omega_g^2 \quad , \quad f < \omega_g < N(z) \tag{7.10}$$

$$= 0 \quad , \quad \text{otherwise.}$$

Noting that

$$\int\overline{q(\alpha,\omega_g)q^*(\alpha,\omega_g')}d\omega_g' = \int E(\alpha,\omega_g')\delta_0(\omega_g'-\omega_g)d\omega_g' = E(\alpha,\omega_g) \quad , \tag{7.11}$$

under conditions of stationarity and homogeneity [7.28] (where $\delta_0$ is the Dirac delta function), the mean-square internal-wave displacement is

$$\overline{\delta^2} = N_0^{-2} \iint Z^2(z)\; E(\alpha,\omega_g)d\alpha\; d\omega_g \quad . \tag{7.12}$$

Defining a displacement spectral density $F_\delta(\omega_g)$, we have

$$\overline{\delta^2} = \int F_\delta(\omega_g)d\omega_g \quad , \tag{7.13}$$

or

$$F_\delta(\omega_g) = N_0^{-2} \int \overline{Z^2(z)}\; E(\alpha,\omega_g)d\alpha \quad , \tag{7.14}$$

Now using (7.6) and (7.10) and integrating yields

$$N(z)F_\delta(\omega_g) = 2\pi^{-1}b^2N_0Ef\omega_g^{-3}\left(\omega_g^2-f^2\right)^{\frac{1}{2}} \quad , \tag{7.15}$$

for the displacement spectrum [7.26]. Note that the spectrum falls of as $\omega_g^{-2}$ for frequencies well above the inertial frequency.

### 7.3.2 Fluctuations in the Index of Refraction

Fluctuations in refractive index and sound velocity, caused by the passage of an internal wave, are primarily a function of temperature,

$$\gamma = \Delta c/c_0 = \varepsilon\delta T_z \quad , \tag{7.16}$$

where $T_z$ is the local temperature gradient, and $\varepsilon$ a constant. PORTER et al. [7.29] gave details of the more general dependence on salinity and potential (or nonadiabatic) temperature gradient. In (7.16), $\delta$ is given by (7.9) except that the vertical wave function Z(z) is replaced by its rms value; $[\overline{Z^2(z)}]^{\frac{1}{2}}$ is found from (7.10). In reality, the local index of refraction oscillates rapidly and randomly due to the small vertical wavelengths and lack of vertical correlation of the internal-wave fields. By using the rms behavior of the vertical wave function we are smoothing over depth but leaving horizontal and slowly varying z dependencies. The justification for this smoothing is that vertical wavenumbers are significantly larger than horizontal ones at a given frequency due to the anisotropy of the oceans. The rms internal-wave displacement can be found by integrating (7.15) over $\omega_g$; it is $(\delta^2)^{\frac{1}{2}} \sim 10$ m at the top of the main thermocline. It should be noted that we are assuming that the horizontal variation in sound velocity is due to the progressing internal wave defined by $\delta$.

MUNK has derived a relationship between the local buoyancy frequency and sound speed fluctuations due to the vertical displacement of an internal wave [7.30,31]. It is shown in [7.29] that

$$\gamma = -N^2(z)\delta\sigma/g_r \quad , \tag{7.17}$$

where $\sigma$ depends on the salinity and temperature, and $g_r$ is the acceleration of gravity. The rms vertical wave displacement increases with depth as $N^{-\frac{1}{2}}$, see (7.9-12). Thus, from (7.17), sound-velocity fluctuations increase with depth as $N^{3/2}$. Therefore, sound-velocity fluctuations induced by internal gravity waves are concentrated where N(z) is greatest, i.e., in the main and seasonal thermoclines.

For a single discrete mode the displacement is given by (7.9). For a continuum of incoherently added modes, a particular member of the ensemble in $\alpha,\omega_g$ space has displacement spectral density

$$F_\delta(\alpha,\omega_g) = N_0^{-2}\,\overline{Z^2(z)}\,E(\alpha,\omega_g) \quad , \tag{7.18}$$

and refraction-index density

$$F_\gamma(\alpha,\omega_g) = \varepsilon^2 T_z^2 N_0^{-2}\,\overline{Z^2(z)}\,E(\alpha,\omega_g) \quad . \tag{7.19}$$

## 7.4 Acoustic Fluctuation Theories and Their Relationship in Ocean Dynamics

Early experiments of SAGAR [7.5], WHITMARSH et al. [7.6], SHVACHKO [7.7], and SHEEHY [7.32] were motivated by an oceanic model consisting of patches of slightly warmer or colder water. The ocean was assumed to be frozen during the transit time of the acoustic signal. The motion of the patches was construed to be a combination of convection and diffusion. Most of the early experiments were at high frequencies (10-50 Hz) for which the patches are large compared to a wavelength. Nearly all of the theoretical work from 1950 to the present assumes that the wavelength is small compared to the spatial scale of the temperature variations.

### 7.4.1 Acoustic Fluctuation Theories and Their Relationship to Ocean Dynamics

BERGMAN [7.33] was motivated by observations during World War II to develop a theory of acoustic fluctuations based on geometric ray theory. This theory failed to explain the experimental results of SHEEHY. MINTZER [7.34] was motivated to use the Born approximation of diffraction theory for weak scattering. His theory did provide a satisfactory fit to the range dependence ($r^{\frac{1}{2}}$) observed by SHEEHY. MINTZER did not concern himself with the physics producing the temperature; he assumed that a correlation function of the refractive index variations exists. He used data obtained by LIEBERMANN [7.35] from a temperature sensor mounted on a submarine.

A comprehensive investigation was conducted by WHITMARSH et al. including acoustic and environmental measurements, as well as theory based on a turbulent ocean model [7.6]. Their measurements of temperature gradients were made with thermistor chains suspended from one of the two ships involved in the acoustic measurements. They found horizontal temperature distributions that fit the Kolmogorov law $r^{2/3}$, r being a spatial scale. Their measurements indicate that the mean square temperature fluctuations increase steadily for thermistor spacings less than 30 m. Either temperature fluctuations occur over bigger spatial scales, or they are not statistically homogeneous. They concluded that the fit between data and theory based on 0.6 m scale size measured by LIEBERMANN is overly simplified. They also observed that the temperature patches are not spherical but spheroidal with the longest dimension in the horizontal (recent theories of internal waves discussed earlier support pancake type patches, circular but thin). Their model of acoustic fluctuations is based on a focusing mechanism emphasizing a given patch size at a given range. The standard deviation of the predicted amplitude fluctuations is $r^{3/2}$ by BERGMANN at short ranges and $r^{\frac{1}{2}}$ at long ranges. This saturation phenomena will be discussed later. The transition range depends on the maximum patch size which depends on depth.

TATARSKI [7.28] and CHERNOV [7.36] have developed the theory of weak scattering through turbulent media to a high level. CHERNOV has developed a unified treatment of weak scattering methods. Starting from a description of homogeneous turbulence, due primarily to Kolmogorov, TATARSKI has developed the statistics of temperature fluctuations. The structure function defined as

$$D_f(\rho) = \overline{[f(r+\rho)-f(r)]^2} \quad , \tag{7.20}$$

where the overbar indicates an ensemble average, serves as a unifying mathematical device. An important intuitive idea states that $f(r+\rho) - f(r)$ is statistically homogeneous even when the correlation function

$$C_f(r,\rho) = \overline{f(r+\rho)f(r)} \tag{7.21}$$

is not. The method of smooth perturbations, due to Rytov, is used to extend the range of validity of the weak-scattering Born approximation. In the Born approximation the magnitude of the phase variations $|\Psi|$ must be small, while in Rytov's method we have the more physically appealing $\lambda|\nabla\Psi| << 2\pi$. This restricts weak scattering approximations to media in which the acoustic phase varies slowly over one wavelength.

The fluctuation of amplitude with range r depends on the Fresnel zone size $\sqrt{\lambda r}$ relative to the inner scale $\ell_0$ and the outer scale $L_0$ of the turbulent inhomogeneities. When $\sqrt{\lambda r} << \ell_0$ ray theory yields a good approximation to the propagation and the amplitude fluctuations for a plane wave grow as

$$\chi^2 \rightarrow r^3 \quad . \tag{7.22}$$

When the Fresnel zones are between inner and outer scales $\ell_0 << \sqrt{\lambda r} << L_0$ the amplitude fluctuations are governed by a focusing mechanism as proposed by WHITMARSH et al. They grow with range as

$$\overline{\chi^2} \rightarrow r^{11/6} \tag{7.23}$$

for homogeneous turbulence. Finally, when the Fresnel zone is large, $\sqrt{\lambda r} >> L_0$, the Born approximation used by MINTZER is valid; we have

$$\overline{\chi^2} \rightarrow r \quad . \tag{7.24}$$

The range dependencies for small and large Fresnel zones hold without regard to the physics of the patches. It is the transition range where the physics becomes truly important.

The theories and experiments of this early period suffered from two major shortcomings. It was unknown whether the ocean could be adequately modeled by homogeneous turbulence; few oceanographic measurements were available to guide those attempting to explain observed sound fluctuations. Consequently the early work concentrated on range and frequency dependencies. No information on the time scale of acoustic or oceanic variations was available. Also, only single point statistics were studied. The coherence of the fluctuations at closely separated points was not widely investigated. Coherence measures are useful for testing various wave theories and are of considerable practical importance.

### 7.4.2 Effect of Large Scale Flows on Acoustic Amplitude and Phase

The establishment of fixed low-frequency acoustic ranges between Miami and Bimini (MIMI) and between Block Island and Fisher's Island (BIFI) [7.37] has yielded observations of the interaction between underwater sound and ocean dynamics. The MIMI data showed that the phase of a continuous tone (420 Hz) contained much energy at the frequency of the semi-diurnal tide. The relationship between tide and acoustic phase remained obscure.

JACOBSON and others [7.38-43] have studied the effect of current, geostrophic, and barotropic flow on acoustic amplitude and phase. These analyses are based entirely on ray theory; they do not consider scattering from random inhomogeneities. They are interested in dynamic events with large, spatial (200 km) and long temporal scales (12 hours) such as Rossby waves. They argue that the acoustic phase can be related to ocean dynamics. They find that phase is linearly dependent on current (interpreted as average current) but that the amplitude exhibits scintillations of as much as 40 dB for very small variations of the current.

STALLWORTH and JACOBSON [7.38] have presented some of the basic results for a channel with constant sound speed and uniform tidal current. Acoustic energy travels along many separate paths; straight lines for this problem. The ray undergoing n surface reflections arrives with signal

$$V_n(t) = A_n \sin(\omega t - \Phi_n) \quad . \tag{7.25}$$

The authors showed that single path amplitude is negligibly influenced by a weak current. The phase of a single path consists of two terms

$$\Phi_n = \Phi_{n0} - \omega TM \quad , \tag{7.26}$$

where T is the straight line travel time between source and receiver, and $M = v/c$ is the Mach number for the current. $\Phi_{n0}$ is the phase for path n in the absence of current flow. This result is accurate for small M. The total acoustic field can then be written as

$$\sum_{n=0}^{N} A_n \exp[i(\omega t - \Phi_n)] = \left\{ \sum_{n=0}^{N} A_{n0} \exp(-i\Phi_{n0}) \times \exp[i(\omega t - \omega TM)] \right\} \quad . \tag{7.27}$$

For a spatially uniform tidal current

$$M = m_0 \sin(2\eta t/\tau) \quad . \tag{7.28}$$

STALLWORTH and JACOBSON associate the acoustic phase difference with the term

$$\Delta\Phi = -\omega T m_0 \sin(2\pi t/\tau) \quad . \tag{7.29}$$

These authors have fit a sinusoidal tidal current to data from the BIFI range and obtained good agreement. Observed phase variations between Miami and Bimini cannot be explained as an explicit current effect. For this geometry the current flow between receivers is small; the Gulf Stream flows through the Florida Straits perpendicular to the transmission path. FRANCHI and JACOBSON [7.39,40] have considered modifications of the sound speed due to the northward current that explain much of the phase fluctuation. The earth's rotation produces a pressure gradient to the east across the straits. The pressure gradient is matched by an east-west density gradient which is related by the equation of state to an east-west temperature gradient. Finally, the speed of sound in sea water depends on pressure, temperature, and salinity. For small Mach numbers M these authors show that the sound speed is a linear function of the northerly current. As in the explicit dependence on current developed by STALLWORTH and JACOBSON, the acoustic phase is a linear function of the current. Theoretical calculations of the phase, based on the measured transport and measured sound speed variations, agree well with phase measurements.

An attempt has been made to explain phase variations, observed between Eleuthera and Bermuda, as resulting from Rossby waves [7.41]. A ray trace model has been developed for a bilinear sound speed profile whose horizontal variations are produced by Rossby waves. Not all the observed phase fluctuations can be accounted for. Also, suitable sound speed measurements for validating this theory are not available. More recently, WATSON et al. [7.42] have developed a general relationship between sound speed and vorticity for quasigeostrophic flow.

### 7.4.3 Effect of Internal Waves on Acoustic Amplitudes and Phases

Ray theory can be used to study acoustic fluctuations induced by internal waves. When there are a few isolated paths, it can be shown that the spectrum of the phase fluctuations is directly related to the spectrum of internal wave displacements. The earliest analysis assumed that the internal wave activity was concentrated at a single depth [7.29]. The rays were considered for the deep sound channel where a single ray path can interact many times with the internal wave field. This phase-rate spectrum can be approximated by

$$\begin{aligned} F_{\dot{\theta}}(\omega_g) &= \nu_{0\ell}/(N_0-f) \quad , \quad f < \omega_g < N_0 \\ &= 0 \quad , \quad \omega > N_0 \quad , \quad \omega < f \quad , \end{aligned} \tag{7.30}$$

where $\nu_{0\ell}$ is the rms phase rate for path $\ell$.

There is considerable theoretical and experimental support for the idea that acoustic phase or travel time are directly related to dynamic events occurring in the ocean. However, detailed comparison of theory and experiment are lacking. Also, a considerable controversy exists because the phase of an actual signal arriving along many paths does not agree with the phase estimated by these simple models. The phase of a measured signal can be determined from

$$\theta = \tan^{-1}\left(\frac{y}{x}\right) + \ell\pi = \tan^{-1}\left(\frac{S_y+n_y}{S_x+n_x}\right) + \ell\pi \quad , \tag{7.31}$$

where S and n are signal and noise, and where $\ell$ is an arbitrary integer. For a multipath signal there are times when near cancellation of the signal takes place. Then the estimated phase is that of the noise which is a random number between $-\pi/2$ and $\pi/2$. Since the phase difference before and after the fade is arbitrary, each segment of the phase time series is independent of the other segments. Figure 7.6 is an illustration of phase and amplitude time series showing these segments and showing the phase reversals occurring at the fading time. Observations show that phase reversals are correlated with the occurrence of amplitude scintillations.

How far must the amplitude drop before a phase reversal occurs? The simple answer is to the noise level. Since the noise is random with some probability distribution, the frequency of occurrence of phase reversals is also random.

Various workers have used a phase random hypothesis to study the multipath problem [7.44-46]. They consider a large number N of equal energy paths where the Cartesian components of the total field are

$$X = \sum_1^N x_n \quad , \quad Y = \sum_1^N y_n \quad . \tag{7.32}$$

By the central limit theorem we can argue that X and Y are Gaussian random variables. We further assume that they are independent, have zero mean, and equal variance. ADAMS, based on results of LAWSON and UHLENBECK [7.47], has found the covariance function of the phase rate

$$C_{\dot\theta}(\tau) = \frac{1}{2}\left(\frac{\ddot\rho}{\rho} - \frac{\dot\rho^2}{\rho^2}\right)\log(1-\rho^2) \quad , \tag{7.33}$$

where $\rho$ is the normalized covariance function of the Cartesian components. Assuming that $\rho(\tau)$ is Gaussian

$$\rho(\tau) = \exp\left[-2\pi^2\nu_0^2\tau^2\right] \quad , \tag{7.34}$$

we find the spectrum of the phase to be approximately

$$F_{\dot\theta}(\omega_g) = \nu_0\left(\omega_g^2 + 1.27\nu_0^2\right)^{-\frac{1}{2}} \quad . \tag{7.35}$$

The characteristic frequency $\nu_0$ is closely related to the mean time between fades. Thus the phase spectrum does not depend directly on ocean dynamics; the single parameter $\nu_0$ characterizes the spectrum.

We have a dilemma. Simple ray theories predict phase directly dependent on space-time scales of ocean events while simple theories of random processes predict a phase that is not dependent on these scales. Both approaches yield a portion of the truth. As a first attempt to reduce the influence of amplitude scintillations on the phase, PORTER and SPINDEL [7.16] have considered the phase rate given by (7.4). This quantity is single-valued but diverges when the amplitude vanishes. A phase reversal, occurring at a fade, has an impulsive phase rate as shown in Fig.7.12. Between impulses we hypothesize that the phase rate is dependent on ocean dynamics; the spectrum of the phase rate contains high frequency energy due to the spikes.

Clearly, we do not want to investigate those portions of the phase data occurring during a fade. The intensity-weighted phase rate $I\dot{\theta}$ is given by

$$I\dot{\theta} = \frac{1}{2}[X(t)\dot{Y}(t) - \dot{X}(t)Y(t)] \quad , \tag{7.36}$$

which is well behaved. For the problem treated by STALLWORTH and JACOBSON with the carrier removed

$$X(t) = A_{n0} \cos(\Phi_{n0} + \omega TM) \tag{7.37}$$

$$Y(t) = A_{n0} \sin(\Phi_{n0} + \omega TM) \tag{7.38}$$

and

$$I\dot{\theta} = \frac{T}{2\pi}\frac{dM}{dt}\sum_{mn} A_{n0}A_{m0} \cos(\Phi_{n0} - \Phi_{m0}) \quad . \tag{7.39}$$

The intensity-weighted phase rate is always proportional to the rate of change of the current even though the phase rate is large during a fade.

In more general problems the phase along individual paths are different time functions. Consider two paths with differing phase rates $\dot{\theta}_1$ and $\dot{\theta}_2$ where

$$I\dot{\theta} = A_1^2\dot{\theta}_1 + A_2^2\dot{\theta}_2 + A_1A_2(\dot{\theta}_1+\dot{\theta}_2) \cos\left[2\pi(\theta_1-\theta_2)\right] \quad . \tag{7.40}$$

Suppose, the phases are proportional and related to time by

$$\theta_1 = a\theta_2 = ba \cos(2\pi t/T) \quad , \tag{7.41}$$

where T can be the tidal period. Then the intensity-weighted phase rate is

$$I\dot{\theta} = -\left(A_1^2 a + A_2^2\right) \frac{2\pi b}{T} \sin\left(\frac{2\pi t}{T}\right) + A_1A_2b(1+a) \cos[2\pi b(1-a) \cos(2\pi t/T)] \quad . \tag{7.42}$$

The second term is the distortion introduced by multipath interference. For large b the phase variation is several cycles. PORTER and SPINDEL [7.48] have shown that the major contribution of the interference term occurs at a frequency in the neighborhood of

$$\omega = \frac{2\pi b(1-a)}{T} \quad . \tag{7.43}$$

Thus we can extract the time scale T even though multipath interference complicates the relationship between the physics and the acoustic phase.

Complicated physical dependencies can be extracted from the intensity-weighted phase. Apparently, the only restriction is to frequencies much less than the root-mean-square fading rate. Consider internal wave fluctuations discussed previously. MUNK and ZACHARIASEN [7.49] and DESAUBIES [7.50] have calculated single-path phase fluctuations induced by the internal wave field. These workers find phase spectra that fall approximately at $\omega_g^{-3}$ for frequencies higher than the inertial frequency. These spectra are adequately approximated by

$$F_{\dot{\theta}\ell}(\omega_g) = \nu_{0\ell}^2/\left[\omega_g \ln(N_0/f)\right] \quad , \quad f < \omega < N_0$$

$$= 0 \quad , \quad \omega > N_0 \quad , \quad \omega < f \quad . \tag{7.44}$$

The mean-square phase rate for path $\ell$ is $\nu_{0\ell}^2$.

## 7.4.4 Multipath Fluctuations for Stable Paths

We now know how the multipath fluctuations are related to the single-path fluctuations. It is assumed that single-path phase fluctuations contribute more than amplitude fluctuations to the variation of the total field,

$$\dot{b}_\ell^2 << 4\pi^2 \, i_\ell \dot{\theta}_\ell^2 \quad . \tag{7.45}$$

For purposes of this simplified discussion, $i_\ell$ is constant and the same for each path. Also the phase statistics are the same for each path. We shall need the covariance of the phase rate, the covariance of the phase, and the structure function of the phase

$$i_0 = \bar{i} \tag{7.46}$$

$$\nu^2(\tau) = \overline{\dot{\theta}_{i\ell}\dot{\theta}_{j\ell}} \tag{7.47}$$

$$\eta^2(\tau) = \overline{\theta_{i\ell}\theta_{j\ell}} \tag{7.48}$$

$$\Delta^2(\tau) = \overline{(\theta_{i\ell}-\theta_{j\ell})^2} \quad , \tag{7.49}$$

where $\theta_{i\ell}$ is the phase at time $t_i$ for path $\ell$ and where $\tau = t_j - t_\ell$. For the phase-rate spectrum of (7.44) the structure function is

$$\Delta^2(\tau) = \nu_0^2\tau^2 \quad . \tag{7.50}$$

In evaluating the multipath statistics, we assume L stationary, independent paths where

$$\eta_0^2 > 1 \quad . \tag{7.51}$$

A detailed analysis by PORTER and SPINDEL [7.48] shows that the covariance of the intensity-weighted phase rate is

$$C_{I\dot{\theta}}(\tau) = Li_0^2\nu^2(\tau) - \frac{L(L-1)i_0^2}{16\pi^2}\frac{d^2}{d\tau^2}\exp\left[-4\pi^2\nu_0^2\tau^2\right] \quad . \tag{7.52}$$

The first term is directly related to the oceanic space-time scales through the single-path phase fluctuations. The second term results from multipath interference similar to the second term in (7.42). The normalized intensity-weighted phase rate spectrum is

$$F_{I\dot{\theta}}/C_I(0) = \frac{1}{2L}F_{\dot{\theta}}(\omega_g) + \frac{\omega_g^2}{16\nu_0\sqrt{\pi}}\exp\left(-\omega_g^2/4\nu_0^2\right) \quad . \tag{7.53}$$

At frequencies $\omega_i < \omega_g < 2\nu_0$, $F_{\dot{\theta}}$ behaves as $\omega_g^{-1}$ while the interference term behaves as $\omega_g^{-2}$. This term is small at frequencies well below $\nu_0$. At non-zero frequencies the normalized intensity spectrum is

$$F_I(\omega_g)/C_I(0) = \frac{1}{4\nu_0\sqrt{\pi}}\exp\left(-\omega_g^2/4\nu_0^2\right) \quad . \tag{7.54}$$

It is flat at frequencies below $2\nu_0$ and then falls rapidly. The mean-square phase rate controls the amplitude fading rate. As long as the number of paths is not very large and, more important, the single-path amplitude fluctuations are small we can extract single-path phase fluctuations from the intensity-weighted multipath phase. In turn, single-path phase fluctuations are directly related to oceanic space-time scales.

The predictions of the multipath analysis can be compared with the data shown in Figs.7.15-20. Intensity data (Figs.7.15,17) agree very well with (7.54). The intensity-weighted phase-rate spectra of Fig.7.20 are described quite well by (7.53) where the single-path $F_{\dot{\theta}}$ is phase rate given by (7.44) based on the GARRETT and MUNK internal-wave spectrum. The spectral break occurs at $\omega_g = 3$ cph giving a $\nu_0$ of 1.5 cph which is in good agreement with internal wave theory. Apparently the sharp roll off in the spectra are due to interference effects not internal-wave buoyancy.

## 7.4.5 Stability of Single Paths

We now study the limits on the stability of single paths. What happens to a single path? We have seen that the statistics are Gaussian for a large number of equal energy paths. At long ranges in the ocean Gaussian statistics should provide an adequate description of the acoustic field [7.44]. How far away from the source can we go and still use simple ray ideas to describe the propagation? MUNK and ZACHARIASEN have shown that the Rytov approximation or eikonal description closely describes a single ray interacting with an internal wave field [7.49]. FLATTE and TAPPERT have performed a simulation study showing that amplitude fluctuations for a single ray are small at a range of 100 km and a frequency of 100 Hz [7.51].

DASHEN has developed a general systematics that permits analysis of the break up of ray paths. Since his work has been covered in a recent publication we only highlight his results here [7.52]. Using a path integral approach with the Markov approximation, he shows that a single ray eventually disintegrates into a large number of paths, each with random amplitude and phase. Total disintegration leads to a limit for amplitude phase fluctuations. The fluctuations are saturated at these ranges and frequencies. He uses two parameters, the diffraction parameter $\Lambda$ and the phase-fluctuation strength $\Phi$, to characterize ray break up. This phenomenon is known as micro-multipath formation.

The diffraction parameter is the relative Fresnel zone size

$$\Lambda = R^{-1} \int_0^R (2\pi)^{-1}\left[R_F(\chi)/L\right]^2 d\chi \quad , \tag{7.55}$$

where $\chi$ is a straight propagation path, $R_F(\chi)$ is the radius of the first Fresnel zone at $\chi$, and L is the correlation length for the medium variations (internal waves) for the ocean. The strength parameter is the expected mean-square phase fluctuation along a single path

$$\Phi^2 = \left\langle\left(k \int_0^R \mu d\chi\right)^2\right\rangle \quad . \tag{7.56}$$

If the range is large compared to the scale size L

$$\Phi^2 \sim k^2 RL \langle\mu^2\rangle \quad . \tag{7.57}$$

Others have used these parameters [7.53].

DASHEN has developed a $\Lambda - \Phi$ plot allowing us to determine when the break-up of a single path into many paths occurs (Fig.7.22). For large $\Lambda$ and small $\Phi$ the Fresnel zone is large compared to the scale size of the medium. The field is diffracted by the cells of warm and cool water. The Rytov extension of the Born approximation can be used to calculate the acoustic fluctuations. When both $\Lambda$ and $\Phi$ are small the per-

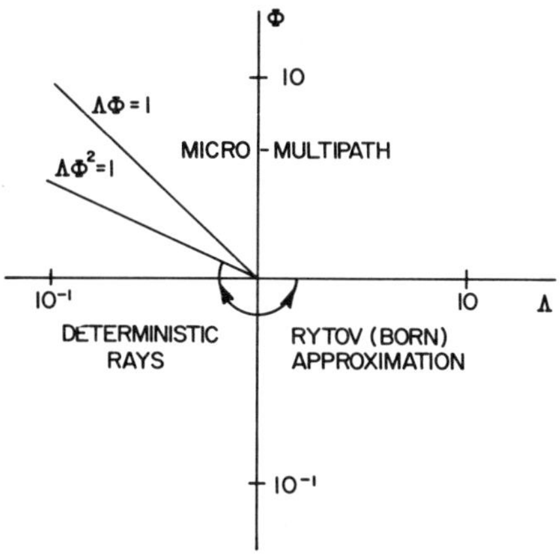


Fig. 7.22. $\Lambda - \Phi$ plot showing regions where fluctuations are saturated. In the unsaturated region propagation obeys rules of geometric optics or diffraction theory

turbations to the ray are small. The behavior developed by TATARSKI and discussed earlier holds. As long as $\Lambda$ is small we can use ray theory. DASHEN has shown that as long as $\Phi^2\Lambda \ll 1$ the fluctuations along a single path will not cause it to disintegrate. Saturation occurs when $\Phi^2\Lambda \gg 1$. In this region the Cartesian components of the acoustic field fluctuate according to Gaussian statistics.

Thus, as long as $\Phi^2\Lambda$ is small the amplitude fluctuations for a single geometric path will be small. The condition of weak fluctuations (7.45) will be met. The statistics of the multipath field are non-Gaussian when each path is weakly fluctuating. The covariances of the phase and amplitude fluctuations have already been presented. FLATTE et al. have calculated the strength and diffraction parameters for real ocean conditions [7.52]. They predict that a single steep path splits into many micro-multipaths at ranges greater than 300 km at all frequencies above 50 Hz.

These workers have substantiated these results from data collected on a fixed acoustic range near the Azores. A complete description of the AFAR experiment and its interpretations is available in their recent book. They show that the fluctuations in acoustic travel time are adequately predicted by the micro-multipath theory.

Deep-ocean long-range data collected by PORTER and SPINDEL show the transition from weak fluctuation to saturated behavior. Phase rate and intensity-weighted phase-rate fluctuations are shown in Figs.7.18 and 7.20. At long range the spectra of the phase rate fluctuations fit the random multipath field precisely ([7.45,46] and (7.35)). The only deviation occurs at 0.08 cph for the 550 km data; this is a tidal peak absent from the other data. At very short ranges, the phase-rate spectra fall steadily as predicted by the weak-fluctuation multipath theory of PORTER and SPINDEL [7.48]. At 280 km, the data at 1500 m, consisting of many paths, produces a nearly saturated spectrum. The data at 2800 m for the same range was produced by few paths; this data is not yet saturated. The transition between saturated and weak fluctuations occurs at 300 km as predicted by FLATTE et al.

## 7.5 Implications for Ocean Probing

All applications of acoustic probing to the measurement of ocean dynamics depends on the stability of propagation paths. Short pulses arriving along resolvable paths can be used to monitor ocean currents. Because the propagation paths cycle vertically as well as travelling horizontally the relationships between acoustic travel time and the current field is complex. Still, the dynamic information is contained in the travel time of a single path.

Energy will propagate along single paths determined by the mean sound-speed profile as long as the strength-diffraction product $\Lambda\Phi^2 \leq 1$. Both theoretical and experimental results support a range of 300 km at frequencies less than 200 Hz. At higher frequencies and greater ranges a single-ray path disintegrates into a collection of multipaths with random amplitudes.

Experimental systems for measuring ocean currents and other oceanographic parameters are difficult to achieve if we are restricted to resolving individual paths. The experiment performed by WORCESTER required pulse lengths of the order of milliseconds; individual paths are separated by 30 ms. Bandwidths of the order of 1 kHz imply high propagation frequencies that limit the range of the acoustic probe and decrease the range of path stability to 50 km. If we reduce the bandwidth, increasing the pulse length, we still must accept some multipath interference.

The phase of a scintillating acoustic signal is not simply related to the phase variations of the individual paths. Principally the phase estimate fails during deep fades; during those instances, the phase rate can be large and random. Weighting the phase rate by the intensity emphasizes phase estimates with the greatest signal to noise. The intensity-weighted phase rate is proportional to the sum of the single-path phase rates at frequencies low compared to the mean fading rate.

Acoustic probes do not require resolution of all paths. Only those ray paths that traverse substantially different portions of the water column need to be resolved. At ranges of 300 km all those ray paths with the same number of lower half cycles (in general 4 rays) will have travel times within 50 ms of each other. We do not require resolution of individual rays but only one family from another. The bandwidth requirements are now reduced substantially; 10 Hz with a carrier frequency of 200 Hz is sufficient.

We can expect to see development of acoustic systems for probing the ocean during the 1980's. A possible configuration, consisting of a narrow-band source surrounded by acoustic receivers at various ranges, can be used for observing deep ocean eddies. A narrow-band source with a transmission life of one year can be deployed in the deep ocean. Digital acoustic buoys with the capability of recording one year of preprocessed data are now feasible. An array of these receivers can be deployed within 300 km of the source and still obtain reliable measurements of oceanographic parameters. A preliminary experiment using 4 to 8 receivers can map an ocean region of

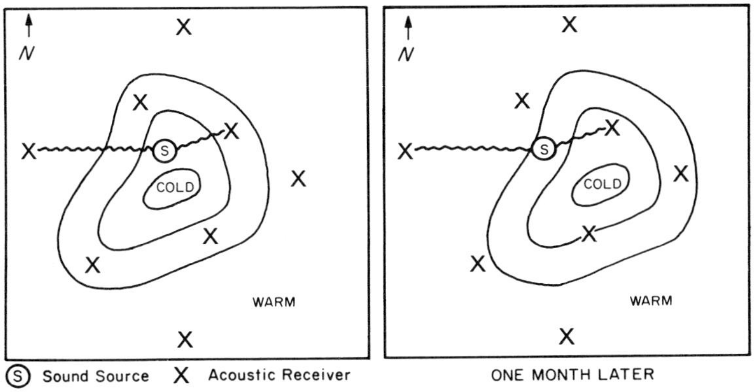


Fig. 7.23. Illustration of an eddy passing through an array of acoustic receivers. Two different locations of the eddy are shown

600 km diameter. This system could be used to map the temperature field of oceanic eddies for periods of several months. The average temperature between source and an individual receiver will change as the eddy passes by. One hundred kilometer spacings between sensors appears sufficient to resolve the temperature field. Figure 7.23 illustrates a cold eddy in two different positions relative to the acoustic array. As the eddy passes, the travel time to the receiver on the left decreases (increased sound speed), while the travel time to the receiver on the right increases. Relatively short pulses can be transmitted, permitting the mapping of the eddy field at several depths.

The experience of recent years has taught us that sound is greatly influenced by the condition of the sea through which it travels. We can now relate sound fluctuations to tides and internal waves. New experiments will utilize this data to monitor the ocean and to extract quantitative information on its energy and momentum. In a few years we can expect that acoustic probes will join the instruments routinely used by the oceanographer in his attempt to understand the movement of water around the globe.

## References

[7.1] J.C. Steinberg, J.G. Clark, H.A. DeFerrari, M. Kronengold, K. Yacoub: J. Acoust. Soc. Am. *52*, 1521-1535 (1972)
[7.2] T. Rossby, A.D. Voorhis, D. Webb: J. Marine Res. *33*, 355-382 (1975)
[7.3] R.E. Cheney, W.H. Gemmill, M.K. Shank, R.L. Richardson, D. Webb: J. Phys. Oceano. *6*, 741-749 (1976)
[7.4] P. Worcester: J. Acoust. Soc. Am. *62*, 895-905 (1977)
[7.5] F.H. Sagar: J. Acoust. Soc. Am. *32*, 112-121 (1960)
[7.6] D.C. Whitmarsh, E. Skudrzyk, R.J. Urick: J. Acoust. Soc. Am. *29*, 1124-1143 (1951)

[7.7] R.F. Shvachko: Sov. Phys.-Acoust. *9*, 280-282 (1964)
[7.8] J.C. Steinberg, T.G. Birdsall: J. Acoust. Soc. Am. *39*, 301-315 (1966)
[7.9] J.G. Clark, M. Kronengold: J. Acoust. Soc. Am. *56*, 1071-1083 (1974)
[7.10] H.A. DeFerrari: J. Acoust. Soc. Am. *54*, 162-180 (1973)
[7.11] G.E. Stanford: J. Acoust. Soc. Am. *55*, 968-977 (1974)
[7.12] T.E. Ewart: J. Acoust. Soc. Am. *60*, 46-59 (1976)
[7.13] R.M. Kennedy: J. Acoust. Soc. Am. *46*, 737-745 (1969)
[7.14] L.H. Larsen, J.D. Irish: J. Geophys. Res. *80*, 1691-1695 (1975)
[7.15] R.C. Spindel, R.P. Porter, R.J. Jaffee: J. Acoust. Soc. Am. *56*, 440 (1974)
[7.16] R.P. Porter, R.C. Spindel: J. Acoust. Soc. Am. *61*, 943-958 (1977)
[7.17] R.C. Spindel, R.P. Porter, W.M. Marquet, J.L. Durham: IEEE J. OE-*1*, 6-13 (1976)
[7.18] R.C. Spindel, R.P. Porter, D.C. Webb: IEEE J. OE-*2*, 331-337 (1977)
[7.19] C. Eckart: *Hydrodynamics of Oceans and Atmosphere* (Pergamon Press, New York 1969) pp.59-63
[7.20] C. Wunsch, R. Hendry: Geophys. Fluid Dynam. *4*, 101-145 (1972)
[7.21] N.P. Fofonoff: "Spectral Characteristics of Internal Waves in the Ocean", Deep-Sea Res. Suppl. *16*, 58-71 (1969a)
[7.22] W. Krauss: Progr. Oceanog. *5*, 95-101 (1969)
[7.23] E.J. Katz: J. Phys. Oceanog. *3*, 448-457 (1973)
[7.24] E.C. LaFond, K.G. LaFond: "Thermal Structure through the California Front", Rpt. NUC TP224, 133 (1971)
[7.25] H.T. Perkins: "Internal Oscillations in the Mediterranean"; Thesis, Mass. Inst. of Technol. and Woods Hole Oceanographic Institution (1970)
[7.26] C. Garrett, W. Munk: Geophys. Fluid Dynam. *2*, 225-264 (1972)
[7.27] O.M. Phillips: *The Dynamics of the Upper Ocean* (Cambridge University Press, Cambridge 1966)
[7.28] I. Tatarski: *Wave Propagation in a Turbulent Medium* (McGraw-Hill, New York 1961)
[7.29] R.P. Porter, R.C. Spindel, R.J. Jaffee: J. Acoust. Soc. Am. *56*, 1426-1436 (1974)
[7.30] W.H. Munk: Personal communication
[7.31] W.H. Munk: J. Acoust. Soc. Am. *55*, 220-226 (1974)
[7.32] M.J. Sheehy: J. Acoust. Soc. Am. *2*, 24 (1950)
[7.33] P.G. Bergmann: Phys. Rev. *70*, 486-492 (1946)
[7.34] D. Mintzer: J. Acoust. Soc. Am. *25*, 922-929, 1107-1111 (1953); *26*, 186-190 (1954)
[7.35] L. Liebermann: J. Acoust. Soc. Am. *23*, 561 (1951)
[7.36] L.A. Chernov: *Wave Propagation in a Random Medium* (McGraw-Hill, New York 1960)
[7.37] L.C. Huff: J. Acoust. Soc. Am. *44*, 650 (1968)
[7.38] L.A. Stallworth, M.J. Jacobson: J. Acoust. Soc. Am. *48* (2), 382-391 (1970)
[7.39] E.R. Franchi, M.J. Jacobson: J. Acoust. Soc. Am. *53*, 835-847 (1973)
[7.40] E.R. Franchi, M.J. Jacobson: J. Acoust. Soc. Am. *54*, 1302-1311 (1973)
[7.41] R.N. Baer, M.J. Jacobson: J. Acoust. Soc. Am. *54*, 80-91 (1973)
[7.42] J.G. Watson, W.L. Seigmann, M.J. Jacobson: J. Acoust. Soc. Am. *60*, 355-364 (1976)
[7.43] K.G. Hamilton, W.L. Siegmann, M.J. Jacobson: J. Acoust. Soc. Am. *62*, 53-62 (1977)
[7.44] I. Dyer: J. Acoust. Soc. Am. *48* (2), 337-345 (1970)
[7.45] F. Dyson, W. Munk, B. Zettler: J. Acoust. Soc. Am. *59*, 1121-1133 (1976)
[7.46] S.L. Adams: J. Acoust. Soc. Am. *60*, 1218-1219 (1976)
[7.47] J.L. Lawson, G. Uhlenbeck: *Threshold Signals* (McGraw-Hill, New York 1950)
[7.48] R.P. Porter, R.C. Spindel: J. Acoust. Soc. Am. *64*, 769-779 (1978)
[7.49] W.H. Munk, F. Zachariasen: J. Acoust. Soc. Am. *59*, 818-830 (1976)
[7.50] Y.S.F. Desaubies: J. Acoust. Soc. Am. *60*, 795-800 (1976)
[7.51] S.M. Flatte, F.D. Tappert: J. Acoust. Soc. Am. *58*, 1151-1159 (1975)
[7.52] S.M. Flatte, R. Dashen, W.H. Munk, F. Zachariasen: "Sound Transmission through a Fluctuating Ocean", Stanford Res. Inst. JSR-76-39 (May 1977)
[7.53] D.A. deWolf: Radio Sci. *10*, 53-57 (1975)

# Subject Index